Intentional and Uni
Contaminants
Food and Fe

ACS SYMPOSIUM SERIES **1020**

Intentional and Unintentional Contaminants in Food and Feed

Fadwa Al-Taher, M.S.
National Center for Food Safety and Technology/Illinois Institute of Technology

Lauren Jackson, Ph.D.
National Center for Food Safety and Technology/Food and Drug Administration

Jonathan W. DeVries, Ph.D.
Medallion Labs

Sponsored by the
ACS Division of Agricultural and Food Chemistry

American Chemical Society, Washington DC

Library of Congress Cataloging-in-Publication Data

Intentional and unintentional contaminants in food and feed / Fadwa Al-Taher, Lauren Jackson, Jonathan W. DeVries ; sponsorship by the ACS Division of Agricultural and Food Chemistry.

p. cm. -- (ACS symposium series ; 1020)

Includes bibliographical references and indexes.

ISBN 978-0-8412-6979-8

1. Food--Microbiology--Congresses. 2. Food contamination--Congresses. 3. Food--Safety measures--Congresses. 4. Food--Safety regulations--Congresses. I. Al-Taher, Fadwa. II. Jackson, Lauren S. III. DeVries, Jonathan W. IV. American Chemical Society. Division of Agricultural and Food Chemistry. V. American Chemical Society. Meeting (235th : 2008 : New Orleans, La.)

QR115.I456 2009

363.19'262--dc22

2009024253

The paper used in this publication meets the minimum requirements of American National Standard for Information Sciences—Permanence of Paper for Printed Library Materials, ANSI Z39.48–1984.

Distributed by Oxford University Press

PRINTED IN THE UNITED STATES OF AMERICA

Foreword

The ACS Symposium Series was first published in 1974 to provide a mechanism for publishing symposia quickly in book form. The purpose of the series is to publish timely, comprehensive books developed from the ACS sponsored symposia based on current scientific research. Occasionally, books are developed from symposia sponsored by other organizations when the topic is of keen interest to the chemistry audience.

Before agreeing to publish a book, the proposed table of contents is reviewed for appropriate and comprehensive coverage and for interest to the audience. Some papers may be excluded to better focus the book; others may be added to provide comprehensiveness. When appropriate, overview or introductory chapters are added. Drafts of chapters are peer-reviewed prior to final acceptance or rejection, and manuscripts are prepared in camera-ready format.

As a rule, only original research papers and original review papers are included in the volumes. Verbatim reproductions of previous published papers are not accepted.

ACS Books Department

Contents

Indexes

PREFACE

Food safety issues have always been of great concern to scientists working in the agriculture and food scientific community. This, along with increased public concern continues into the new millennium. Significant scientific and traditional media attention resulted from the discovery of acrylamide in heat processed carbohydrate rich foods in 2002; the presence of furan in foods that undergo thermal treatments such as canned or jarred foods in 2004; the illness outbreak resulting from *E. coli* O157:H7 in leafy greens in 2006; imported farm-raised seafood contaminated with banned antimicrobial reagents from 2006 to 2007; the deliberate contamination of pet food (along with coincidental contamination of feed for animals intended for food) with melamine in 2007; and the *Salmonella* Litchfield outbreak from cantaloupe imported from Honduras in 2008.

To understand the research completed and ongoing and how the health risks to consumers are being addressed with regard to these food safety-related issues, we organized a two-day symposium on the "Intentional and Unintentional Contaminants in Food and Feed." The symposium took place in New Orleans, Louisiana on April 9-10, 2008 at the 235th American Chemical Society Spring National Meeting and brought 28 expert speakers from the U.S., Canada, and Europe together. Ideas and viewpoints were shared particularly via interaction amongst the experts. The proceedings from this symposium are published in this volume.

The content of this book, *Intentional and Unintentional Contaminants in Food and Feed,* includes a variety of topics under the following headings: overview of chemical contaminants- intentional and unintentional; emerging analytical methods for food contaminants; potential strategies to prevent contamination of food; and regulatory policy and risk assessment for intentional and unintentional contaminants. We hope that readers will examine these articles in depth so that they will benefit from a breadth of expert viewpoints. The most important purpose of this volume is the exchange of ideas.

This book focuses on the science of undesirable chemicals found in food resulting from either natural occurrence or deliberate addition. Initial chapters contain overviews on contaminants of concern. Heat-produced chemicals (acrylamide and furan) and other possible carcinogens have been detected in staple foods and have caused alarm in recent years. Several chapters cover the factors affecting formation of furan and acrylamide and approaches for mitigating their formation in food.

Microbiologist authors discuss microbial contamination of fresh produce and methods to prevent contamination. Illness outbreaks resulting from *E. coli* O157:H7 in spinach and lettuce had microbiologists scrambling to determine the cause of this outbreak and how to prevent future microbiological contamination.

Research continues on mycotoxins, secondary metabolites of molds and fungi that can cause disease and death in humans. A few chapters outline strategies for reducing mycotoxin levels in food. The events of September 11, 2001 have also made us more attentive to the possible potential deliberate contamination of food with chemical agents such as ricin or abrin, or microbial pathogens. Food defense programs have been developed to protect the safety of our nation's food supply. Several chapters highlight research being done to understand the stability of chemical and biological threat agents and plans being developed for reducing the risk of contamination in food with these chemical and microbial agents. Intentional contamination of pet food using melamine to falsely boost the measured protein content has also drawn attention and is addressed. This book contains valuable information on novel analytical technologies, toxicology of food contaminants, strategies to reduce or eliminate food contaminants, and regulatory policy and risk assessment for intentional and unintentional foods. A chapter focuses on the US Food and Drug Administration's tightened regulations on imports to maintain control of the safety of the U.S. food supply. Finally, risk assessment and worldwide food safety regulations are discussed.

We hope that *Intentional and Unintentional Contaminants in Food and Feed* will be a valuable reference and resource for further research in the effort of continuing to make our food supply even safer. We are confident that the effort of all contributors will be worthwhile. Food scientists, food processors, food technologists, chemists, microbiologists, industry, government and academia will all be interested in reading this important volume. Graduate students will find this book an excellent reference to aid them in their research. Anyone who performs research in the area of food and beverages in industry or government agencies such as the United States Food and Drug Administration, United States Department of Agriculture, and the Environmental Protection Agency will find this book a great asset. The book will certainly gain some international interest as far away as Canada, Europe, and China.

Finally, we want to express our gratitude to all the contributors of this volume for their help in bringing together the important issues of food safety, food defense and food protection both on a domestic level and internationally. We are also thankful to the ACS Division of Agricultural and Food Chemistry for their approval and financial support of the symposium. We also express our gratitude to the following sponsors: Frito-Lay, PepsiCo, Silliker Laboratories, and Waters. Without their financial support, this symposium would not have been possible.

Fadwa Al-Taher, M.S.
National Center for Food Safety and Technology/Illinois Institute of Technology
6502 S. Archer Rd.
Summit-Argo, IL 60501
(708)563-8167
(708)563-1873 fax
altaher@iit.edu

Lauren Jackson, Ph.D.
National Center for Food Safety and Technology/Food and Drug Administration
6502 S. Archer Rd.
Summit-Argo, IL 60501
(708)728-4162
(708)728-4177 fax
lauren.jackson@fda.hhs.gov

Jonathan W. DeVries, Ph.D.
Medallion Labs
9000 Plymouth Avenue
Minneapolis, MN 55427
(763)764-2774
(763)764-4010 fax
jon.devries@genmills.com

Chapter 1

Overview Chapter

Fadwa Al-Taher

Although Americans have one of the safest food supplies in the world, the new millennium has brought challenges to the realm. The globalization of the food supply, the discoveries of intentional contamination of the food supply, the growing volumes of imports, and advances in production and distribution methods require updated approaches to protecting our food supply from unintentional and deliberate contamination.

The U.S. Centers for Disease Control and Prevention (CDC) estimates 76 million food-related illnesses, 325,000 food-related hospitalizations and 5,000 food-related deaths occur in the U.S. each year.. The number of food borne illnesses associated with fresh produce is on the rise. Leafy greens have become one of the most common sources of food borne outbreaks in the U.S. Increased attention has been focused on pathogenic microorganisms

The chapter "Microbial contamination of fresh produce" looks at *E. coli* and *Salmonella* contamination of lettuce, spinach and other greens occurring in the fields or at processing plants. The increase of bacterial contamination in produce has caused microbiologists, epidemiologists, and chemists to work quickly to find methods for the rapid detection and control of the causative agents and the prevention of illnesses associated with these pathogens. Research is being conducted on new methods to control bacteria on produce and in biofilms without compromising the quality of food. This is discussed in the chapter "Inactivation of microbial contaminants in fresh produce." Development of antimicrobial chemicals that can help ensure the safety of produce is underway. Immunoassays and PCR detection methods have been improved in terms of accuracy and speed to determine the cause of an outbreak. In-line rapid detection of microorganisms in produce wash and/or rinse solutions, irradiation of produce, surface pasteurization treatments of produce and cold plasma technologies are some novel technologies that have been developed to prevent or reduce pathogens from produce while preserving eating quality.

With advances in analytical detection, chemists are now capable of analyzing for and detecting more chemicals at much lower concentrations in more foods due mostly to hyphenated techniques as is pointed out in "When philosophies collide: Dealing with very low levels of chemicals in food." Questions arise as to whether a chemical detected at parts per billion levels poses more health problems and concerns about the food supply for consumers than those chemicals detected at the parts per million level. The "U.S. Food and Drug Administration's program for chemical contaminants in food" provides an overview of the chemicals of concern to the agency. Chemicals that may enter the food supply are food additives (colors and preservatives), food contact chemicals (food packaging materials, additives, Bisphenol A), agrichemicals (pesticides and veterinary drugs), mycotoxins, thermally-processed induced chemicals (acrylamide, furan, heterocyclic aromatic amines), environmental contaminants (heavy metals), and food allergens. Two chapters, "Effect of heat-processed foods on acrylamide formation" and "Furan in thermally processed foods," address specific chemicals that form as a result of heat treatment of food.

It is difficult to determine a zero threshold for chemicals in food and difficult to explain acceptable tolerances to consumers. Toxicologists are in demand to supply more animal data and better ways to extrapolate to the low levels humans might be exposed to. Although safe levels of human exposure have been set for many chemicals, most people are concerned when new chemicals have been identified in their food and those chemicals have demonstrated potentially adverse chronic effects such as neurotoxic effects (acrylamide) and reproductive effects (dioxins) at high levels, but the effects of low level chronic exposure are uncertain.

As of yet, there has been no known treatment for life-threatening food allergens, only prevention of allergic reaction by avoidance of exposure to the allergen. Food allergy affects about two percent of the population in the U.S. Food labeling of the presence of allergens educates sensitive consumers regarding products they should avoid consuming. Unintentional cross-contamination during manufacturing can occur and thus, it is important to discover novel approaches to monitoring food for undeclared allergens. Enzyme-Linked Immunosorbent Assay (ELISA) has been used traditionally for screening for undeclared allergens in foods, but it often gives ambiguous results due to cross-reactivity. Alternative methods are needed for confirmation. Mass spectrometric methods, such as what is discussed in the chapter "Detection and confirmation of food allergens using mass spectrometric techniques: Characterization of allergens in hazelnut using ESI and MALDI mass spectrometry," have recently been introduced as detection and confirmatory tools for many potential allergenic contaminants. Mass spectrometric methods offer specificity, sensitivity and multi-target identification and quantification. Some allergens that have been investigated using mass spectrometry include ovalbumin in egg; tropomyosin in shellfish; Ara h1, Ara h2, and Ara h3 in peanut; and α-S1 casein and β-lactoglobulin in milk.

Mycotoxins, secondary metabolites produced by certain fungi, occur naturally, and can contaminate food during growth in the field, processing, transportation or storage. "Mycotoxins of concern in imported grains" is an example of the occurrence of mycotoxins and the characteristics and formation of the various mycotoxins is discussed. Mycotoxins are considered both poisonous and chronic hazards and are best avoided by implementing good agricultural and manufacturing practices. Many mycotoxins are stable to heat and food processing procedures. Consumers may exhibit various toxicological outcomes from ingestion. Historically, efforts have been undertaken to minimize mycotoxins in the food supply by establishing guidelines and action levels and monitoring the food supply for mycotoxins. Regulatory action will be taken against products that exceed the action levels for a mycotoxin. Some mycotoxins of greatest concern are aflatoxins in corn, peanuts, tree nuts, rice and cottonseed; fumonisins in corn, wheat, barley, and rice; and ochratoxin in wheat, barley, oats, rye, sorghum peanuts, wine, beer, and raisins. The European Union (EU) action limits for mycotoxins are lower compared to those established action limits in the United States. This can cause problems for international trade. For example, the EU has set an action level for Ochratoxin A in imported grains to be 5.0ppm. The U.S. has not established a limit as yet.

Food imports to the U.S. have almost doubled in the past decade, from $36 billion in 1997 to more than $70 billion in 2007. Because of reduced budgets, the number of FDA inspectors at the Office of Regulatory Affairs dropped from 1,642 in 2003 to 1,389 in 2005, while food imports rose from 9.3 million shipments per year to more than 13.8 million shipments annually. The FDA inspectors sample just 1.3 percent of all imported food shipments entering the country and perform few on-site inspections of foreign farms and food processing plants. Since there are not enough inspectors at the borders, there have been incidents of food imports entering the U.S. unapproved, The FDA is responsible for inspecting all imported foods except for meat and egg products, which are regulated by the Food Safety and Inspection Service, part of the U.S. Department of Agriculture. "Dealing with intentional and unintentional contaminants in meat and poultry products regulated by the USDA/FSIS" demonstrates the economic concerns and what happens when there is a recall.

In the time frame of October 2006 through May 2007, the FDA found that farm-raised seafood imported from China was contaminated with antimicrobial agents unapproved in the U.S. (nitrofuran, malachite green, gentian violet, and fluoroquinolone). Nitrofuran, malachite green, and gentian violet have been shown to be carcinogenic based on long-term exposure studies in lab animals. Residues of fluoroquinolones may increase antibiotic resistance to this class of antibiotics. These drugs had been used to treat the seafood to inhibit growth of bacteria and fungi or to prevent or treat parasitic infestation. However, they are not approved for use in farm-raised seafood in the U.S. As a result of these findings, in June 2007, the FDA restricted imports of five types of farmed seafood (catfish, basa, shrimp, dace, and eel) from China because of concerns regarding the unapproved drug residues.

Lead-based and other heavy metal-based inks used for labeling candy wrappers have historically been a regulatory issue. The history of lead contamination is elucidated in Michael Kashtock's chapter "Lead in Food: The Neo-classical contaminant." Although the U.S. and EU have banned the use of heavy metal-based inks in food wrappers, lead has been found in the wrappers of candy imported from Mexico. Lead-based inks have been found on both the exterior and interior surfaces. If lead derived from a lead-based printing ink is found on the portion of the package that directly contacts the food or, if the lead is expected to migrate into the packaged food, the product would likely be regarded as being in violation of the Federal Food, Drug, and Cosmetic Act. Also, certain ingredients such as chili powder and certain types of salt, often used in Mexican candy products may contain lead and these are an avoidable source of lead in the food supply. In 2006, FDA issued updated guidance regarding lead with the express purpose of lowering children's exposure to small traces of lead present in certain candies. This occurred after testing and finding that certain types of Mexican candy products showed levels of lead above 0.5 ppm. FDA has reduced the allowable level of lead in candy to 0.1 ppm.

In 2007, pet food manufacturers recalled more than 150 brands of dog and cat food contaminated with melamine, amelide, amiline, and/or cyanuric acid. Animals consuming the food developed symptoms of kidney failure, including loss of appetite, vomiting, lethargy, frequent urination, increased thirst and in some cases, ultimately death. At the exposure levels experienced by the affected cats and dogs, melamine, in combination with amelide, amiline, and/or cyanuric acid appears toform highly insoluble crystals in the animal's kidney systems, resulting in kidney damage. A chapter on the "Renal toxicity of pet foods contaminated with melamine and related compounds" discusses the implications.

The pet foods contained wheat gluten and rice protein as sources of protein for the animals diets. The wheat gluten and rice protein contained melamine, amelide, amiline, and/or cyanuric acid.. The Chinese suppliers had added these compounds to the pet food to increase the measured level of protein. These adulterants are high in nitrogen on a weight/weight basis, and artificially provide elevated protein levels when the wheat gluten/rice protein samples are analyzed by conventional methodology. Subsequently, it was discovered that manufacturers of pet foods convert the scrap and rework of the dog and cat foods into food for hogs, chickens, and fish. This low levels of melamine were also present in food given to hogs, chicken, and fish. The US FDA's Forensic Chemistry Center first detected and identified the melamine adulterant in pet food. The Center made a preliminary identification of the melamine using a mass spectrometric technique that relies on an open-air ionization method known as direct analysis in real time (DART). The group further confirmed the finding with additional DART analysis coupled with GC-MS analysis. The DART technique has the advantage of being more rapid than traditional GC-MS primarily because it requires no sample preparation, however, the instrumentation is not widely available.

China has recently reacted to international pressure by agreeing to tighten food safety standards. In December 2007, the United States and China signed an agreement to place new registration and inspection requirements on 10 food products exported by Chinese companies. These products include some preserved foods, pet foods, and farm-raised fish, all of which have been found to be contaminated in the past.

A major concern with regard to food safety is that the U.S. food supply might be vulnerable to attack. The U.S. FDA has worked with Sandia National Laboratories to develop a tool for defending the food production systems. "CARVER + Shock: Risk Assessment Tool" shows how this food-defense software can be used to increase protection of the food supply.

To further enhance consumers safety with regard to contaminants, FDA developed the FDA Food Protection Plan which was made public in November 2007. This tool is designed to address both unintentional and deliberate contamination of the nation's food supply. The Food Protection Plan proposes the use of science and a risk-based approach to ensure the safety of domestic and imported foods eaten by American consumers. This plan implements a strategy of prevention, intervention and response to build safety into every step of the food supply chain. The Food Protection Plan, which focuses on both domestic and imported food, complements the Presidential Initiative: Import Safety Action Plan that recommends how the U.S. can improve the safety of all imported products. It is estimated that $2 trillion worth of goods were imported into the U.S. in 2007, and it is expected that will increase to over $6 trillion by 2015. The Import Safety Action Plan lays out a road map with short- and long-term recommendations to increase product safety at every step of the import life cycle. Together, these plans will improve efforts by the public and private sectors to enhance the safety of wide array of products used by American consumers. The plan is based on preventing harm before it can occur, intervening at key points in the food production system, and responding immediately when problems are identified. These efforts will provide a food protection framework that ensures that the U.S. food supply remains safe.

Chapter 2

U.S. Food and Drug Administration's Program for Chemical Contaminants in Food

M. Bolger, S. Egan, P. South, C. Murray, L. Robin, G. Wood, H. Kim and N. Beru

Office of Food Safety
Center for Food Safety and Applied Nutrition
U.S. Food and Drug Administration
5100 Paint Branch Parkway
College Park, MD 20740

The Food and Drug Administration (FDA) enforces the Federal Food, Drug and Cosmetic Act, a law intended among other things, to ensure that the U.S. food supply is safe for human consumption particularly when it may contain poisonous or deleterious substances. FDA conducts a broad range of activities to identify emerging food-borne natural and anthropogenic chemical contaminants. The primary focus is the development of analytical methods for testing food for the presence of these contaminants and a fairly extensive monitoring program that has been expanded considerably since its inception in the 1960s. The FDA uses science-based risk analysis including risk assessment to determine if the chemical contaminants may pose a health risk, and risk management and risk communication to protect the public health. Generally, the risk assessment involves a safety assessment, which is the identification of a level of exposure deemed to represent negligible risk. In the case of dietary chemical contaminants, this generally results in the derivation of a Tolerable Daily Intake (TDI). The TDI is derived from a no observed adverse dose level or the lower confidence level of a benchmark dose from an animal or human study and the application of one or more 10 fold safety/uncertainty factors to account for intra- and interspecies sensitivity differences. In some circumstances, dose-response information may be used to describe the differential response of sensitive populations. These methods are then used to evaluate the merits of alternative public health and risk management programs that are designed to mitigate exposure and risk.

Introduction

The Federal Food, Drug, and Cosmetic Act (FFDCA) provides the U.S. Food and Drug Administration (FDA) with broad regulatory authority over food that is introduced or delivered for introduction into interstate commerce. Section 402(a)(1) of the FFDCA provides that a food is deemed to be adulterated if it contains any poisonous or deleterious substances, such as chemical contaminants, which may or ordinarily render it harmful to health. Under this provision of the FFDCA, FDA oversees the safety of the U.S. food supply (domestic and imports), in part, through its monitoring programs for natural toxicants (e.g., mycotoxins), pesticides, and anthropogenic (e.g., industrial chemicals, such as dioxins; cooking or heating related chemicals, such as acrylamide; trace elements, such as lead) contaminants in food and the assessment of potential exposure and risk.

This statute allows the FDA to control the presence of chemical contaminants in the U.S. food supply by means of the following general strategies : a) establishing guidelines (e.g., action levels, guidance levels in foods, (b) monitoring the food supply to ensure compliance with established levels, (c) initiating appropriate enforcement action, (d) conducting science-based safety/risk assessments and analyses and (e) cooperating with state and other federal agencies in regards to the safety of foods. Action levels are used as a guide by FDA field staff to determine when it may be necessary to take enforcement action against a food producer, processor or distributor. Guidance levels are provided to food producers and represent the best estimation of the negligible risk level associated with a particular chemical contaminant based on available exposure and toxicological information. Although guidance levels are not inherently enforceable, FDA reserves the right to take appropriate actions when warranted under given circumstances.

Foods are monitored routinely by FDA through its compliance and surveillance programs. The objectives of these programs are to collect and analyze samples of foods and feeds to determine the occurrence and levels of chemical contaminants; to remove from interstate commerce those foods and feeds that contain chemical contaminants at concentrations judged to be of regulatory significance to protect public health; and to determine the awareness of potential problems and control measures employed by distributors, manufacturers and/or processors. The objective of the exploratory surveillance program is to obtain background exposure data that may be used in conjunction with toxicological data to conduct risk assessments that will characterize the potential risk that the presence of a certain chemical contaminant in food poses to the U.S. population. This information can be used to inform risk managers to evaluate potential risk reduction options (e.g., establishing guidelines) if a regulatory control program is warranted. The monitoring efforts are directed at regions and commodities that historically have a high level of contamination or in response to new information on contamination problems developing in regions or commodities not normally affected. All FDA district laboratories involved in monitoring activities are provided with a list of commodities susceptible to contamination, a sampling plan (including product sample size), and a quota of the number of samples to be collected. The collected samples are analyzed by official collaboratively studied methods specific for each product.

In assessing the risks of dietary chemical contaminants, the public health risk question must be unambiguous. Generally, a risk assessment is safety assessment which is the identification of a level of exposure deemed to represent negligible risk. In the case of dietary chemical contaminants this results in the derivation of a Tolerable Daily Intake on the basis of a single no adverse dose level from an animal or human study and the application of one or more 10 fold safety/uncertainty factors to account for intra- and interspecies sensitivity differences. These default extrapolation factors account for the uncertainty regarding extrapolation of dose response information between species and/or the sensitivity of members of a population. An example is the a prior position that children are more sensitive than adults. This is based on the generic pre- and post-natal sensitivity of the developing organism and higher exposure of children versus adults on a body weight basis because of higher caloric consumption per body weight. An example is the observed sensitivity of the fetus and children to lead. Sensitivity may also occur because of unique exposure patterns (e.g., subsistence consumers), nutritional deficiencies, genetic polymorphisms, reduced clearance capacity and pre-existing health conditions. The safety assessment paradigm is useful for determining exposures which are of no public health concern. The determination that an exposure is "unsafe" or a population is "at risk" should then lead to a consideration of risk as a matter of degree. This will allow for the inclusion of other issues such as the avoidability of exposure and competing dietary risks which is particularly relevant for ubiquitous environmental contaminants. Dose-response information may be used quantitatively to describe the differential response of sensitive populations by deriving estimates that account for both variability and uncertainty of the risk. These methods can then be generalized to evaluate the merits of other public health and risk management programs that involve trade-offs between food borne chemical contaminant risks.

The following are succinct summaries of individual program areas that comprise the FDA's dietary chemical contaminants program. The activities and results of the chemical contaminants program area is described and maintained with up-to-date residue results and descriptions of the various program areas on FDA's web site (http://www.cfsan.fda.gov/~lrd/pestadd.html).

Total Diet Study

FDA's Total Diet Study (TDS) provides another approach to monitoring levels of contaminants in foods. The study involves the periodic purchase and chemical analysis of foods to determine the levels of pesticide residues, anthropogenic contaminants and nutrients. The major objective of the study is to monitor over time the concentrations of these substances in foods and to estimate dietary intake by the US population. FDA initiated the TDS in 1961 primarily in response to public concerns about the potential for radionuclide contamination of foods resulting from atmospheric nuclear testing. Initially, the study analyzed for two radionuclides, several organochlorine and organophosphate pesticides, and dietary intakes estimated only for teenage boys were evaluated since they represented the greatest potential dietary exposure per grams of food consumed per day. Since 1961, the TDS has been expanded to include many more foods and analytes. TDS samples are also used in other dietary contaminant program work. The study has also seen improvement in analytical techniques and the expansion of the population subgroups for which dietary intakes are estimated *(1)*. TDS samples are also used in other dietary contaminant program work, particularly for determining background levels of those contaminants in the wide range of foods collected in the TDS.

In the early studies, foods were composited for analysis into 11 or 12 major commodity groups (e.g., meat and eggs, grains, and fruits). In 1982, the TDS was revised and foods are now analyzed individually rather than as commodity composites. In the current program, samples of about 280 "core foods" are collected and analyzed for about 230 analytes (elements, selected nutrients, pesticide residues, industrial chemicals, and radionuclides). These core foods represent the foods (including beverages) most commonly consumed and those consumed in the greatest quantities by the US population based upon dietary data collected by the United States Department of Agriculture (USDA). In 1991, 20 additional infant and toddler foods were included in the TDS food list specifically to provide more information on levels of pesticides and lead in the diets of young children. TDS samples are generally collected four times a year, once in each of four regions of the country (West, North Central, South and Northeast). Different cities are selected for sample collections each year to allow for more geographic representation. The foods are prepared table-ready. Samples from the three cities are composited to form a single analytical sample. The TDS is unique among the monitoring programs within FDA in that it determines levels of the analytes in foods as they would be consumed. This is particularly important for estimating the dietary intake of substances which may be reduced as a result of washing, peeling and cooking. The primary goal of the TDS is to determine concentrations and dietary intake of analytes that are usually present in foods at very low levels. For this reason, the analytical methods used in the TDS are five to ten times more sensitive than those used in other monitoring and surveillance programs.

Dietary intakes of analytes measured in the TDS are estimated by multiplying their concentrations in foods by the amounts of foods that are consumed. For estimating intakes, model diets have been constructed for 14 age-sex groups based on results of national food consumption surveys including the USDA 1987-88 Nationwide Food Consumption Survey, the Continuing Surveys of Intakes by Individuals (CSFII) conducted in the 1990s, and the National Health and Nutrition Examination Survey (NHANES). In addition to using the model diets to calculate intakes, FDA routinely estimates dietary intakes by linking the analytical results from the TDS to the detailed results from the consumption surveys, which allows for more targeted dietary intake assessments for other age-sex groups and for those who may be at the upper percentiles of intake.

The capability of the TDS to measure very low levels of contaminants in food provides an invaluable baseline reference for ensuring the continuing safety of the food supply. And due to the ongoing nature of the study, the TDS results allow for an assessment of the impact of efforts to reduce dietary exposure to contaminants. Lead provides a clear example of this. Beginning in the early 1970s, a number of regulations and voluntary actions by industry resulted in substantial decreases in lead levels in food. The TDS showed a steady decrease in dietary intake of lead from the late 1970s through the late 1980s and into the 1990s. During that time period, the dietary intake of lead by young children decreased from 30 μg/day in 1976 to about 1 μg/day in 1995.

Acrylamide

In 2002, Swedish researchers reported finding acrylamide in a variety of fried and oven-baked foods. Research to date indicates that acrylamide formation is particularly associated with traditional high temperature cooking processes for certain carbohydrate-rich foods.

Acrylamide forms in food due to a reaction between the amino acid asparagine and reducing sugars such as glucose and fructose *(2,3)*. The formation of acrylamide is part of the Maillard reaction, which leads to browning and flavor changes in cooked foods. Acrylamide formation occurs primarily at elevated cooking temperatures used when frying or baking (above 120 °C) and in low moisture conditions. Also, formation occurs primarily in plant-based foods, notably potato products such as French fries and potato chips; cereal foods such as cookies, crackers, breakfast cereals, and toasted bread; and coffee.

The discovery of acrylamide in food is a concern because it is a potential human carcinogen and genotoxicant, based on high-dose animal studies, and a known human neurotoxicant. A major international evaluation of acrylamide by the Joint FAO/WHO Expert Committee on Food Additives *(4)* identified margins of exposure (MOEs) for acrylamide of 300 for average consumers and 75 for persons consuming large amounts of acrylamide in the diet (high consumers). JECFA considered these MOEs (ratios between the lowest amount of acrylamide causing cancer in animal studies and the amount of acrylamide found in food) to be low for a compound that is genotoxic and carcinogenic and concluded that the levels of acrylamide in food were of concern. Acrylamide is also a known human neurotoxicant and animal reproductive and developmental toxicant. However, acrylamide is considered unlikely to have neurological, reproductive, or developmental effects at the levels encountered in human foods *(4,5),* although more research is being carried out in these areas.

After the discovery of acrylamide in food in 2002, FDA initiated a broad range of activities on acrylamide. FDA accomplishments include development of an analytical method: analysis of acrylamide levels in more than 2600 food samples; development of exposure assessments that indicate the amount and primary sources of exposure to acrylamide for U.S. consumers; research on acrylamide formation and mitigation; research on acrylamide toxicology, including toxicokinetic, bioavailability, mutagenicity, carcinogenicity, and neurodevelopmental studies; and participation in international work on acrylamide.

One notable aspect of FDA's work is the development of a large database (over 2600 samples) of acrylamide levels in food, from analyzing both individually purchased food products and samples from the TDS. This database includes samples representative of food from food processors, restaurants, and home cooking, as well as foods from three cuisines common in the American diet: Asian, Hispanic, and Southern/Creole/Cajun.

FDA has estimated that the average U.S. consumer's intake of acrylamide is 0.4 microgram/kilogram body weight/day (μg/kg-bw/d), *(6)* while international estimates for the average consumer range from 0.2 to 1.4 μg/kg-bw/d *(7).* FDA exposure estimates have remained stable over the past several years, even as more acrylamide results were added to the database. Therefore, as of summer 2006, no new sampling was planned, although more sampling can be added in the future if needed.

FDA's National Center for Toxicological Research (NCTR) completed new long-term rat and mouse carcinogenicity bioassays of acrylamide in 2007. Pathology working group results from these studies are expected in early 2009 and review by the National Toxicology Program (NTP) is scheduled for 2010. The bioassays address deficiencies in earlier carcinogenicity studies and should provide more reliable data on potential carcinogenic risks of acrylamide exposure. NCTR has also completed development of a physiologically based pharmacokinetic (PBPK) model for acrylamide. The new potency data and PBPK model will be used along with food exposure data in a risk assessment intended to estimate the risk of cancer from acrylamide in food.

Much work has been done internationally on potential ways to reduce acrylamide in food, such as developing alternative cooking profiles, changing ingredients, and using the enzyme asparaginase to break down the acrylamide precursor asparagine. The Confederation of Food and Drink Industries of the European Union (EU) has prepared a summary of strategies for acrylamide mitigation in its Acrylamide "Toolbox" *(8)* and the Codex Committee on Contaminants in Foods is also preparing a Draft Code of Practice for the Reduction of Acrylamide in Food *(9).* By 2007, FDA had responded to two notices confirming no objection to the Generally Recognized as Safe status of asparaginase for acrylamide reduction.

Currently, FDA is still in the information gathering stage on acrylamide, and has not instituted any regulatory action. The planned risk assessment on acrylamide will be critically important to FDA's risk management decision makers. FDA's best advice for acrylamide and eating is that consumers adopt a healthy eating plan, consistent with the Dietary Guidelines for Americans, that emphasizes fruits, vegetables, whole grains, and fat-free or low-fat milk and milk products; includes lean meats, poultry, fish, beans, eggs, and nuts; and is low in saturated fats, *trans* fats, cholesterol, salt (sodium) and added sugars. FDA also added optional information to the Acrylamide: Questions and Answers section of its website in 2008 for consumers who want to reduce their acrylamide exposure from food now.

Dioxin-like Compounds (DLCs)

Dioxin and chemically-related compounds (referred to collectively as dioxin-like compounds or DLCs) are a group of environmental contaminants found throughout the world. Studies suggest that human exposure to DLCs may lead to a variety of adverse health effects including reproductive and developmental problems, diabetes, as well as increased risk of cancer *(10)*.

DLCs are members of three closely related families: polychlorinated dibenzo-*p*-dioxins (PCDDs), polychlorinated dibenzofurans (PCDFs) and certain polychlorinated biphenyls (PCBs). PCDDs and PCDFs are not created intentionally, but produced inadvertently during combustion and by certain chemical processes. In contrast, PCBs were commercially produced and widely used as a coolant and insulator in electrical equipment until production was banned by the United States and other countries in the late 1970s and early 1980s. The potential for PCB exposure exists, however, because PCBs, like all DLCs, are extremely persistent compounds that are still found in the environment due to past releases. In addition, PCBs are contained in older transformers, capacitors and other electrical equipment still in use.

PCDDs and PCDFs are found in most food-producing animals and animal feeds and are suspected to be the primary pathway of exposure *(11)*. Because PCDDs and PCDFs tend to accumulate in the fat of food-producing animals, consumption of animal-derived foods (e.g., meat, poultry, eggs, fish, and dairy products) is considered to be the major route of human exposure. FDA has been monitoring specific foods with the goal of identifying ways to reduce dietary exposure. For example, since 1995, FDA has been monitoring dioxin levels in finfish, shellfish, and dairy products. In 1999, FDA initiated dioxin analysis of foods collected under its TDS survey.

DLC levels are reported as toxicity equivalents (TEQs) which are used to estimate the relative toxicity of DLC congeners. The TEQ is calculated by multiplying the concentration of each DLC congener by the corresponding Toxic Equivalency Factor (TEF) established by the World Health Organization (WHO) in 1998 and revised in 2006 *(12, 13)*. In 1998, the WHO also established a tolerable daily intake range of 1-4 pg TEQ/kg bw/d for DLCs *(14)*.

In 2001, JECFA established a provisional tolerable monthly intake of 70 pg TEQ/kg bw/mo for DLCs *(15)*.

Because DLC analysis is costly and time consuming, available data on background levels in most foods and feeds are limited. In fact, for many foods and most feeds, there are no data. Consequently, it is difficult to determine where and how DLC levels in foods and feeds can be further reduced. Therefore, in 2001, FDA developed a strategy for DLCs (www.cfsan.fda.gov/~lrd/dioxstra.html) that substantially expanded its dioxin monitoring program to obtain more comprehensive data on background levels of DLCs in specific food and feed samples, as well as to identify and reduce pathways of DLC contamination and improve assessments of human DLC exposure. Specific goals for the program are as follows:

- Increasing sampling and analysis of human foods and animal feeds that contribute most significantly to human dietary exposure to DLCs,
- Expanding the capability of FDA field staff to collect and analyze increased numbers of food and feed samples for DLCs,
- Performing trace-back investigations of unusually high levels in food and feeds to determine if the source of contamination can be reduced or eliminated, and
- Enhancing research into new or modified methods for DLC analysis so that less time consuming and less expensive methods become available to the public.

Since 2001, FDA's dioxin monitoring program has analyzed food collected under its TDS and food and feed from non-TDS sampling. FDA non-TDS sampling collects and analyzes foods suspected of having both higher DLC levels and more variability in those levels than other foods *(16)*. This sampling provides additional estimates of the various distributions of DLC levels in specific foods. FDA non-TDS sampling also identifies foods with elevated DLC levels to allow investigation of potential sources and pathways for DLC contamination of the food supply. FDA has posted on its web site data for DLC levels in both TDS and non-TDS food samples as well as exposure estimates from samples analyzed *(17)*.

Elemental Contaminants

Dietary exposures to elemental contaminants, particularly inorganic arsenic (tri- and pentavalent forms), cadmium, lead, and methylmercury have been a source of public health concern and a major focus of the monitoring and safety/risk assessment activities of FDA for many years. A major part of FDA activities in this program area have consisted of targeted sampling and the TDS.

Like that for other chemical contaminants, these efforts have allowed the FDA to identify and mitigate major sources or pathways of dietary exposure.

Methylmercury and lead in particular have been on-going concerns for many years. Methylmercury is most specifically a contaminant found in fish that arises from natural (e.g., volcanic) and anthropogenic (e.g., coal-fired power generation) activities and tends to be found at the highest levels in long-lived species at the top of the marine food chain. Methylmercury produces pronounced adverse effects on the nervous system which is of particular concern to the developing nervous system. The same is the case for lead which also adversely affects other organs depending on the level of exposure *(18, 19, 20)*. While dietary exposures to methylmercury and cadmium *(1, 21, 22, 23, 24)* have remained fairly constant over the years, those to lead have been substantially reduced (e.g., 90% or greater) since 1980 *(25, 26)*.

While arsenic levels in the diet and resulting exposures can be significant, this occurs as an organic form of arsenic which demonstrates little inherent toxicity *(27, 28)*. However, high levels of inorganic arsenic in water used for food processing can result in elevated levels in foods which can pose a significant hazard. The best example of such a case is rice. Rice can also have inherently elevated levels of inorganic arsenic *in situ* such that consumption of rice can result in elevations in the daily exposure to inorganic arsenic.

While cadmium has been shown to produce pronounced chronic effects on the cardiovascular system and particularly on the kidneys, this has occurred in only a few specific areas of the world and with a lifetime of exposure *(23)*. Exposures in the United States have always been much lower than those associated with adverse cardiovascular effects and remained unchanged for many years.

Furan

Furan is an industrial chemical used in production of other chemicals, including tetrahydrofuran, resins, lacquers, and agricultural products. Before 2004, the presence of furan due to heating had been reported in a limited number of foods *(29)*. In Spring 2004, FDA scientists announced that they had found furan in a wide range of foods, particularly foods subjected to retorting in cans and jars *(30)*. Furan was subsequently identified in certain low moisture foods as well, including crackers, potato chips, and tortilla chips. The levels detected by FDA range from < 0.2 parts per billion (ppb) to over 170 ppb (http://www.cfsan.fda.gov/~dms/furandat.html), and FDA has estimated the average intake for U.S. consumers as 0.3 μg/kg-bw/d *(31)*.

The formation of furan in food during thermal processing is not as well understood as that of acrylamide. Multiple mechanisms have been proposed, including oxidation of polyunsaturated fatty acids, breakdown of ascorbic acid derivatives, breakdown of carbohydrates, and breakdown of amino acids in the presence or absence of reducing sugars *(32, 33)*. Research into the mitigation of furan formation is limited, but the fact that furan appears to form by different mechanisms may complicate mitigation attempts.

Furan is both carcinogenic and cytotoxic in rodents. In a bioassay conducted by the National Toxicology Program (NTP), furan administered by gavage to Fisher 344 rats (2, 4, or 8 milligram per kilogram per body weight (mg/kg-bw)) and B6C3F1 mice (8 or 15 mg/kg-bw) 5 days a week for up to 2 years produced hepatic cholangiocarcinoma, hepatocellular adenoma and carcinoma, and mononuclear cell leukemia in rats, and hepatocellular adenoma and carcinoma and benign pheochromocytoma of the adrenal gland in mice *(34)*. In both the 2-year NTP bioassay and a 13-week NTP study, furan also caused cell proliferation, inflammation, biliary tract fibrosis, hyperplasia, hepatocellular cytomegaly, degeneration, necrosis, and vacuolization in rats and mice *(34)*. Furan has also been shown to induce apoptosis in mice at hepatocarcinogenic doses *(35)* and uncoupling of hepatic mitochondrial oxidative phosphorylation *(36)*. In addition, furan has been shown to be a mutagen and clastogen in *in vitro* mammalian systems *(36, 37, 38)*, and to induce chromosomal aberrations in *in vivo* mammalian systems *(38)*.

Although no human studies have been reported on furan, it is considered possibly carcinogenic to humans by the IARC *(39)* and is listed in the Department of Health and Human Services Report on Carcinogens, based on animal tests *(40)*. A European Food Safety Authority (EFSA) review *(41)* concluded that there is a relatively small difference between possible human exposures and doses that caused cancer in animals, although emphasizing that both toxicity and exposure data were limited and that more data were needed to draw conclusions. FDA stated in 2004 that its preliminary estimate of consumer exposure was well below what FDA expects would cause harmful effects.

FDA's work on furan includes development of an analytical method, analysis of furan levels in more than 650 samples, cooperation with the Alcohol and Tobacco Tax and Trade Bureau to gather data on furan levels in alcoholic beverages, and assessment of furan exposures from food *(42)*. FDA is currently conducting research on the effects of ionizing radiation and consumer cooking practices on furan levels in foods, as well as analyzing additional foods for furan levels. In the toxicology area, FDA has proposed a new chronic cancer bioassay of furan in rats to the NTP. The proposed study would use lower doses of furan than the previous rat bioassay, in which cancers occurred at all doses tested. Other work proposed to the NTP includes short-term studies of cell proliferation; dose-response studies of the formation of liver adducts of *cis*-2-butene-1,4-dial, the putative genotoxic metabolite of furan; studies of the use of hemoglobin adducts as biomarkers of furan exposure in rats and humans; and subchronic *in vivo* mutagenesis assays with Big Blue rats to determine mutant frequencies and mutation types. All these proposed studies are intended to define the liver cancer risk in humans from low-dose furan exposure. As with acrylamide, FDA has not initiated any regulatory action on furan or recommended dietary change beyond adoption of a healthy eating plan.

Mycotoxins

Mycotoxins are naturally occurring toxic metabolites produced by certain fungi that can infect and proliferate on various agricultural commodities in the field and/or during storage or processing. The occurrence of these toxins on grains, nuts and other commodities is influenced by environmental factors such as temperature, humidity, and extent of rainfall during the pre- and post-harvesting periods. Some mycotoxins are teratogenic, mutagenic and/or carcinogenic in certain susceptible animal species and are associated with various diseases in animals and humans in many parts of the world. The occurrence of mycotoxins in foods is not entirely avoidable, therefore small amounts of these toxins may be legally permitted in foods, provided the amounts involved are not considered to be injurious to human health *(43)*. The monitoring data obtained over the years reveal that human exposure to these mycotoxins is relatively low.

The food industry is monitored routinely through formal compliance programs *(44)*. The objectives of these programs are to collect and analyze food samples to determine compliance with FDA regulatory levels, and to remove from interstate commerce those commodities that contain specific mycotoxins at levels judged to be of regulatory significance. The collected samples are analyzed chemically in FDA District laboratories by official, collaboratively studied methods. FDA's efforts are complemented by control programs carried out by other federal departments, agricultural state departments, and various trade associations.

Mycotoxins that are currently being monitored by the FDA include the aflatoxins, patulin, deoxynivalenol, fumonisins and ochratoxin A. These mycotoxins are relatively stable and are not completely destroyed by normal cooking and other processing procedures *(45)*.

Aflatoxins (B_1, B_2, G_1, and G_2), that are produced by *Aspergillus flavus* and *A. parasiticus* fungi, may infect many food commodities including peanut/peanut products, corn/corn products, and tree nuts. Aflatoxin M_1 is a toxic metabolite that may be produced in the liver of mammals that have ingested and metabolized high levels of aflatoxin B_1; it is excreted in the milk. Aflatoxins are potent liver toxins and are carcinogenic to humans and susceptible animal species. The current action level established by FDA for total aflatoxins in human food is 20 micrograms per kilogram (20 ppb); the action level for aflatoxin M_1 in fluid milk and fluid milk products is 0.5 micrograms per kilogram (0.5 ppb).

Patulin is produced by *Penicillium, Aspergillus,* and *Byssochylamys* fungi that may grow on apples and other fruits. Patulin is not destroyed by normal heat processing, therefore it can occur at various levels in apple juice if rotten, moldy or damaged apples were used to make the juice. Animal feeding studies have demonstrated that high levels of patulin in apple juice can pose a health risk to humans if contaminated juice is consumed over an extended period of time. FDA has established an action level of 50 micrograms per kilogram (50 ppb) for patulin in apple juice, apple juice concentrates, and apple juice products.

Deoxynivalenol (DON), commonly referred to as vomitoxin, is produced by several fungi of the genus *Fusarium*, especially *F. graminearum*, which is a common contaminant of grains such as wheat, corn, rye and barley. DON has been associated with a number of adverse health effects in humans and animals. The current guidance level for DON in finished wheat products, e.g., flour, bran, and germ, that may be consumed by humans is 1 microgram per gram (1 ppm).

Fumonisins (FB_1, FB_2 and FB_3) are produced by *Fusarium verticillioides* (previously known as *F. moniliforme)* and other *Fusarium* species that are common natural contaminants of corn and are found world-wide in corn and corn-based products. The carcinogenicity of fumonisins has been demonstrated in rodent species. Epidemiological studies have demonstrated that the fumonisins are toxic to humans but a direct relationship with human esophageal cancer has not been definitely established. The FDA has established guidance levels for fumonisins (FB_1, FB_2+FB_3) in various milled corn products that range from 2 to 4 ppm. These levels are considered adequate at this time to protect human health and are achievable with the implementation of good agricultural and good manufacturing practices.

Ochratoxin A is a nephrotoxic metabolite produced by certain species of *Aspergillus and Penicillium* fungi and is mainly a contaminant of cereals (corn, barley, wheat and oats). It is also found in various beans (coffee, soya, and cocoa), dried vine fruit, wine and cheese. An association between the intake of ochratoxin A and nephropathy in humans has been postulated, but causality has not been established. The results from surveys of grains and processed food products for ochratoxin A in the U.S. suggest that ochratoxin A contamination is not a major problem in this country; therefore, no regulatory limits have been established by FDA at this time.

The data obtained over the years from monitoring programs are used to provide: (a) estimates of the incidence and levels of contamination by various mycotoxins in affected areas of the country, (b) dietary exposure data (estimates) for use in making risk assessments, (c) background data for use in establishing guidance levels, (d) an estimate of the economic impact of the enforcement of regulatory guidelines on foods during a given year, and (e) information that can be used by U.S. delegates at international meetings involving mycotoxin issues.

Perchlorate

In recent years, perchlorate (ClO_4^-) has received a fair amount of attention in the scientific literature. Perchlorate is used as an oxidizing agent in solid rocket propellant. In addition, it is found in other items (e.g., explosives, road flares, fireworks, car airbags, herbicides, etc.). Perchlorate is also found to occur naturally in Chilean nitrate fertilizer, which has been used in the United

States *(46)*. In recent years, scientific evidence suggests that perchlorate is possibly generated under certain climatic conditions *(47, 48)*. Also, it has been detected in surface and groundwater and in foods. As a response to the environmental concerns, the U.S. Environmental Protection Agency (EPA), along with other government agencies and academia, has sought to understand and assess the potential health effects of perchlorate levels in soil, groundwater, drinking water, and agricultural commodities around the country. Greer et al. *(49)* reported that perchlorate at high pharmacological doses (0.02, 0.1, and 0.5 mg/kg-day) interfere with iodide uptake into the thyroid gland, disrupting its function. Also, the National Academy of Sciences (NAS) has identified that the fetuses of pregnant women who might have hypothyroidism or iodide deficiency as the most sensitive population .

FDA has recognized the potential for perchlorate contamination in food through the use of some fertilizers, contaminated irrigation water, processing water, and source waters for bottle water. During 2004 and 2005, the FDA conducted exploratory surveys to monitor perchlorate levels in 28 types of foods and beverages consisting of bottled water, milk, fruits and fruit juices, vegetables, grain products, and seafood. In 2005, FDA began testing all samples from the TDS to determine whether perchlorate is found in a broader range of foods. In 2008, Murray et al. *(50)* provided intake estimates of perchlorate and iodine, a precursor to iodide, using the analytical results from the TDS. Estimated average perchlorate and iodine daily intakes as well as the contribution of specific food groups to total intakes were estimated for 14 age-sex subgroups of the U.S. population. The estimated smallest lower bound to the largest upper bound average perchlorate intakes by the 14 age-sex groups range from 0.08 to 0.39 micrograms per kilogram body weight per day (μg/kg bw/day), compared with the EPA Reference Dose (RfD) of 0.7 μg/kg bw/day. Infants and children demonstrated the highest estimated intakes of perchlorate on a body weight basis. Also, Murray et al. *(50)* compared the TDS perchlorate results with perchlorate levels in selected foods that have been reported previously in the literature *(51, 52, 53, 54, 55, 56)* and results from FDA conducted exploratory surveys in 2004 and 2005. The perchlorate residue results show fairly good agreement for seven of the 12 commodities (milk, infant formula-milk based, infant formula-soy based, iceberg lettuce, green leaf lettuce, oranges, and grapefruit) with results form these other limited surveys. In conclusion, the recent perchlorate results from the TDS increases substantially the available data for characterizing dietary exposure and provide a useful basis for the beginning to evaluate overall perchlorate estimated dietary intakes in the US population.

Pesticides

The responsibility of FDA in the regulation of pesticides is to enforce tolerances established by the EPA in foods and feeds, except for meat, poultry, and certain egg products which fall under the statutory authority of the USDA.

FDA's pesticide monitoring program consists of 3 approaches: regulatory monitoring, incidence/level monitoring, and the TDS. This monitoring program is designed to track compliance with U.S. tolerances in domestically produced foods in interstate commerce and in imported foods offered for entry into the United States. FDA establishes monitoring priorities through development of an annual National Sampling Plan, which is a compilation of Regional Sampling Plans prepared by FDA personnel throughout the United States. Factors considered by FDA in planning the types and numbers of samples to collect include review of recently generated state and FDA residue data, regional intelligence on pesticide use, dietary importance of foods, foods eaten by infants and children, information on the amount of food entering commerce, chemical characteristics and toxicity of the pesticide.

Domestic samples are collected as close as possible to the point of production in the distribution system and imported samples are collected at the point of entry into the U.S. commerce. The emphasis of the program is on the raw agricultural product, which is analyzed as unwashed, whole, unpeeled, raw commodity. In general, fresh fruits and vegetables account for the largest proportion of commodities, comprising 75 - 80% of the total number of samples. However, some processed foods are also included. Most samples collected by FDA are of the surveillance type; that means there is no prior knowledge or evidence that a specific food shipment contains illegal pesticide residues. Compliance samples are taken as follow-up to the finding of an illegal residue or when other evidence indicates that a pesticide residue may be problematic.

To analyze the large number of samples whose pesticide treatment history is usually unknown, FDA uses analytical methods capable of simultaneously determining a number of pesticide residues. These multi-residue methods (MRMs) can determine about half of the pesticides with EPA tolerances, and many others that have no tolerances. The most commonly used MRMs can also detect many metabolites, impurities, and alteration products of pesticides. The lower limit of residue measurement of a specific pesticide is usually well below tolerance levels, which generally range from 0.1 to 50 ppm. In general, no residues are found in 60 - 70% of the domestic and imported samples analyzed under the regulatory monitoring. Violation rates for domestic samples ranges from 1 to 2 %, while for import samples, it ranges from 2 to 4 % each year.

Conclusion

The FDA has had a fairly extensive chemical contaminants program in place since the 1960s, that over time has expanded as new contaminants of concern arose. It encompasses a broad based monitoring approach, as well as targeted sampling monitoring work, and an ongoing consideration of potential risks of exposure. The chemical contaminants of concern include natural toxins, mycotoxins, pesticides, and chemical compounds of anthropogenic origin. This work ultimately serves the public health mission of the FDA to identify elevated levels of dietary exposure, and sources and pathways of exposure that can be mitigated through appropriate risk management activities.

References

1. Egan, S.K; Tao, S.S; Pennington, J.A; Bolger, P.M. U.S. Food and Drug Administration's Total Diet Study: intake of nutritional and toxic elements, 1991-96. *Food Addit Cont* **2002**, *19,* 103-25.
2. Mottram, D.S.; Wedzicha, B.L.; Dodson, A.T. 2002. Acrylamide is formed in the Maillard reaction. *Nature* **2002**, *419*, 448-449.
3. Stadler, R.H.; Blank, I.; Varga, N.; Robert, F.; Hau, J.; Guy, P.A.; Robert, M.C.; Riediker, S. Acrylamide from Maillard reaction products. *Nature* **2002,** *419*, 449-450.
4. JECFA (Joint FAO/WHO Expert Committee on Food Additives JECFA). Summary and conclusions: sixty-fourth meeting. 2005 http://www.who.int/ipcs/food/jecfa/summaries/summary_report_64_final.pdf.
5. NTP (National Toxicology Program), Center for Evaluation of Risks to Human Reproduction (CERHR). NTP-CERHR monograph on the potential human reproductive and developmental effects of acrylamide. 2005. http://cerhr.niehs.nih.gov/chemicals/acrylamide/Acrylamide_Monograph.pdf.
6. DiNovi, M. The 2006 exposure assessment for acrylamide. U.S. FDA. 2006. http://www.cfsan.fda.gov/~dms/acryexpo.html.
7. CAST (Council for Agricultural Science and Technology). Acrylamide in food. Issue paper 32. CAST, Ames, Iowa, 2006.
8. CIAA (Confederation of Food and Drink Industries of the EU). The CIAA acrylamide toolbox. Revision 11: December 2007. http://www.ciaa.be/asp/documents/brochures_form.asp?doc_id=41.
9. CCCF (Codex Committee on Contaminants in Foods). 2008. Proposed draft code of practice for the reduction of acrylamide in food (N06-2006). CX/CF 08/2/8. ftp://ftp.fao.org/codex/cccf2/cf02_08e.pdf.
10. IARC (International Agency for Research on Cancer). 1997. IARC Monographs on the evaluation of carcinogenic risks to humans. Volume 69. Polychlorinated dibenzo-para-dioxins and polychlorinated dibenzofurans. Geneva: WHO.
11. Institute of Medicine (IOM) and National Research Council (NRC), 2003. Dioxin and Dioxin-Like Compounds in the Food Supply: Strategies to Reduce Exposure. National Academy Press, Washington, DC.
12. van den Berg, M.; Birnbaum, L.; Bosveld, A.T.,; Brunstrom, B.; Cook, P.; Feeley, M.; Giesy, J.P; Hanberg, A.; Hasegawa, R.; Kennedy, S.W.; Kubiak, T.; Larsen, J.C.; van Leeuwen, F.X.; Liem, A.K.; Nolt, C.; Peterson, R.E.; Poellinger, L.; Safe, S.; Schrenk, D.; Tillitt, D.; Tyskind, M.; Younes, M.; Waern, F.; Zacharewski, T. Toxic equivalency factors (TEF) for PCBs, PCDDs, PCDFs for humans and wildlife. *Environ Health Perspect* **1998**, *106*, 775-792.

13. van den Berg, M.; Birnbaum, L.; Denison, M.; De Vito, M.; Farland, W.; Feeley, M.; Fiedler, H.; Hakansson, H.; Hanberg, A.; Haws, L.; Rose, M.; Safe, S.; Schrenk, D.; Tohyama, C.; Tritscher, A.; Tuomisto, J.; Tysklind, M.; Walker, N.; Peterson, R.E. The 2005 World Health Organization reevaluation of human and mammalian toxic equivalency factors for dioxins and dioxin-like compounds. *Toxicol Sci* **2005,** *93*, 223-241.
14. WHO (World Health Organization). 1998. Assessment of the health risks of dioxins: re-evaluation of the tolerable daily intake (TDI), Geneva: WHO.
15. JECFA (Joint FAO/WHO Expert Committee on Food Additives). 2001. Summary and conclusions of the fifty-seventh meeting. http://whqlibdoc.who.int/trs/WHO_TRS_909.pdf
16. FDA (Food and Drug Administration). 2001. Dioxins: FDA strategy for monitoring, method development, and reducing human exposure. February 7, 2002. www.cfsan.fda.gov/~lrd/dioxstra.html
17. FDA (Food and Drug Administration). 2007. Dioxin analysis results/exposure estimates. Updated November 2007. www.cfsan.fda.gov/~lrd/dioxdata.html
18. Carrington, C.; Bolger, M. An assessment of the hazards of lead in food. *Regulatory Toxicology and Pharmacology* **1992**, *16*, 265-272.
19. Carrington, C.; Sheehan, D.; Bolger, M. Hazard assessment of lead. *Food Addit Cont* **1993**, *10*, 325-335.
20. Carrington, C.; Bolger, M.; Scheuplein, R. Risk analysis of dietary lead exposure. *Food Addit Cont* **1996**, *13*, 61-76.
21. Carrington, C.; Bolger, M. A risk assessment for methylmercury in tuna. *Water, Air Soil Poll.* **1997,** *97*, 273-283.
22. Egan, S.K.; Bolger, P.M.; Carrington, C.D. Update of U.S. FDA's Total Diet Study food list and diets. *J Expo Sci Environ Epidemiol.* **2007**, *17*, 573-82.
23. JECFA (Joint Expert Committee on Food Additives), 2006 http://whqlibdoc.who.int/publications/2006/9241660554_eng.pdf.
24. JECFA (Joint Expert Committee on Food Additives), 2007 http://whqlibdoc.who.int/publications/2007/9789241660587_eng.pdf
25. Bolger, P.M.; Carrington, C.; Capar, S.; Adams, M. Reductions in dietary lead exposures in the United States. *Chemical Speciation Bioavailability* **1991**, *3*, 31-36.
26. Bolger, M.; Yess, N.; Gunderson, E.; Troxell, T.; Carrington, C. Identification and reduction of sources of dietary lead in the United States. *Food Addit Cont* **1996**, *13*, 53-60.
27. Adams, M.; Bolger, M. Dietary intake and hazards of arsenic. In *Arsenic - exposure and health, Science Technology Letters*, 1995, pp 41-49.
28. Tao, S.; Bolger, M. Dietary arsenic intakes in the United States: FDA Total Diet study, September 1991-1996, *Food Addit Cont* **1998**, *16*, 465-472.
29. Maga, J.A. Furans in foods. *Crit Rev Food Sci Nutr* **1979,** *11*, 355-400.

30. DHHS (Department of Health and Human Services), Food and Drug Administration. Furan in Food, Thermal Treatment; Request for Data and Information. *Fed Regist* **2004,** *69,* 25911-25913.
31. DiNovi, M.; Mihalov, J. 2007. An Updated Exposure Assessment for Furan from the Consumption of Adult and Baby Foods. U.S. FDA. http://www.cfsan.fda.gov/~dms/furanexp.html
32. Becalski, A.; Seaman, S. Furan precursors in food: a model study and development of a simple headspace method for determination of furan. *J AOAC Intl* **2004,** *88,* 102-106.
33. Perez-Locas, C.; Yaylayan, V.A. Origin and mechanistic pathways of formation of the parent furan--a food toxicant. *J Agric Food Chem* **2004**, *52*, 6830-6.
34. NTP (National Toxicology Program), ''Toxicology and carcinogenesis studies of furan (CAS No. 110–00–9) in F344/N rats and B6C3F1 mice (gavage studies),'' NTP Technical Report No. 402., U.S. Department of Health and Human Services, Public Health Service, National Institutes of Health, Research Triangle Park, NC, 1993.
35. Fransson-Steen, R.; Goldsworthy, T. L.; Kedderis, G. L.; Maronpot, R.R. ''Furan induced liver cell proliferation and apoptosis in female B6C3F1 mice,'' *Toxicol* **1997**, *118,* 195–204.
36. Mugford, C. A.; Carfagna, M. A.; Kedderis, G.L. ''Furan-mediated uncoupling of hepatic oxidative phosphorylation in Fischer-344 rats—an early event in cell death,'' *Toxicol Appl Pharmacol* **1997**, *144*, 1–11.
37. Mugford, C.A.; Kedderis, G.L. ''Furan-mediated DNA double strand breaks in isolated rat hepatocytes,'' *Fundamental Applied Toxicology*, **1996,** *30*, 128.
38. Wilson, D. M.; Goldsworthy, T.L.; Popp, J.A.; Butterworth, B.E. ''Evaluation of genotoxicity, pathological lesions, and cell proliferation in livers of rats and mice treated with furan,'' *Environmental and Molecular Mutagenesis*, **1992**, *19*, 209–222.
39. International Agency for Research on Cancer (IARC), Summaries and evaluations, Furan (Group 2B) (http://www.inchem.org/documents/iarc/vol63/furan.html). 1995.
40. DHHS Report on Carcinogens, Eleventh Edition, U.S. Department of Health and Human Services, Public Health Service, National Toxicology Program. http://ntp.niehs.nih.gov/ntp/roc/toc11.html., 2005.
41. EFSA (European Food Safety Authority). Report of the Scientific Panel on Contaminants in the Food Chain on provisional findings of furan in food. *The EFSA Journal* **2004**, *137*, 1-20.
42. FDA (Food and Drug Administration) . Exploratory Data on Furan in Food: Individual Food Products, August 2, 2007 update. Website: http://www.cfsan.fda.gov/~dms/furandat.html#table1.
43. Wood, G.E.; Trucksess, M.W. Regulatory control programs for mycotoxin-contaminated food. In: *Mycotoxins in Agriculture and Food Safety,* Sinha, K.K., Bhatnagar, D., Eds.; Marcel Dekker, Inc.: New York, 1998, pp 459- 481.

44. Wood, G.E.; Trucksess, M.W. and Henry, S.H. Major fungal toxins of regulatory concern. In: *International Handbook of Foodborne Pathogens,* ed. M.D. Miliotis and J.W. Bier, Marcel Dekker, Inc.: New York, 2003, pp. 423-443.
45. Wood, G.; Lee, Y.; Egan, K. and Bolger, M. US Food and Drug Administration's monitoring and surveillance programs for mycotoxins, pesticides and contaminants in food. *J. Environ. Monit.* **2003**, *3*, 79N-83N.
46. Dasgupta, P.K.; Dyke, J.V.; Kirk, A.B.; Jackson, W.A. Perchlorate in the United States. Analysis of relative source contributions to the food chain. *Environ Sci. Technol.* **2006**, *40*, 6608–6614.
47. Dasgupta, P.K.; Martinelango, P.K.; Jackson, W.A.; Anderson, T.A.; Tian, K.; Tock, R.W.; Rajagopalan, S. The Origin of Naturally Occurring Perchlorate: The Role of Atmospheric Processes. *Environ. Sci. Technol.* **2005**, *39,* 1569 – 1575.
48. Rao, B.; Anderson, T.A.; Orris, G.J.; Rainwater, K.A.; Rajagopalan, S.; Sandvig, R.M.; Scanlon, B.R.; Stonestrom, D.A.; Walvoord, M.A.; Jackson, W.A. Widespread Natural Perchlorate in Unsaturated Zones of the Southwest United States. *Environ Sci. Technol.* **2006,** *41,* 4522–4528.
49. Greer, M.A.; Goodman, G.; Pleus, R.C.; Greer, S.E. Health Effects Assessment for Environmental Perchlorate Contamination: The Dose Response for Inhibition of Thyroidal Radioiodine Uptake in Humans. *Environmental Health Perspectives* **2002**, *110*, 927-937.
50. Murray, C.W.; Egan, S. K.; Kim, H.; Beru, N.; Bolger, P.M. U.S. Food and Drug Administration's Total Diet Study: Dietary intake of perchlorate and iodine. *J Expo Sci Environ Epidemiol* **2008**, 1-10
51. Jackson, W.A.; Joseph, P.; Laxman, P.; Tan, K.; Smith, P.N.; Yu, L.; Anderson, T.A. Perchlorate Accumulation in Forage and Edible Vegetation. *J. Agric. Food Chem.* **2005**, *53,* 369 – 373
52. Kirk, A.B.; Martinelango, P.K.; Dutta, A.; Smith, E.E.; Dasgupta, P.K. Perchlorate and Iodide in Dairy and Breast Milk. *Environ. Sci. Technol.* **2005**, *39,* 2011 – 2017
53. Pearce, E.N.; Leung, A.M.; Blount, B.C.; Bazrafshan, H.R.; He, X.; Pino, S.; Valentin-Blasini, L.; Braverman, L.E. Breast Milk Iodine and Perchlorate Concentrations in Lactating Boston-area Women *J. Clin. Endocrinol. Metab.,* **2007**, *92*, 1673–1677
54. Sanchez, C.A.; Crump, K.S.; Krieger, R.I.; Khandaker, N.R.; Gibbs, J.P. Perchlorate and Nitrate in Leafy Vegetables of North America. *Environ. Sci. Technol.* **2005a**, *39,* 9391 – 9397.
55. Sanchez, C.A., Krieger, R.I.; Khandaker, N.; Moore, R.C.; Holts, K.C.; Neidel, L.L. Accumulation and Perchlorate Exposure Potential of Lettuce Produced in the Lower Colorado River Region. *J. Agric. Food Chem.* **2005b**, *53,* 5479 – 5486.
56. Sanchez, C.A.; Krieger, R.I.,; Khandaker, N.R.; Valentin-Blasini, L.; Blount, B.C. Potential perchlorate exposure from Citrus sp. irrigated with contaminated water. *Analytica Chimica Acta* **2006**, *57*, 33-38.

Chapter 3

Microbial Contamination of Fresh Produce

Susanne E. Keller

CFSAN/FDA
NCFST, Summit-Argo IL 60540

Fresh produce is typically viewed as an important part of a balanced and healthy diet. However, fresh produce can be particularly vulnerable to contamination since there is no final kill-step involved priot to consumption to ensure the complete destruction of any foodborne pathogens present. Outbreaks have involved typical foodborne pathogens such as the bacteria *Salmonella, Shigella, Listeria, Escherichia coli* O157:H7, parasites such as *Cryptosporidium* and viruses particularly Norovirus. Once pathogens are introduced, they become almost impossible to remove, consequently, an understanding of where contamination can occur and its prevention is critical to ensure the safety of the food supply.

Introduction

Fresh fruits and vegetables are a critical part of a balanced and healthy diet. The US dietary guidelines recommend the consumption of a variety of fruits and vegetables each day to meet the requirements of a healthy diet *(1)*. Fruits and vegetables are nutrient rich, particularly when consumed fresh; and have not generally been considered a high risk for foodborne illness. Therefore, it may come as a surprise that an examination of foodborne illness data reported by the CDC from 2000 to 2005 reveals that in outbreaks where the microbiological cause was known and the food source was identified, an average of 20% involved either minimally processed or fresh produce (Table 1)*(2-7)*. Consequently, to provide a safe supply of fresh produce and to reduce the rate of associated foodborne illnesses it is critical that the source as well as the cause and possible remedial steps are understood.

Table 1. Foodborne outbreaks reported to the CDC from 2001 through 2006.

Year	Total outbreak caused by microorganisms	Outbreaks with known etiology and identified vehicle	Produce related outbreaks
2001	440	264	49
2002	448	268	42
2003	355	187	41
2004	472	249	54
2005	370	233	59
2006	623	280	73
2001 to 2006	2671	1201	318

Data taken from http://www.cdc.gov/foodborneoutbreaks/outbreak_data.htm

Organisms of concern

Any foodborne pathogen including *Campylobacter* sp., *Salmonella* sp., *Shigella* sp., *Listeria monocytogenes,* enterovirulent *Escherichia coli*, *Clostridium* sp., *Yersinia* sp., parasites such as *Cryptosporidium parvum* and the noroviruses could theoretically be acquired through produce, particularly if unhygienic or other poor practices are used anywhere in their production. This includes the fields in which they are grown, their transport to and through processing, and their eventual sale and consumption. Characteristics of each pathogen and their associated foodborne illnesses can be found both on the CDC and the FDA websites. In particular, the "Bad Bug Book" located at http://www.cfsan.fda.gov/~mow/intro.html provides a good overview of foodborne pathogens in general. However, of the currently known foodborne pathogens, some have rarely or never been associated with fresh produce while others represent the bulk of the foodborne illnesses reported to the CDC each year.

The microorganisms causing the bulk of the foodborne illnesses associated with fresh produce include *Salmonella* sp., enterovirulent *Escherichia coli*, *Cryptosporidium parvum* and *Norovirus* (Norwalk or Norwalk-like). From 2001 to 2005, over 50% of outbreaks caused by fresh produce for which the etiology was identified were attributed to foodborne virus. Approximately 25% was attributed to *Salmonella* sp and another 10% was attributed to *E. coli* O157:H7. In addition, although not associated with any major outbreaks of illness associated with fresh produce in the last 10 years, *Listeria monocytogenes* contamination has resulted in considerable economic loses to the industry due to product recalls. Although none of these microorganisms is considered normal epiflora of fresh fruits or vegetables, they all share the ability to survive for fairly long periods in relatively hostile environments.

Foodborne viruses

In the years 2001 through 2005, viruses were the cause of over 40% of foodborne outbreaks for which etiology was known that were reported to the CDC. In 2006, viral outbreaks rose to 54% of those outbreaks with a confirmed etiology *(7)*. This continues a trend already apparent from 1998 to 2002 were viral pathogens, predominately norovirus increased from 16% in 1998 to 42% in 2002 *(8)*. Viruses transmitted by food or water fall into three groups, hepatovirus, enterovirus, and norovirus. Of these, the hepatovirus and norovirus (Norwalk and Norwalk-like, NLV) appear to be of greatest concern with fresh produce. The incidence of NLV occurrence in foods is unknown as detection is difficult. As with all foodborne virus, NLV does not grow in foods. Viral outbreaks are frequently the result of poor sanitation or poor worker hygiene *(9, 10)*.

Salmonella

The genus *Salmonella* contains over 2000 serotypes which can cause illness *(11)*. Of these serotypes, approximately half of all foodborne cases are caused by either Enteritidis or Typhimurium *(11)*. Although more commonly associated with animal derived foods, outbreaks associated with fresh produce have occurred with regular frequency. There are numerous animal reservoirs for *Salmonella* including domestic birds, swine, and cattle. They have also been isolated from reptiles, amphibians, fish, and insects *(12)*. *Salmonella* is resistant to desiccation and once introduced to the environment may remain viable for long periods of time. This potential for long term environmental survival was demonstrated by Danyluk et al., 2007 and may have contributed to outbreaks of Salmonellosis attributed to the consumption of almonds in late fall of 2000 to the spring of 2001 and September 2003 until April of 2004 *(13, 14)*.

Salmonella are also acid resistant and capable of growing at a pH as low as 3.8. Survival of *Salmonella* may occur even at a pH where growth is inhibited. Survival in fresh orange juice by various *Salmonella* serovars was studied by Parish et al *(15)*. Salmonella serovars Gaminara, Hartford, Rubislaw, and

Typhimurium were inoculated at log 6 cfu/ml into orange juices at various pH levels. Survival was recorded at pH 3.5 from a low of 14.3 ± 0.9 days for S. Typhimurium to a high of 26.7 ± 4.0 days for S. Hartford. Not surprisingly, Salmonellosis has been linked to the consumption of fresh juices.

Enterovirulent Escherichia coli

Escherichia coli is a thoroughly studied microorganism that is part of the normal bacterial flora resident in the intestines of many animals and is commonly used as a non-pathogenic indicator of recent fecal contamination *(16)*. However, not all *E. coli strains* are benign. Pathogenic *E. coli* all produce toxins of various types and have been described previously *(17)*. Diarrheagenic *E. coli* are further subdivided into six classes based on the virulence factors they possess and on the symptoms they produce *(18)*. Of these the group of greatest concern is the enterohemorrhagic (EHEC) class, due to its low infectious dose, and severity of symptoms produced. One of these, hemolytic uremic syndrome (HUS), occurs primarily in children under 10 years of age and has a mortality of 3 to 5%.*(18)* The severity of this disease in children led the FDA to issue a warning in November of 2001 to the public concerning the health risk of consuming untreated juices *(19)*.

Several serotypes of EHEC are known, however, the most common serotype, particularly in the United States, Canada, Great Britain and parts of Europe, is *E. coli* O157:H7 *(18)*. In the years from 2001 to 2006 approximately 33 outbreaks caused by enterohemorrhagic *E. coli* were linked to the consumption of produce *(2-7)*.

Like *Salmonella*, EHEC strains of *E. coli* are not normal endogenous microflora of produce. Their presence on fresh produce is believed to be the consequence of some form of fecal contamination prior to consumption. The major reservoir for this microorganism is believed to be cattle,*(20-23)* however, wild animals such as deer, feral pigs, and birds may be an additional sources of the organism *(24)*. Animal droppings and contaminated ground water were implicated in a 2006 outbreak that occurred due to the consumption of raw spinach *(25)*. This outbreak resulted in 205 illnesses, 31 cases of kidney failure and 3 deaths *(25)*. It also became the single largest recall of its type resulting in significant financial losses to the industry.

E. coli O157:H7 can also survive well enough at low pH to result in serious illness. Although the pH of most fruit juice is low enough to either significantly slow or inhibit growth of *E. coli*, EHEC strains have tolerance to high levels of acid allowing for extended survival time *(26, 27)*. This acid tolerance is a complex, induced response, involving several mechanisms and is enhanced in stationary phase. *E. coli* O157:H7 was the cause of the infamous apple cider outbreak that occurred in the Western states during October of 1996, resulting in 66 cases of illness and one death *(28, 29)*.

Other microbiological causes

Although not as frequent as foodborne outbreaks caused by *Salmonella* and enterohemorrahgic *E. coli*, *Shigella* sp, *Yersinia pseudotuberculosis*, *Listeria monocytogenes* and *Campylobacter jejuni* have also resulted in foodborne outbreaks related to fresh produce. *Shigella* sp., like its related microorganisms, *Salmonella* and enterohemorrahgic *E. coli*, is resistant to acidic pH and can survive at a pH as low as 2.5 *(30)*. This characteristic no doubt aids in its survival on and in many types of fresh produce. *Shigella* sp. have been isolated from oranges in surveys of fruit and juices in Mexico *(31)*. Its presence was attributed to poor hygiene. *Shigella* sp. have been the cause of outbreaks associated with the consumption of lettuce, tomatoes, and orange juice *(32-35)*.

Less is known about the foodborne pathogen *Y. pseudotuberculosis*, and it is considered by some to be a recently emerging foodborne pathogen *(36)*. *Y. pseudotuberculosis* is closely related to two other pathogenic *Yersinia* species, *Y. pestis* and *Y. enterocolitica* for which more information is available particularly related to meat and dairy contamination. *Y. pseudotuberculosis* has resulted in outbreaks due to the consumption of lettuce and raw carrots *(36, 37)*.

L. monocytogenes and *C. jejuni* are two organisms that have been common causes of foodborne outbreaks world wide. However, outbreaks that are associated with the consumption of fresh produce are rare, despite their common association with fresh produce *(38-41)*. The frequent association, particularly of *L. monocytogenes* with produce, may stem in part from its ubiquitous nature. On the other hand, *L. monocytogenes* at least, has been the subject of numerous recalls related to its presence in fresh produce *(42-45)*. The lack of confirmed outbreaks related to both *L. monocytogenes* and *C. jejuni* and associated with fresh produce may be related to difficulties involved with detection and reporting rather then a real absence of foodborne illness tied to fresh produce. On the other hand, they are clearly not as prevalent a cause of outbreaks associated with produce as are *Salmonella* and enterohemorrahgic *E. coli.*

In addition to bacterial causes, several protozoan parasites have resulted in foodborne illnesses in fresh produce. These includc *Cryptosporidium parvum, Cyclospora* and *Giardia.* *Cryptosporidium parvum* is one that is highly infectious. Common reservoirs include cattle, deer and sheep *(46, 47)*. Cryptosporidium is more commonly associated with contaminated water. The largest waterborne outbreak in U.S. history occurred in Milwaukee, WI in 1993 and affected an estimated 403,000 people *(48)*. Cryptosporidium cannot replicate in the environment; however, its oocysts are thick-walled, resistant to chlorine and persist in the environment. Presumably, the thick-wall also confers some acid resistance, as outbreaks of cryptosporidiosis have also occurred from fresh-pressed cider *(46, 47)*. Apple cider-associated outbreaks were reported in 1993, 1996 and 2003.

Cyclospora was unknown prior to 1977 *(49)*. It has since been linked to several outbreaks mostly involving imported fresh raspberries, basil, and mesclun *(49-52)*. Unlike *Cryptosporidium, Cyclospora* oocysts are not generally infectious when first excreted but become so after approximately 2 days. Consequently, human to human transfer is unlikely and the most probable route of infection is through water or food consumption. Although reported in many countries, *Cyclospora* appears most common in tropical and subtropical areas.

Less commonly reported with outbreaks linked to produce but more geographically distributed are *Giardia*. Although more commonly associated with waterborne illness, *Giardia* can be acquired through food, particularly through cross-contamination. In a study of food-handlers in Turkey, 24.6% were found to have *Giardia* in stool samples *(53)*.

Sources of Contamination

Without a final "kill" step during processing that can destroy any pathogen resident on or in the harvested produce the only means of protecting the public is prevention. Consequently, a through understanding of contamination sources and their control, particularly at the farm level, is critical to the production of safe produce.

Many of the foodborne pathogens are zoonotic. The ultimate sources for these pathogens are both wild and domestic animals. As such, access to farm fields should be restricted in so far as it is possible. Buffer zones around fields can work to discourage wildlife. Fencing can also restrict access to a limited extent, particularly for domestic livestock. The Good Agricultural Practices (GAPS) jointly issued by the FDA and USDA recommends keeping grass and weeds trimmed and removing debris to further discourage wild animals *(54)*. In addition to the GAPs, various produce industry groups have developed their own product specific guidelines incorporating many of these points *(55-57)*.

Domestic animal waste can be a serious concern if not handled properly. Raw manure can contain high levels of human pathogens and should be treated prior to use to reduce the level of pathogens *(58)*. The use of non-composted manure should not occur close to harvest. The United States Department of Agriculture National Organic Program regulations require a minimum of 120 days between the application of non-composted manure and the harvest of crops that may be exposed to the soil. However, according to a study by Ingram et al 2004 of soil fertilized with *E. coli* inoculated manure, even this time interval may be inadequate to ensure the safety of the harvested crop *(59)*. The actual length of survival of foodborne pathogens in soil is unknown. However, the potential for long term survival has been demonstrated. Uesugi et al 2007 studied the survival of *Salmonella* Enteritidis phage type 30 in an almond orchard over a five year period *(60)*. This was the same *Salmonella* strain responsible for outbreaks associated with the consumption of almonds in 2000 and 2001. *Salmonella* Enteritidis phage type 30 was recovered in each of the five years of the study. Greater recover was associated with harvest periods. No attempt was made to determine if an animal vector was in part responsible, however the recovery of the same *Salmonella* strain over the five year period would argue for its continued persistence in the environment once initially introduced. The persistence of *Salmonella* Newport in the environment was demonstrated when the same strain caused foodborne outbreaks traced to tomatoes in 2002 and in 2005 *(61)*. The tomatoes were traced to the eastern shore of Virginia where the outbreak strain was isolated from pond water used to irrigate the tomato fields. This suggests a persistent contamination of the fields. Possible animal vectors contributing to the persistence were not identified.

Composting of raw waste can successfully reduce pathogen levels prior to application to the soil as fertilizer. According to one recent study, both the temperature and moisture content of the waste is critical to the successful composting of waste *(62)*. Despite reductions of pathogens in manure prior to its use, some risk of contamination may remain. The extent of the risk is unknown. The use of manure has resulted in an increased risk of contamination with E. coli on farms located in Wisconsin and Minnesota *(63)*. The risk was increased further when raw manure was used rather then composted manure.

Foodborne pathogens can easily be transferred from their original source to water, particularly surface water which can then become a source of contamination. Rain can spread pathogens from manure throughout soil regardless of soil type, although soil type may affect the extent of spread *(64)*. Survival of foodborne pathogens in water and pond sediment can be extensive *(65)*. Numerous studies have confirmed the contamination of produce through irrigation with contaminated water. These studies were reviewed by Steele and Odumeru 2004 *(66)*. Among factors believed to increase the risk of contamination are the source of the irrigation water, with ground water being generally of higher quality then surface water. The use of waste water was tied to the highest risk of contamination. The method of irrigation can also influence the risk of contamination. Spray irrigation can result in greater contamination then drip irrigation methods *(67)*. In a study by Choi et al 2004, furrow irrigation resulted in a reduced level of contamination when compared to a subsurface irrigation system *(68)*. Moreover, contact with leaves or other visible parts of the plant may not be required for contamination to occur. *E. coli* O157:H7 has been shown to contaminate lettuce plant tissue through migration from the roots *(69)*. Similarly, *S. enterica* serovar Newport was able to contaminate Romaine lettuce through the root system in a controlled laboratory setting *(70)*. Jablasone et al 2005 demonstrated the uptake of foodborne pathogens during growth from contaminated seed in a variety of vegetables including lettuce, tomato, spinach, and carrots *(71)*. The rate at which such uptake through a root system will occur under field conditions has yet to be determined. Nonetheless, these findings illustrate the importance of preventing the initial contamination.

Irrigation water is not the only possible route of transfer from contaminated water. One additional source of contamination has been pesticide solutions made with contaminated water. Contaminated pesticide was suspected as the probable cause of contamination of Mandarin oranges *(72)*. Several different foodborne pathogens were tested for survival and growth in different pesticide solutions by Guan et al *(73)*. Although most solutions were somewhat inhibitory, *Salmonella* sp did show growth in some pesticides. *E. coli* O157:H7 and *Salmonella* mixed with pesticide and sprayed on tomato plants were recovered after the 1 day recommend pre-harvest interval. Similar results were also obtained by Ng et al *(74)*.

Water is also frequently used during harvesting to wash and/or cool produce. GAPs recommend the use of potable water for any water used in direct contact with the edible portion of the plant harvested. Furthermore, an antimicrobial agent should be used in wash water when possible to prevent the spread of any contamination. Contamination has been shown to spread through washing and/or cooling procedures. In a study examining the potential sources of contamination during the production of fresh unpasteurized apple cider, washing was cited as a major contributor to the final microbial load *(75)*. In an experiment following cross-contamination in washing lettuce, experiments were conducted to simulate washing in a restaurant setting *(76)*. Water, without agitation, was shown to allow the spread of contamination from the point of inoculation throughout a batch of over 500 grams uncontaminated lettuce. Once contaminated, three washings could not remove the pathogen.

While the quality of water used during production represents a major concern with regards to contamination of fresh produce, it is not the only concern. Worker hygiene can also impact contamination both in the field and in subsequent processing. Worker hygiene has been implicated in numerous foodborne outbreaks including Shigellosis from orange juice, norovirus on cruise ships, and Salmonellosis from several sources *(9, 35, 77, 78)*.

Hand washing facilities and toilets should be available to field workers as well as to workers located in processing facilities. Training should be provided to all food service workers. The benefit of such training was noted by Kassa et al who found higher sanitation scores for restaurants with trained employees *(79)*. An association between foodborne outbreaks and food service operations with food health code violations has been recognized *(80)*.

Interventions

Treatments/procedures to destroy or remove foodborne pathogens from produce can be divided into two categories: those that target only the surface of the fruit or vegetable and those that will be effective both at the surface and in the interior. Both categories include treatments that can be extremely effective in meeting the objectives of their designed purpose.

In the first category (surface-directed agents or methods) reside procedures such as washing and brushing, treatment with chemical sanitizers such as chlorine or chlorine dioxide, ozone and many other sanitizing agents *(81-116)*. Typical reductions reported range from 2 log to greater then 5 log. Differences in reported ranges may be more related to the method by which they were evaluated rather then any real difference in efficacy. Differences in reported efficacy related to methods used were clearly demonstrated by Fleischman et al., who used two different methods to inoculate apples *(117)*. In the first method apples were inoculated by submerging fruit in a liquid culture and dried. This resulted in a maximum of 2-log reductions when the fruit was treated by briefly submerging the fruit into hot water. Using a second procedure where drops of inoculum were placed on the surface and allowed to dry, the same treatment resulted in as much as an 8-log reduction. Clearly with the first procedure, substantial infiltration of the pathogen inside the fruit occurred, thus protecting the organism from the lethal effect of the surface heat treatment.

Internalization of foodborne pathogens, or their residence in some other protected state, is well-documented, although the extent of such occurrences in vivo may be unknown *(71, 118-131)*. Although many surface treatments will provide efficacy against loosely attached or free bacteria in solution, efficacy against biofilms and internalized microorganisms will be greatly reduced or absent. Fatemi and Knabel *(96)* examined sanitizer penetration and destruction of *E. coli* O157:H7 in Golden Delicious apples. They concluded that although some increases in pathogen reductions could be achieved with agitation, *E. coli* O157:H7 continued to survive and may have been protected by crevices or hydrophobic interactions. Consequently, the success or failure of any surface treatment for the decontamination of produce depends upon contact of the treatment with the pathogen, not upon contact of the treatment with the produce. When the pathogen resides in a protected niche, either through attachment in crevices, in biofilms, or by becoming internalized into the produce, surface treatment will be ineffective.

Because of inflitration into fruit and vegetables and/or the presence of biofilms, the only completely effective intervention treatment for produce is one that treats the entire product, rather than just the surface. Currently, two such treatment types exist; thermal treatment and irradiation. Unfortunately, thermal treatments will render the produce "cooked" and no longer fresh, therefore they are not useful for fresh products. Although irradiation could leave many types of produce with a "fresh" appearance and "fresh" organoleptic properties, as of this writing, irradiation is not approved as a food additive for use on fresh produce (21 CFR 179.26). [Didn't this change this past summer? I thought it resulted in a lot of controversy?] Irradiation at up to 1 kiloGray (kGy) may be used for growth and maturation inhibition of fresh foods and for deinfestation of arthropod pests, but this level alone is unlikely to prove efficacious in the disinfection of fresh produce *(132, 133)*. However, low level irradiation may be combined with other methods with some success for some types of produce (134). However, it should be noted that acceptance by the public may be limited.

In contrast to commonly applied intervention methods, post-harvest preventive measures, can be effective. Removal of damaged produce has been associated with lower levels of bacterial microflora in general as well as lower levels of pathogens *(135-141)*. In addition, numerous treatments have shown efficacy in preventing the spread of contamination *(84, 86, 96, 142-144)*. Temperature control and modified atmosphere packaging have also proved efficacious in preventing the multiplication of foodborne pathogens in produce *(128)*. While none of these methods alone may provide complete protection, when taken as a whole they can significantly impact the risk of contamination and subsequent outbreaks. It was for this reason that the GAPs guidance was jointly formulated by the USDA and the FDA *(54)*. More recently, various produce industry groups have also provided their own additional commodity specific guidelines *(55-57)*. It is projected that with adherence of producers and processors to these guidelines the risk of foodborne illness from the consumption of fresh produce will be reduced.

Summary

Currently available intervention technologies are largely ineffective in the complete removal of foodborne pathogens from fresh produce. Consequently, the sole means to ensure safe fresh produce for the public is through the prevention of contamination.

Unfortunately, the primary sources of foodborne pathogens are frequently zoonotic. The complete removal of all wild life from farm fields is clearly not possible. Even where restriction of animal activity is more easily accomplished, such as in greenhouses, risk remains, particularly if water quality is not strictly monitored. In addition, human hygiene must be controlled to further reduce the risk of contamination. A through understanding of the source and various routes of contamination can lead to better control of these risks. To this end the US government and the industry itself has provided guidance to allow producers and processors greater understanding and hence control of the risks involved. Through adherence to the best and most current standards provided by these guidances the risk involved with the consumption of fresh produce should be reduced to the point where foodborne outbreaks become a rare occurrence.

References

1. U. S. Health and Human Services; U. S. Department of Agriculture Dietary guidelines for Americans. www.healthierus.gov/dietaryguidelines (6/11/2008).
2. Centers for Disease Control and Prevention. 2001 Summary Statistics. http://www.cdc.gov/foodborneoutbreaks/us_outb/fbo2001/2001linelists.pdf (June 19, 2001).
3. Centers for Disease Control and Prevention. 2002 Summary Statistics. http://www.cdc.gov/foodborneoutbreaks/us_outb/fbo2002/2002linelist.pdf (June 19, 2008).
4. Centers for Disease Control and Prevention. 2003 Summary statistics. http://www.cdc.gov/foodborneoutbreaks/documents/2006_line_list/2006_line_list.pdf (June 19, 2008).

5. Centers for Disease Control and Prevention. 2004 Summary statistics. http://www.cdc.gov/foodborneoutbreaks/us_outb/fbo2004/Outbreak_Linelist_Final_2004.pdf (27 Nov 2007).
6. Centers for Disease Control and Prevention. 2005 Annual listing of foodborne disease outbreaks, United States. http://www.cdc.gov/foodborneoutbreaks/us_outb/fbo2005/2005_Linelist.pdf (Dec 6 2007).
7. Centers for Disease Control and Prevention. 2006 Annual listing of Foodborne disease outbreaks, United States. http://www.cdc.gov/foodborneoutbreaks/documents/2006_line_list/2006_line_list.pdf (June 19, 2008).
8. Lynch, M.; Painter, J.; Woodruff, R.; Braden, C. Centers for Disease Control and Prevention, Surveillance for foodborne-disease outbreaks--United States, 1998-2002. *MMWR CDC Surveillance Summaries* **2006,** 55, 1-42.
9. Herwaldt, B. L.; Lew, J. F.; Moe, C. L.; Lewis, D. C.; Humphrey, C. D.; Monroe, S. S.; Pon, E. W.; Glass, R. I. Characterization of a variant strain of norwalk virus from a foodborne outbreak of gastroenteritis on a cruise ship in Hawaii. *J. Clin. Micro.* **1994,** 32, 861-866.
10. Einstein, A. B.; Jacobsohn, W.; Goldman, A. An epidemic of infectious hepatitis in a general hospital: probable transmission by contaminated orange juice. *JAMA* **1963,** 185, 171-174.
11. Centers for Disease Control and Prevention. Salmonellosis. http://www.cdc.gov/ncidod/dbmd/diseaseinfo/salmonellosis_t.htm (Oct 23).
12. ICMSF. Salmonellae. In *Microorganisms in Foods: Microbiological specifications of food pathogens,* Roberts, T. A.; Baird-Parker, A. C.; Tompkin, R. B., Eds. Chapman and Hall: New York, 1996; Vol. 5, 217-264.
13. Centers for Disease Control and Prevention. Outbreak of Salmonella Serotype Enteritidis Infections Associated with Raw Almonds --- United States and Canada, 2003--2004. *Morbid. Mortal. Weekly Rep.* **2004,** 53, 484-487.
14. Isaacs, S.; Aramini, J.; Ciebin, B.; Farrar, J. A.; Ahmed, R.; Middleton, D.; Chandran, A. U.; Harris, L. J.; Howes, M.; Chan, E.; Pichette, A. S.; Campbell, K.; Gupta, A.; Lior, L. Y.; Pearce, M.; Clark, C.; Rodgers, F.; Jamieson, F.; Brophy, I.; Ellis, A. An international outbreak of salmonellosis associated with raw almonds contaminated with a rare phage type of *Salmonella enteritidis. J. Food Prot.* **2005,** 68, 191-198.
15. Parish, M. E.; Narcisco, J. A.; Fredrich, L. M. Survival of Salmonellae in orange juice. *J. Food Saf.* **1997,** 17, 273-281.
16. Edberg, S. C.; Rice, E. W.; Karlin, R. J.; Allen, M. J. *Escherichia coli*: the best biological drinking water indicator for public health protection. *J. Appl. Micro.* **2000,** 88, 106S-116S.
17. Chart, H. Toxigenic *Escherichia coli. J. Appl. Micro.* **1998,** 84, (Symposium Supplement), 77S-86S.

18. Buchanan, R. L.; Doyle, M. E. Foodborne disease significance of *Escherichia coli* O157:H7 and other enterohemorrhagic *E. coli*. *Food Tech.* **1997,** 51, 69-76.
19. U. S. Food and Drug Administration. Untreated juice may pose health risk to children. http://www.cfsan.fda.gov/~dms/juicsaf2.html (Oct 17).
20. Whipp, S. C.; Rasmussen, M. A.; Cray, W. C. Public veterinary medicine: Food safety and handling. Animals as a source of *Escherichia coli* pathogenic for human beings. *J. Amer. Veter. Med. Assoc.* **1994,** 204, 1168-1175.
21. Zhao, T.; Doyle, M. E.; Shere, J.; Garber, L. Prevalence of enterohemorrhagic *Escherichia coli* O157:H7 in a survey of dairy herds. *Appl. Environ. Micro.* **1995,** 61, 1290-1293.
22. Hancock, D. D.; Rice, D. H.; Herriott, D. E.; Besser, T. E.; Ebel, E. D.; Carpenter, L. V. Effects of farm manure-handling practices on *Escherichia coli* O157:H7 prevalence in cattle. *J. Food Prot.* **1997,** 60, 363-366.
23. Tauxe, R. V. Emerging foodborne dieseases:An evolving public health challenge. *Emerging Infectious Diseases* **1997,** 3, 425-434.
24. Rice, D. H.; Besser, T. E. Verotoxigenic E. coli O157 colonisation of wild deer and range cattle. *Veterinary Record* **1995,** 137, 524.
25. U. S. Food and Drug Administration. Case study--Summary chronology Escherichia coli O157:H7 outbreak in spinach United States September 2006. http://www.fda.gov/oia/Mepi/Handouts/Spinach-Outline.pdf (June 20, 2008).
26. Benjamin, M. M.; Datta, A. R.. Acid tolerance of enterohemorrhagic *Escherichia coli*. *Appl. Environ. Micro.* **1995,** 61, 1669-1672.
27. Buchanan, R. L.; Edelson, S. G. Culturing enterohemorrhagic *Escherichia coli* in the presence and absense of glucose as a simple means of evaluating the acid tolerance of stationary-phase cells. *Appl. Environ. Micro.* **1996,** 62, 4009-4013.
28. Centers for Disease Control and Prevention. Outbreaks of *Escherichia coli* O157:H7 infections associated with drinking unpasteurized commercial apple juice-October, 1996. *MMWR* **1996,** 45, 975.
29. Deliganis, C. V. Death by apple juice: The problem of foodborne illness, the regulatory response, and further suggestions for reform. *Food and Drug Law Journal* **1998,** 53, 681-728.
30. Lin, J.; Lee, I. S.; Frey, J.; Slonczewski, J. L.; Foster, J. W. Comparative analysis of extreme acid survival in *Salmonella typhimurium*, *Shigella flexneri*, and *Escherichia coli*. *J. Bacter.* **1995,** 177, 4097-4104.
31. Castillo, A.; Villarruel-Lopez, A.; Navarro-Hidalgo, V.; Martinez-Gonzalez, N. E.; Torres-Vitela, M. R. *Salmonella* and *Shigella* in freshly squeezed orange juice, fresh oranges, and wiping cloths collected from public markets and street booths in Guadalajara, Mexico: Incidence and comparison of analytical routes. *J. Food Prot.* **2006,** 69, 2595-2599.

32. Davis, H.; Taylor, J. P.; Perdue, J. N.; Stelma, G. N. J.; Humphreys, J. M. J.; Rowntree, R. r.; Greene, K. D. A shigellosis outbreak traced to commercially distributed shredded lettuce. *Amer. J. Epidem.* **1988,** 128, 1312-1321.
33. Kapperud, G.; Rørvik, L. M.; Hasseltvedt, V.; Høiby, E. A.; Iversen, B. G.; Staveland, K.; Johnsen, G.; Leitao, J.; Herikstad, H.; Andersson, Y.; Langeland, G.; Gondrosen, B.; Lassen, J. Outbreak of *Shigella sonnei* infection traced to imported iceberg lettuce. *J. Clin. Micro.* **1995,** 33, 609-614.
34. Reller, M. E.; Nelson, J. M.; Molbak, K.; Ackman, D. M.; Schoonmaker-Bopp, D. J.; Root, T. P.; Mintz, E. D. A large, multiple-restaurant outbreak of infection with *Shigella flexneri* Serotype 2a traced to tomatoes. *Clin.Infect.Dis.* **2006,** 42, 163-169.
35. Thurston, H.; Stuart, J.; McDonnell, B.; Nicholas, S.; Cheasty, T. Fresh orange juice implicated in an outbreak of *Shigella flexneri* among visitors to a South African game reserve. *J. Infect. Dis.* **1998,** 36, 350.
36. Jalava, K.; Hakkinen, M.; Valkonen, M.; Nakari, U.; Palo, T.; Hallanvuo, S.; Ollgren, J.; Siitonen, A.; Nuorti, J. P. An outbreak of gastrointestinal illness and erythema nodosum from grated carrots contaminated with *Yersinia pseudotuberculosis*. *J. Infect. Dis.* **2006,** 194, 1209-1216.
37. Nuorti, J. P.; Niskanen, T.; Hallanvuo, S.; Mikkola, J.; Kela, E.; Hatakka, M.; Fredriksson-Ahomaa, M.; Lyytikainen, O.; Siitonen, A.; Korkeala, H.; Ruutu, P. A widespread outbreak of Yersinia pseudotuberculosis O:3 infection from iceberg lettuce. *J. Infect. Dis.* **2004,** 189, 766-774.
38. Crepet, A.; Albert, I.; Dervin, C.; Carlin, F. Estimation of microbial contamination of food from prevalence and concentration data: Application to *Listeria monocytogenes* in fresh vegetables. *Appl. Environ. Micro.* **2007,** 73, 250-258.
39. Gombas, D. E.; Chen, Y.; Clavero, R. S.; Scott, V. N. Survey of *Listeria monocytogenes* in ready-to-eat foods. *J. Food Prot.* **2003,** 66, 559-569.
40. Kärenlampi, R.; Hänninen, M. L. Survival of *Campylobacter jejuni* on various fresh produce. *Intern. J. Food Micro.* **2004,** 97, 187-195.
41. Park, C. E.; Sanders, G. W. Occurrence of thermotolerant campylobacters in fresh vegetables sold at farmers' outdoor markets and supermarkets. *Can. J.Micro.***1992,** 38, 313-316.
42. Anonymous. Duck Delivery Produce, Inc. recalls apple snack trays because of possible health risk. http://www.fda.gov/oc/po/firmrecalls/duck3_01.html (27 Nov 2007).
43. Anonymous. Duck Delivery Produce recalls cut honeydew and cut cantaloupe melon for possible health risk. http://www.fda.gov/oc/po/firmrecalls/duck07_03.html (27 Nov 2007).
44. Anonymous. Georgia Ag department finds contaminated mushrooms. http://www.fda.gov/oc/po/firmrecalls/southmill08_03.html (27 Nov 2007).

45. Anonymous. Monterey Mushrooms recalls fresh sliced white and baby bella mushrooms in PA, MD, NC, NJ, NY, OH, and VA because of possible health risk. http://www.fda.gov/oc/po/firmrecalls/monterey09_08.html (27 Nov 2007).
46. Centers for Disease Control and Prevention. Outbreaks of *Escherichia coli* O157:H7 infection and cryptosporidiosis associated with drinking unpasteurized apple cider-Connecticut and New York, October 1996. *MMWR* **1997,** 46, 4-8.
47. Millard, P. S.; Gensheimer, K. F.; Addiss, D. G.; Sosin, D. M.; Beckett, G. A.; Houck-Jankoski, A.; Hudson, A. An outbreak of Cryptosporidiosis from fresh-pressed apple cider. *JAMA* **1994,** 272, 1592-1596.
48. Guerrant, R. L. Cryptosporidiosis: An emerging highly infectious threat. *Emerging Infect. Dis.* **1997,** 3, 51-57.
49. Herwaldt, B. L. *Cyclospora cayetanensis*: A review, focusing on the outbreaks of cyclosporiasis in the 1990s. *Clin. Infect.Dis.* **2000,** 31, 1040-1057.
50. Centers for Disease Control and Prevention. Update: Outbreaks of Cyclosporiasis--United States and Canada, 1997. *MMWR.* **1997,** 46, 521-523.
51. Centers for Disease Control andPrevention. Outbreak of cyclosporiasis associated with snow peas --- Pennsylvania, 2004. *MMWR.* **2004,** 53, 1-3.
52. Döller, P. C.; Dietrich, K.; Filipp, N.; Brockmann, S.; Dreweck, C.; Vonthein, R.; Wagner-Wiening, C.; Wiedenmann, A. Cyclosporiasis outbreak in Germany associated with the consumption of salad. *Emerging Infect. Dis.* **2002,** 8, 992-994.
53. Gündüz, T.; Limoncu, M. E., .; Cümen, S.; Ari, A.; Serdağ, E.; Tay, Z. The prevalence of intestinal parasites and nasal *S. aureus* carriage among food handlers. *J. Environ. Health* **2008,** 70, 64-65, 67.
54. U. S. Food and Drug Administration. Guide to minimize microbial food safety hazards for fresh fruits and vegetables. http://vm.cfsan.fda.gov/~dms/prodguid.html (6 Dec 2007).
55. International Fresh-cut Produce Association; Produce Marketing Association; United Fresh Fruit and Vegetable Association; Western Growers Commodity specific food safety guidelines for the lettuce and leafy greens supply chain. http://www.cfsan.fda.gov/~dms/lettsup.html
56. North American Tomato Trade Work Group Commodity specific food safety guidlines for the fresh tomato supply chain. http://www.cfsan.fda.gov/~dms/tomatsup.html (July 2, 2008).
57. Produce Marketing Association; United Fresh Fruit and Vegetable Association Commodity specific food safety guidlines for the melon supply chain. http://www.cfsan.fda.gov/~dms/melonsup.html (July 2, 2008).
58. Cole, D. J.; Hill, V. R.; Humenik, F. J.; Sobsey, M. D. Health, safety and environmental concerns of farm animal waste. *Occupational Medicine* **1999,** 14, 423-428.

59. Ingham, S. C.; Losinski, J. A.; Andrews, M. P.; Breuer, J. E.; Breuer, J. R.; Wood, T. M.; Wright, T. H. *Escherichia coli* contamination of vegetables grown in soils fertilized with noncomposted bovine manure: Garden-scale studies. *Appl. Environ. Micro.* **2004,** 70, 6420 - 6427.
60. Uesugi, A. R.; Danyluk, M. D.; Mandrell, R. E.; Harris, L. J. Isolation of *Salmonella* Enteritidis Phage Type 30 from a single almond orchard over a 5-year period. *J. Food Prot.***2007,** 70, 1784-1789.
61. Greene, S. K.; Daly, E. R.; Talbot, E. A.; Demma, L. J.; Holzbauer, S.; Patel, N. J.; Hill, T. A.; Walderhaug, M. O.; Hoekstra, R. M.; Lynch, M. F.; Painter, J. A. Recurrent multistate outbreak of Salmonella Newport associated with tomatoes from contaminated fields, 2005. *Epidemiology and Infection* **2008,** 136, 157-165.
62. Ceustermans, A.; DeClercq, D.; Aertsen, A.; Michiels, C.; Coosemanns, J.; Ryckeboer, J. Inactivation of *Salmonella* Senftenberg strain W 775 during composting of biowastes and garden wastes. *J. Appl. Micro.* **2007,** 103, 53-64.
63. Mukherjee, A.; Speh, D.; Dyck, E.; Diez-Gonzales, F. Preharvest evaluation of coliforms, *Escherichia coli*, *Salmonella*, and *Escherichia coli* O157:H7 in organic and conventional produce grown by Minnesota farmers. *J. Food Prot.* **2004,** 67, 894-900.
64. Gagliardi, J. V.; Karns, J. S. Leaching of Escherichia coli O157:H7 in diverse soils under various agricultural practices. *Appl. Environ Micro.* **2000,** 66, 877-883.
65. Moore, B. C.; MArtinez, E.; Gay, J. M.; Rice, D. H. Survival of *Salmonella enterica* in freshwater and sediments and transmission by the aquatic midge *Chironomus tentans* (Chironomidae: Diptera). *Appl. Environ. Micro.* **2003,** 69, 4556-4560.
66. Steele, M.; Odumeru, J. Irrigation water as source of foodborne pathogens on fruit and vegetables. *J. Food Prot.* **2004,** 67, 2839-2849.
67. Solomon, E. B.; Potenski, C. J.; Matthews, K. R. Effect of irrigation method on transmission to and persistence of *Escherichia coli* O157:H7 on lettuce. *J. Food Prot.* **2002,** 65, 673-676.
68. Choi, C.; Song, I.; Stine, S.; Pimentel, J.; Gerba, C. Role of irrigation and wastewater reuse: comparison of subsurface irrigation and furrow irrigation. *Water Science and Technology* **2004,** 50, 61-68.
69. Solomon, E. B.; Yaron, S.; Matthews, K. R. Transmission of *Escherichia coli* O157:H7 from contaminated manure and irrigation water to lettuce plant tissue and its subsequent internalization. *Appl.Environ. Micro* **2002,** 68, 397-400.
70. Bernstein, N.; Sela, S.; Neder-Lavon, S. Assessment of contamination potential of lettuce by *Salmonella enterica* serovar Newport added to the plant growing medium. *J. Food Prot.* **2007,** 70, 1717-1722.
71. Jablasone, J.; Warriner, K.; Griffiths, M. Interactions of *Escherichia coli* O157:H7, *Salmonella typhimurium* and *Listeria monocytogenes* plants cultivated in a gnotobiotic system. *Int. J. Food Micr.* **2005,** 99, 7-18.

72. Poubol, J.; Tsukada, Y.; Sera, K.; Izumi, H. On-the-farm contamination of Satsuma mandarin fruit at different orchards in Japan. *Acta Hort.* **2006,** 712, 551-560.

73. Gaun, T. T. Y.; Blank, G.; Holley, R. A. Survival of pathogenic bacteria in pesticide solutions and on treated tomato plants *J. Food Prot* **2005,** 68, 296-304.

74. Ng, P. J.; Fleet, G. H.; Heard, G. M. Pesticides as a source of microbial contamination of salad vegetables. *Int J Food Micr* **2005,** 101, 237-250.

75. Garcia, L.; Henderson, J.; Fabri, M.; Moustapha, O. Potential sources of microbial contamination in unpasteurized apple cider. *J Food Prot* **2006,** 69, 137-144.

76. Wachtel, M. R.; Charkowski, A. O. Cross-contamination of lettuce with *Escherichia coli* O157:H7. *J Food Prot* **2002,** 65, 465-470.

77. Birkhead, G. S.; Morse, D. L.; Levine, W. C.; Fudala, J. K.; Kondracki, S. F.; Chang, H.; Shayegani, M.; Novick, L.; Blake, P. A. Typhoid fever at a resort hotel in New York: A large outbreak with an unusual vehicle. *J. Infect Dis* **1993,** 167, 1228-12232.

78. Todd, E. C. D.; Greig, J. D.; Bartleson, C. A.; Michaels, B. S. Outbreaks where food workers have been implicated in the spread of foodborne disease. Part 2. Description of outbreaks by size, severity, and settings. *J. Food Prot.* **2007,** 70, 1975-1993.

79. Kassa, H.; Harrington, B.; Bisesi, M.; Khuder, S. Comparison of the microbiological evaluation of selected kitchen areas with visual evaluation for preventing potential risk of foodborne outbreaks in food service operations. *J. Food Prot.* **2001,** 64, 509-513.

80. Kassa, H. An outbreak of Norwalk-like viral gastroenteritis in a frequently penalized food service operation: a case for mandatory training of food handlers in safety and hygiene. *J. Environ. Health* **2001,** 64, 9-12, 33.

81. Sengun, I. Y.; Karapinar, M. Effectiveness of household natural sanitizers in the elimination of *Salmonella typhimurium* on rocket (*Eruca sativa* Miller) and spring onion (*Allium cepa* L.). *Int. J. Food Micro* **2005,** 98, 319-323.

82. Sengun, I. Y.; Karapinar, M. Effectiveness of lemon juice, vinegar and their mixture in the elimination of *Salmonella typhimurium* on carrots (*Daucus carota* L.). *Int. J. Food Micro* **2004,** 96, 301-305.

83. Pirovani, M. E.; Guemes, D. R.; Di Pentima, J. H.; Tessi, M. A. Survival of *Salmonella hadar* after washing disinfection of minimally processed spinach. *Lett. Appl. Micro* **2000,** 31,143-148.

84. Parnell, T. L.; Harris, L. J.; Suslow, T. V. Reducing *Salmonella* on cantaloupes and honeydew melons using wash practices applicable to postharvest handling, foodservice, and consumer preparation. *Inter. J. Food Micro.* **2005,** 99, 59-70.

85. Lanciotti, R.; Belletti, N.; Patrignani, F.; Gianotti, A.; Gardini, F.; Guerzoni, M. E. Application of hexanal, (E)-2-hexenal, and hexyl acetate to improve the safety of fresh-sliced apples. *J Agric Food Chem* **2003,** 51, 2958-2963.

86. Ongeng, D.; Devlieghere, F.; Debevere, J.; Coosemans, J.; Ryckeboer, J. The efficacy of electrolysed oxidising water for inactivating spoilage microorganisms in process water and on minimally processed vegetables. *Int. J. Food Micro* **2006,** 109, 187-197.
87. Pierre, P. M.; Ryser, E. T. Inactivation of *Escherichia coli* O157:H7, *Salmonella* Typhimurium DT104, and *Listeria monocytogenes* on inoculated alfalfa seeds with a fatty acid–based sanitizer. *J. Food Prot* **2006,** 69, 582-290.
88. Kumar, M.; Hora, R.; Kostrzynska, M.; Waites, W. M.; Warriner, K. Inactivation of Escherichia coli O157:H7 and Salmonella on mung beans, alfalfa, and other seed types destined for sprout production by using an oxychloro-based sanitizer. *J. Food Prot* **2006,** 69, 1571-1578.
89. Kreske, A. C.; Ryu, J.-H.; Beuchat, L. R. Evaluation of chlorine, chlorine dioxide, and a peroxyacetic acid–based sanitizer for effectiveness in killing Bacillus cereus and Bacillus thuringiensis spores in suspensions, on the surface of stainless steel, and on apples. *J. Food Prot* **2006,** 69, 1892-1903.
90. Koseki, S.; Isobe, S. Effect of ozonated water treatment on microbial control and on browning of iceberg lettuce (*Lactuca sativa* L.). *J Food Prot.* **2006,** 69, 154-160.
91. Kim, H.; Ryu, J.-H.; Beuchat, L. R. Survival of *Enterobacter sakazakii* on fresh produce as affected by temperature, and effectiveness of sanitizers for its elimination. *Int. J. Food Micro* **2006,** 111, 134-143.
92. Kenney, S. J.; Beuchat, L. R. Comparison of aqueous commercial cleaners for effectiveness in removing *Escherichia coli* O157:H7 and *Salmonella muenchen* from the surface of apples. *Int. J. Food Micro* **2002,** 74, 47-55.
93. Jaquette, C. B.; Beuchat, L. R.; Mahon, B. E. Efficacy of chlorine and heat treatment in killing *Salmonella stanley* inoculated onto alfalfa seeds and growth and survival of the pathogen during sprouting and storage. *Appl. Environ. Micro* **1996,** 62, 2212-2215.
94. Hellstrom, S.; Kervinen, R.; Lyly, M.; Ahvenainen-Rantala, R.; Korkeala, H. Efficacy of disinfectants to reduce Listeria monocytogenes on precut iceberg lettuce. *J Food Prot* **2006,** 69, 1565 - 1570.
95. Gandhi, M.; Matthews, K. R. Efficacy of chlorine and calcinated calcium treatment of alfalfa seeds and sprouts to eliminate *Salmonella*. *Int. J. Food Micr.* **2003,** 87, 301-306.
96. Fatemi, P.; Knabel, S. J. Evaluation of sanitizer penetration and its effect on destruction of *Escherichia coli* O157:H7 in golden delicious apples. *J. Food Prot* **2006,** 69, 548-555.
97. Adams, M. R.; Hartley, A. D.; Cox, L. J. Factors affecting the efficacy of washing procedures used in the production of prepared salads. *Food Micro* **1989,** 6, 69-77.
98. Venkitanarayanan, K. S.; Lin, C.; Bailey, H.; Doyle, M. E. Inactivation of *Escherichia coli* O157:H7, *Salmonella* Enteritidis and *Listeria monocytogenes* on apples, oranges, and tomatoes by lactic acid and hydrogen peroxide. *J. Food Prot.* **2002,** 65, 100-105.

99. Wright, J. R.; Sumner, S. S.; Hackney, C. R.; Pierson, M. D.; Zoecklein, B. W. Reduction of *Escherichia coli* O157:H7 on apples using wash and chemical sanitizer treatments. *Dairy Food Environ. Sanit* **2000,** 20, 120-126.

100. Kenney, S. J.; Beuchat, L. R. Survival of *Escherichia coli* O157:H7 and *Salmonella* Muenchen on apples as affected by application of commercial fruit waxes. *Int. J. Food Micro* **2001,** 77, 223-231.

101. Pao, S.; Davis, C. L. Maximizing microbiological quality of fresh orange juice by processing sanitation and fruit surface treatments. *Dairy Food Environ. Sanit* **2001,** 21, 287-291.

102. Sapers, G. M.; Simmons, G. F. Hydrogen peroxide disinfection of minimally procesed fruits and vegetables. *Food Tech* **1998,** 52, 48-52.

103. Zhang, S.; Farber, J. M. The effects of various disinfectants against *Listeria monocytogenes* on fresh-cut vegetables. *Food Micro* **1996,** 13, 311-321.

104. Velazquez, L. D.; Escudero, M. E.; Digenaro, M. S.; Decortinez, Y. M.; Deguzman, A. M. S. Survival of *Aeromonas hydrophila* in fresh tomatoes (*Lycopersicum esculentum* mill) stored at different temperatures and treated with chlorine. *J. Food Prot* **1998,** 61, 414-416.

105. Fallik, E.; Aharoni, Y.; Copel, A.; Rodov, V.; Tuvia-Alkalai, S.; Horev, B.; Yekutieli, O.; Wiseblum, A.; Regev, R. Reduction of postharvest losses of Galia melon by a short hot-water rinse. *Plant Path.* **2000,** 49, 333-338.

106. Wisniewsky, M., A. ; Glatz, B. A.; Gleason, M. L.; Reitmeier, C. A. Reduction of *Escherichia coli* O157 : H7 counts on whole fresh apples by treatment with sanitizers. *J. Food Prot.* **2000,** 63, 703-708.

107. Pao, S.; Davis, C. L.; Kelsey, D. F. Efficacy of alkaline washing for the decontamination of orange fruit surfaces inoculated with *Escherichia coli. J.Food Prot.* **2000,** 63, 961-964.

108. Han, Y.; Linton, R. H.; Nielsen, S. S.; Nelson, P. E. Inactivation of Escherichia coli O157 : H7 on surface-uninjured and -injured green pepper (Capsicum annuum L.) by chlorine dioxide gases demonstrated by confocal laser scanning microscopy. *Food Micro.* **2000,** 17, 643-655.

109. Takeuchi, K.; Frank, J. F. Direct microscopic observation of lettuce leaf decontamination with a prototype fruit and vegetable washing solution and 1% NaCl-$NaHCO_3$. *J. Food Prot.* **2001,** 64, 1235-1239.

110. Smid, E. J.; Hendriks, L.; Boerrigter, H. A. M.; Gorris, L. G. M. Surface disinfection of tomatoes using the natural plant compound trans-cinnamaldehyde. *Postharvest Biology and Technology* **1996,** 9, 343-350.

111. Ukuku, D. O.; Pilizota, V.; Sapers, G. M., Influence of washing treatment on native microflora and *Escherichia coli* population of inoculated cantaloupes. *J. Food Saf.* **2001,** 21, 31-47.

112. Sy, K. V.; Murray, M. B.; Harrison, M. D.; Beuchat, L. R. Evaluation of gaseous chlorine dioxide as a sanitizer for killing *Salmonella*, *Escherichia coli* O157:H7, *Listeria monocytogenes*, and yeasts and

molds on fresh and fresh-cut produce. *J. Food Prot.* **2005,** 68, 1176-1187.

113. Deza, M. A.; Araujo, M.; Garrido, M. J. Inactivation of *Escherichia coli* O157:H7, *Salmonella enteritidis* and *Listeria monocytogenes* on the surface of tomatoes by neutral electrolyzed water. *Lett. Appl. Micro.* **2003,** 37, 482-487.

114. Beuchat, L. R.; Nail, B. V.; Adler, B. B.; Clavero, M. R. S. Efficacy of spray application of chlorinated water in killing pathogenic bacteria on raw apples, tomatoes, and lettuce. *J. Food Prot.* **1998,** 61, 1305-1311.

115. Weissinger, W. R.; Chantarapanont, W.; Beuchat, L. R. Survival and growth of Salmonella baildon in shredded lettuce and diced tomatoes, and effectiveness of chlorinated water as a sanitizer. *Int. J. Food Micro.* **2000,** 62, 123-131.

116. Beuchat, L. R.; Nail, B. V.; Clavero, M. R. S. Efficacy of chlorinated water in killing pathogenic bacteria on raw apples, tomatoes, and lettuce. *J. Food Prot.* **1998,** 61, 1305-1311.

117. Fleischman, G. J.; Bator, C.; Merker, R.; Keller, S. E. Hot water immersion to eliminate *Escherichia coli* O157:H7 on the surface of whole apples: thermal effects and efficacy. *J. Food Prot.* **2001,** 64, 451-455.

118. Penteado, A. L.; Eblen, B. S.; Miller, A. J. Evidence of Salmonella internalization into fresh mangos during simulated postharvest insect disinfestation procedures. *J. Food Prot.* **2004,** 67, 181-184.

119. Eblen, B. S.; Walderhaug, M. O.; Edelson-Mammel, S.; Chirtel, S. J.; De Jesus, A.; Merker, R. I.; Buchanan, R. L.; Miller, A. J. Potential for internalization, growth and survival of *Salmonella* spp. and *Escherichia coli* O157:H7 in oranges. *J. Food Prot.* **2004,** 67, 1578-1584.

120. Cherry, J. P. Improving the safety of fresh produce with antimicrobials. *Food Tech.* **1999,** 53, 54-57.

121. Burnett, S. L.; Beuchat, L. R. Human pathogens associated with raw produce and unpasteurized juices, and difficulties in decontamination. *J. Indus. Micro.Biotech.* **2001,** 27, 104-110.

122. Riordan, D. C. R.; Sapers, G. M.; Hankinson, T. R.; Magee, M.; Mattrazzo, A. M.; Annous, B. A. A study of U. S. orchards to identify potential sources of *Escherichia coli* O157:H7. *J. Food Prot.* **2001,** 64, 1320-1327.

123. Han, Y.; Sherman, D. M.; Linton, R. H.; Nielsen, S. S.; Nelson, P. E. The effects of washing and chlorine dioxide gas on survival and attachment of *Escherichia coli* O157:H7 to green pepper surfaces. *Food Micro.* **2000,** 17, 521-533.

124. Burnett, S. L.; Chen, J.; Beuchat, L. R. Attachment of *Escherichia coli* O157:H7 to the surfaces and internal structures of apples as detected by confocal scanning laser microscopy. *Appl. Environ. Micro.* **2000,** 66, 4679-4687.

125. Gorski, L.; Palumbo, J. D.; Mandrell, R. E. Attchment of *Listeria monocytogenes* to radish tissue is dependent upon temperature and flagellar motility. *Appl. Environ. Micro.* **2003,** 69, 258-266.

126. Aruscavage, D.; Lee, K.; Miller, S.; LeJeune, J. T. Interactions affecting the proliferation and control of human pathogens on edible plants. *J. Food Sci.* **2006,** 71, R89-R99.
127. Guo, X.; Chen, J. R.; Brackett, R. E.; Beuchat, L. R. Survival of Salmonella on tomatoes stored at high relative humidity, in soil, and on tomatoes in contact with soil. *J. Food Prot.* **2002,** 65, 274-279.
128. Das, E.; Gurakan, G. C.; Bayindirli, A. Effect of controlled atmosphere storage, modified atmosphere packaging and gaseous ozone treatment on the survival of *Salmonella* Enteritidis on cherry tomatoes. *Food Micro.* **2006,** 23, 430-438.
129. Guo, X.; Chen, J.; Brackett, R. E.; Beuchat, L. R. Survival of Salmonellae on and in tomato plants from the time of inoculation at flowering and early stages of fruit development through fruit ripening. *Appl. Environ. Micro.* **2001,** 67, 4760-4764.
130. Manios, T.; Papagrigoriou, I.; Daskalakis, G.; Sabathianakis, I.; Terzakis, S.; Maniadakis, K.; Markakis, G. Evaluation of primary and secondary treated and disinfected wastewater irrigation of tomato and cucumber plants under greenhouse conditions, regarding growth and safety consideration. *Water Environ. Res.* **2006,** 78, 797-804.
131. Fatemi, P.; LaBorde, L. F.; Patton, J.; Sapers, G. M.; Annous, B. A.; Knabel, S. J. Influence of punctures, cuts, and surface morphologies of golden delicious apples on penetration and growth of *Escherichia coli* O157:H7. *J. Food Prot.* **2006,** 69, 267-275.
132. Mintier, A. M.; Foley, D. M. Electron beam and gamma irradiation effectively reduce *Listeria monocytogenes* populations on chopped romaine lettuce. *J. Food Prot.* **2006,** 69, 570-574.
133. Niemira, B. A. Relative efficacy of sodium hypochlorite wash versus irradiation to inactivate *Escherichia coli* O157:H7 internalized in leaves of Romaine lettuce and baby spinach. *J. Food Prot.* **2007,** 70, 2526-2532.
134. Fan, X.; Annous, B. A.; Sokorai, K. J. B.; Burke, A.; Mattheis, J. P. Combination of hot-water surface pasteurization of whole fruit and low-dose gamma irradiation of fresh-cut cantaloupe. *J. Food Prot.* **2006,** 69, 912-919.
135. Keller, S. E. Chapter 16: Effect of quality sorting and culling on the microbiological quality of fresh produce. In *Microbiology of Fruits and Vegetables*, ed.; Sapers, G. M.; Gorny, J. R.; Yousef, A. E., Eds. CRC Press: Boca Raton, Fl, 2006; pp 365-373.
136. Blender, R. J.; Sargent, S. A.; Brecht, J. K.; Bartz, J. A. Effect of tomato grade on incidence of decay during simulated shipping. *Proc. Fla. State Hortic. Soc.* **1992,** 105, 119-121.
137. Jackson, L. S.; Beacham-Bowden, T.; Keller, S. E.; Adhikari, C.; Taylor, K. T.; Chirtel, S. J.; Merker, R. I. Apple quality, storage, and washing treatments affect patulin levels in apple cider. *J. Food Prot.* **2003,** 66, 618-624.
138. Riordan, D. C. R.; Sapers, G. M.; Annous, B. A. The survival of *Echerichia coli* O157:H7 in the presence of *Penicillium expansium* and

Glomerella cingulata in wounds on apple surfaces. *J. Food Prot.* **2000,** 63, 1637-1642.

139. Keller, S. E.; Chirtel, S. J.; Merker, R. I.; Taylor, K. T.; Tan, H. L.; Miller, A. J. Influence of fruit variety, harvest technique, quality sorting, and storage on the native microflora of unpasteurized apple cider. *J. Food Prot.* **2004,** 67, 2240-2247.
140. Wells, J. M.; Butterfield, J. E. *Salmonella* contamination associated with bacterial soft rot of fresh fruits and vegetables in the marketplace. *Plant Dis.* **1997,** 81, 867-872.
141. Richards, G. M.; Beuchat, L. R. Infection of cantaloupe rind with *Cladosporium cladosporioides* and *Penicillium expansum*, and associated migration of *Salmonella poona* into edible tissues. *Inter. J. Food Micro.* **2005,** 103, 1-10.
142. Ukuku, D. O.; Bari, M. L.; Kawamoto, S.; Isshiki, K. Use of hydrogen peroxide in combination with nisin, sodium lactate and citric acid for reducing transfer of bacterial pathogens from whole melon surfaces to fresh-cut pieces. *Inter.J. Food Micro.* **2005,** 104, 225-233.
143. Keller, S. E.; Merker, R. I.; Taylor, K. T.; Tan, H. L.; Melvin, C. D.; Chirtel, S. J.; Miller, A. J. Efficacy of sanitation and cleaning methods in a small apple cider mill. *J. Food Prot.* **2002,** 65, 911-917.
144. Bartz, J. A. Washing fresh fruits and vegetables: Lessons from treatment of tomatoes and potatoes with water. *Dairy, Food Environ. Sanit.* **1999,** 19, 853-864.

Chapter 4

Nanoscale materials in foods: existing and potential sources

Bernadene A. Magnuson, Ph.D.

Cantox Health Sciences International
Mississauga, ON L5N 2X7
Email: bmagnuson@cantox.com

Advances in nanotechnology are resulting in numerous promising applications for improved food production, processing, packaging, and storage. The safety of nanoscale materials in foods has become an increasingly important issue, both in the US and worldwide. Nanoscale particles in foods can be naturally occurring, intentionally added engineered nanomaterials derived from naturally occurring food components, or may be engineered using materials that are not endogenous to foods. In addition, the presence of nanomaterials in foods may be the result of contamination. To assess the health risk of use of these materials to the consumer, both the potential hazard of the materials and the likely exposure must be considered. Although oral exposure to nanomaterials has not been as intensely investigated as other routes of exposure, recent studies using nanotechnology to improve uptake of nutrients and bioactive components illustrate that pharmacokinetics, such as absorption and distribution, can be altered as compared to microscale materials, thereby potentially changing the potential hazard profile associated with the nutrient or bioactive. Efforts to facilitate international collaboration and information exchange are underway to ensure acceptance and utilization of the many benefits of nanotechnology.

Introduction

The growing interest in the potential use of nanomaterials in food applications is evident by the increasing number of scientific publications in food related journals, and food-specific conferences on nanotechnologies (*1*). As the science develops, and more and more potential applications of nanomaterials are being discovered, there are also issues and challenges arising regarding adoption of this new technology by the food and food-related industries.

What are nanoscale materials?

Nanoscale materials include any materials that contains a structure with at least one dimension in the range of 1-100 nanometers (nm). How small is a nanometer? Really, really small – too small to see, even with a light microscope or basic electron microscope. One comparison is that a nanometer is to a meter, as the diameter of a dime is to the diameter of the earth. Another is that a nanometer is 1/100,000 the width of a human hair. Nanoscale materials include free nanoparticles (such as Buckey balls, carbon nanotubes and quantum dots), nanostructured solid materials that have a nanosized microstructure, and nanoliquids that contain nanosized microstructures. Nanomaterials can also be categorized as naturally occurring (such as nanoparticulates from volcanic eruptions), man-made (such as nanoparticulates resulting from industrial emissions) and engineered nanomaterials (such as single- and double-walled carbon nanotubes).

What is nanoscience? Nanotechnology?

Nanoscience involves research to discover and understand the new behaviors and properties of materials with dimensions at the nanoscale level. Nanotechnology is the ability to manipulate and control nanomaterials in a useful way, and to utilize the unique behaviors and properties of materials at the nanoscale level to enable novel applications (*2*). Nanotechnology has produced a variety of nanomaterials, including nanocomposites used in building materials, nanomedicines and medical imaging compounds containing quantum dots, nanoparticles used in stain-repellent clothing and sunscreens, nanotubes used in tennis rackets, and the list continues to grow (*3*). The rapid development of new nanomaterials and potential applications is at least partially the result of the research support from the National Nanotechnology Initiative (NNI), a federal R&D program established to coordinate the multiagency efforts in nanoscale science, engineering, and technology (*2*). These nanomaterials are often composed of compounds that we have used for many years, such as carbon, gold, lead, and various metals in larger structured form. However, once particles of these materials are reduced in size to below 100 nanaometers (nm), they begin to display novel characteristics based on the quantum mechanical forces that are exhibited at this level. These quantum mechanical forces may make the material stronger, more conducting, better able to transfer heat, absorb light, and/or to have altered solubility properties, altered reactivity, and so on. The change in properties that occurs with the change in size is creating new potential applications for materials, but also may create new health and environmental concerns.

What are sources of nanoscale materials in foods?

Nanoscale particles in foods can be naturally occurring, intentionally added engineered nanomaterials derived from naturally occurring food components, or may be engineered using materials that are not endogenous to foods. In addition, the presence of nanomaterials in foods may be the result of contamination. To assess the trends of development of consumer products using nanotechnology, the Woodrow Wilson Project on Emerging Nanotechnologies compiled an international inventory of commercially available nanobased consumer products (*4*). Numerous food and beverage products are listed in this inventory as consumer products claiming to be produced with nanotechnology, including nanoteas, oils, vitamin supplements and nutraceuticals. In many cases, claims are made that the nanosizing of the components improves absorption and effectiveness of the supplement. However, rarely is evidence presented to demonstrate that the products, as consumed, would contain nanomaterials and/or that absorption or effectiveness is improved.

Naturally occurring nanomaterials in foods

There are many examples of naturally occurring nanomaterials in the foods we have been eating for centuries. The major constituents of milk are nanomaterials, including casein micelles (50-300 nm), whey proteins (4-6 nm), and lactose (0.5 nm) (*5*). As food scientists begin to utilize the advanced imaging technologies for examining nanostructure of materials, more reports on nanomaterials in foods are emerging. Zhang et al. (*6*) explored the nanostructure of pectin in fruits and demonstrated that the nanostructural characteristics of pectins were closely related with fruit firmness. The nanostructure of starch molecules varies with the type of starch, and with cooking and other processing (*7*). Understanding the nanostructure of proteins in meats and fish (*8*) provides the promise of future alternative means of producing highly desirable foods from non-animal protein sources.

Engineered nanomaterials in food and food-related products

Nanomaterials are being developed for a variety of applications including, but not limited to, improved nutrient delivery systems (i.e. fat-soluble vitamin microemulsions), flavor encapsulation, improved microbiological control, improved food packaging, and specific highly sensitive sensors which can be used to detect possible problems with food quality and safety. A short discussion on applications of engineered nanomaterials will be presented below, and readers are referred to more detailed reviews (*1*, *9*, *10*).

Engineered nanomaterials for addition to food and food-related products

Use of nanomaterials for ingredient technologies offer improved control of and sustained release of bioactive compounds (*e.g.* biopolymer based hydrogels and encapsulation technologies); reduced interaction between ingredients within a food system; excellent dispersion and suspension of water insoluble ingredients using liposomes, nanodispersions, and nanoemulsions; improved bioavailability; and improved stability. For example, Aquanova (*11*), a company in Germany, uses nanotechnology to produce nanomicelles to improve solubility of insoluble bioactives, and change water/fat solubility of nutrients including vitamins A, C, D, E, and K, Coenzyme 10, β-carotene, isoflavones, α-lipoic acid, and omega fatty acids.

The ability to alter the solubility of functional lipids is an attractive application, as the poor water solubility of lipids makes them problematic in food formulations. However, how the application of nanotechnology to nutrients and food compounds will alter their chemical and biological properties is not well known at this time. Tan and Nakajima (*12*) described the preparation of beta-carotene nanodispersions for improved solubility and bioavailability. However, the beta-carotene in the nanodispersions was chemically unstable, and the authors showed that the degradation was dependent upon the mean particle diameter. Thus a change in size altered the chemical stability, and further research is needed to develop optimal formulations.

Engineered nanomaterials for food processing, packaging and storage

Potential applications of indirect use of engineering nanomaterials in food processing, food packaging and storage include monitoring of food quality, safety, and biosecurity; improved food packaging; and improved food processing (*10*, *13*). Examples include nanosensors for detecting foodborne pathogens and contaminants; adhesion-specific nanoparticles for selective binding and removal of pathogens and contaminants (non antibiotic approach to disease prevention); active antimicrobials such as metal oxides; and tracers that could help determine sources of contamination (*14*). Nanobiosensors have several advantages such as high sensitivity, high selectivity, near real time detection, low cost, and portability. For example, scientists are using nanotechnology to develop rapid and accurate diagnostics and detection methods for pathogens such as *Staphylococcus aureus*, *Escherichia coli*, *Listeria monocytogenes, Campylobacter jejuni* and *Salmonella*. Latour et al. (*15*) have been investigating the potential for synthesized adhesion-specific nanoparticles to irreversibly bind to targeted bacteria thus inhibiting them from binding to and infecting their host. An outcome of this project is the development of adhesion-specific nanoparticles for removal of *Campylobacter jejuni* from poultry (*15*). The research is aimed at reducing the infective capability of foodborne enteropathogens in poultry products. Clearly, the opportunities for advancement in this arena are very great, but require significant research investment.

Potential food packaging benefits include: high performance packaging with enhanced mechanical and barrier properties; antimicrobial packaging infused with antimicrobial nanoparticles (*e.g.* silver nanoparticles); intelligent packaging technologies that could prevent or respond to spoilage (*e.g.* polymer opal films that change color to indicate spoilage) or DNA biochip nanosensors that detect toxins, contaminants and pathogens; and water and dirt repellant packages. In a study by McHugh et al. (*16*), the tensile, water vapor and oxygen permeability properties of edible films were significantly improved through the application of nanoscience. Each of these technologies demonstrates the many opportunities for nanotechnology to enhance the safety and quality of the food supply.

Potential sources of nanomaterial contaminants

Potential sources of nanomaterial contaminants in foods include unintentional release from nanomaterial-containing food packaging materials, food processing aides or surface coatings on food equipment. In addition, new formulations of agricultural chemicals for use in food production may contain nanomaterials that could end up in or on the food products. Environmental contamination of nanomaterials resulting from use in a growing number of industries is a major area of concern and research. The impact of environmental contamination by other industry has the potential to affect the food supply as has been illustrated by the presence of many existing industrial contaminants being found in foods.

Are nanomaterials safe for food applications?

The answer to this question is not going to be a simple one. In other words, it will depend on the situation, on the type of nanomaterial used and how the nanomaterial is used. The safety of any situation or agent is assessed by determining the likelihood of risk of harm. The risk that is posed is assessed by considering the hazard (inherent toxicity or danger) and the degree of exposure. High risk occurs when there is the combination of inherent high toxicity of a compound, and sufficient exposure such that a dose resulting in toxicity may be consumed. If there is either low hazard (as when the compound has low or no evidence of toxicity) or the exposure is low, then the overall risk will also be low. Thus when considering the safety of food applications of nanomaterials, one must assess both the inherent toxicity of the nanomaterial, and the likelihood of exposure to the nanomaterial.

There is a large body of data on the toxicity of inhaled nanomaterials, due to the presence of nanoparticles in air from industrial pollution and natural sources (*17*). Development of nanomaterials for medical applications is also providing information on the toxicity of nanomaterials injected directly into the body, and in partular, into the blood stream, and the use of nanomaterials in sunscreens and cosmetics has provided data on dermal toxicity. There are, however, fewer studies that have evaluated oral exposure and the ability of nanomaterials to be absorbed from the gastrointestinal tract (*1*). Even fewer have evaluated the bioavailability of nanomaterials from a food product. Thus, the effect of the food matrix on the bioavailability of nanomaterials within those foods is a fertile area of research.

Factors that affect the toxicity of nanomaterials include chemical composition, size and shape, and surface characteristics of the particles or materials (*18-20*). General concerns that apply to nanoscale materials as compared to the same chemical composition in microscale dimensions include: greater exposure to the surface of particles per unit mass; potential differences in exposure route due to small size (*i.e.* may aerosolize more readily); potential differences in distribution due to increased ability to cross cell membranes; and the potential for a new mode of action of the compound based on the novel properties resulting from the nanosizing of the material (*19*). Some of the challenges in the evaluation of nanomaterials are the need to adequately characterize the nanomaterials in terms of size, shape, uniformity of size and shape, agglomeration properties, chemical purity, and stability in the test organism and under the specific test conditions. Use of standards and validated methods for the test material is highly desirable. The National Institute of Standards and Technology (NIST) is leading the efforts to develop reference standards for nanoscale research and the Nanotechnology Characterization Laboratory (NCL) of the National Cancer Institute is developing validated methods for measurement of characteristics of nanomaterials for nanomedicinal uses.

The potential exposure to nanomaterials will greatly depend upon their use in the food and food-related industries. Incorporation of nanomaterials into multi-composite packaging materials, in which the nanomaterial layer is coated with other materials, for example, is likely to result in minimal transfer to the food and thus present exceedingly low exposures to the nanomaterial by the food consumer. In contrast, use of nanomaterials as carriers of nutrients or bioactive compounds that will be added directly to the food product, will result in higher levels of exposure and will depend on the concentration in the food and the amount of that food consumed. In these cases, however, the compositions of the nanomaterials are likely to be modifications of compounds found in food, such as proteins, lipids, *etc*., and thus have low hazard.

What challenges do nanoscale material contaminants in food pose?

When one considers the question of how to assess the presence and possible health implications of nanoscale contaminants in foods, a flurry of questions and challenges arise. Firstly, what would be the appropriate means of assessment of exposure to nanoscale contaminants? How would it be possible to detect nanomaterials in food, if they are composed of similar materials that could or would be present in a microscale form? Are there methods of detection that are sufficiently sensitive to detect low levels of nanomaterials in a food matrix?

Assuming that we are able to overcome these challenges and accurately quantify nanomaterial contaminants, then how do we asses the significance of such information for human health when we have so much yet to learn regarding the biological activity and toxicity of these materials? A critical research need is the development of validated assays for assessment of the toxicity of oral exposure to nanomaterials. For example, validation of *in vitro* gastrointestinal absorption assays, and *in vitro* cytotoxicity assays that take into consideration possible effects of food matrix, food processing, storage *etc.* on the nanomaterials is critical due to the tendency of nanomaterials to agglomerate and adhere to protein and other components that could be present in food. Questions regarding appropriate dose metrics, dose delivery methods, critical endpoints and detection methods for *in vivo* assays also must be addressed. Such issues are currently under intense investigation for other exposure routes, including dermal, inhalation and direct injection of nanomaterials, but these studies may not be relevant to food-related exposures.

Conclusions

Understanding nanomaterials in our food is an exciting new area of food research, as we discover the role of naturally occurring nanomaterials in food and develop new materials for a wide variety of applications in food production, food product development, food safety control and monitoring, and food packaging. At the current time, risk assessment of food-related nanomaterials is very challenging due to insufficient data regarding both hazard and exposure of these materials. However, advances in nanotechnology are occurring in a wide range of industries, and it is recognized that there is the potential for research to simultaneously address cross-industry issues. Gathering and reviewing research from other industries that can be applied to food-related materials will speed progress and reduce the needed investment in research. For example, the 3-phase tiered approach for characterization and safety evaluation of nanomaterials for cancer chemotherapeutics developed by the NCL is a potential model to consider for evaluation of food-related nanomaterials. Environmental, health and safety (EHS) programs designed for other industries may also be appropriate starting points for the food industry. An example of a generic initial EHS program for companies initiating work with nanomaterials has been recently published by Lekas et al. (*21*) on the Woodrow Wilson project website (*21*). A report on cross-industry issues in nanotechnology resulting from a recent workshop organized by NIST will soon be published and is expected to highlight many of these cross-industry opportunities.

The future is exciting for food nanotechnology as evidence mounts that many current problems in food safety, nutrient delivery, food product development and food packaging may be addressed with advances in nanomaterial science and applications. Additional research, international cooperation and cross-industry sharing of information will speed the acquisition of the data needed to realize the potential that nanotechnology holds for the food and food-related sciences.

References

1. Chaudhry, Q.; Scotter, M.; Blackburn, J.; Ross, B.; Boxall, A.; Castle, L.; Aitken, R.; Watkins, R. Applications and implications of nanotechnologies for the food sector. *Food Addit. Contam.* **2008**, *25*, 241-258.
2. NNI. National Nanotechnology Initiative (NNI), Coordination Office; Arlington, VA, 2008; Available from: http://www.nano.gov/. Accessed August 31, 2008.
3. Nanotechnology Now. Current Nanotechnology Applications. http://www.nanotech-now.com/. Accessed August 15, 2008.
4. Project on Emerging Nanotechnologies. Consumer Products. Project on Emerging Nanotechnologies, Woodrow Wilson International Center for Scholars; Washington, DC, 2008; Available from: http://www.nanotechproject.org/inventories/consumer/.
5. Tuinier, R.; de Kruif, C.G. Stability of casein micelles in milk. *J. Chem. Phys*. **2002**, *117*, 1290-1295.
6. Zhang, L.; Chen,F.; An, H.; Yang, H.; Sun, X.; Guo, X.; Li, L. Physicochemical properties, firmness, and nanostructures of sodium carbonate-soluble pectin of 2 Chinese cherry cultivars at 2 ripening stages. *J. Food Sci.* **2008**, *73*, N17-N22.
7. Dang, J.M.; Braet, F.; Copeland, L. Nanostructural analysis of starch components by atomic force microscopy. *J. Microsc*. **2006**, *224*, 181-186.
8. Yang, H.; Wang, Y.; Regenstein, J.M.; Rouse, D.B. Nanostructural characterization of catfish skin gelatin using atomic force microscopy. *J. Food Sci*. **2007**, *72*, C430-C440.
9. Weiss, J.; Takhistov, D.; McClements, J. Functional materials in food nanotechnolog. J *Food Sci*. **2006**, *71*, R107-R116.
10. Chen, H.; Weiss, J.; Shahidi, F. Nanotechnologies in nutraceuticals and functional foods. *Food Technol.* **2006**, *60*, 30-36.
11. Aquanova AG. [Website]. Darmstadt, Germany, 2008; Available from: http://www.aquanova.de/. Accessed August 31, 2008.
12. Tan, C.P.; Nakajima, M. β-Carotene nanodispersions: preparation, characterization and stability evaluation. *Food Chem*. **2005**, *92*, 661-671.
13. Baeumner, A. Nanosensors identify pathogens in food. *Food Tech*. ***2004,*** *58*, 51-55.
14. Scott, N.; Chen H. (Co-Chairs). Nanoscale Science and Engineering for Agriculture and Food Systems. A Report Submitted to Cooperative State Research, Education, and Extension Service, The U.S. Department of Agriculture National Planning Workshop, Nov. 18-19, 2002; 2003; Available from: http://www.nseafs.cornell.edu/roadmap.draft.pdf.
15. Latour, R.A.; Stutzenberger, F.J.; Sun, Y.P.; Rodgers, J.; Tzeng, T.R. Adhesion-specific Nanoparticles for Removal of *Campylobacter jejuni* From Poultry. CSREES Grant, Study Performed by Clemson Univ.; Clemson, SC. for Cooperative State Research, Education, and Extension Service (CREES), U.S. Department of Agriculture (USDA), 2004; Available from: http://cris.csrees.usda.gov/cgi-bin/starfinder/0?path=fastlink1.txt&id=anon&pass=&search=R=15653&format=WEBLINK.

16. McHugh T.H., Avena-Bustillos, R., Dogan, N., Aouada, M., Mattoso, L. Applications of Nanoscience to Edible Films. Submitted to: Recent Progress in Food Science and Engineering International Symposium Proceedings. 2007; pp. 16-21. Available from: http://www.ars.usda.gov/research/publications/publications.htm?seq_no_115=213580.
17. Borm, P.J.; Müller-Schulte, D. Nanoparticles in drug delivery and environmental exposure: same size, same risks? *Nanomed.* **2006**, *1*, 235-249.
18. Oberdörster, G.; Oberdörster, E.; Oberdörster, J. Nanotoxicology: an emerging discipline evolving from studies of ultrafine particles. *Environ. Health Perspect.* **2005**, *113*, 823-939.
19. Nel, A.; Xia, T.; Mädler, L.; Li, N. Toxic potential of materials at the nanolevel. *Science* **2006**, *311*, 622-627.
20. Stern, S.T.; McNeil, S.E. Nanotechnology safety concerns revisited. *Toxicol. Sci.* **2008**, *101*, 4-21.
21. Lekas, D. How to Reduce Your Firm's Risk and Increase Revenues Related to Nanotechnology. An 8-Step Program for Small Firms. Woodrow Wilson International Center for Scholars, 2008; Available from: http://www.nanotechproject.org/process/assets/files/6703/nano_researchbrief_em.pdf.
22. Borm, P.J.; Müller-Schulte, D. Nanoparticles in drug delivery and environmental exposure: same size, same risks? *Nanomed.* **2006**, *1*, 235-249.
23. Oberdörster, G.; Oberdörster, E.; Oberdörster, J. Nanotoxicology: an emerging discipline evolving from studies of ultrafine particles. *Environ. Health Perspect.* **2005**, *113*, 823-939.
24. Nel, A.; Xia, T.; Mädler, L.; Li, N. Toxic potential of materials at the nanolevel. *Science* **2006**, *311*, 622-627.
25. Stern, S.T.; McNeil, S.E. Nanotechnology safety concerns revisited. *Toxicol. Sci.* **2008**, *101*, 4-21.
26. Lekas, D. How to Reduce Your Firm's Risk and Increase Revenues Related to Nanotechnology. An 8-Step Program for Small Firms. Woodrow Wilson International Center for Scholars, 2008; Available from: http://www.nanotechproject.org/process/assets/files/6703/nano_researchbrief_em.pdf.

Chapter 5

Renal toxicity of pet foods contaminated with melamine and related compounds

William H. Tolleson

Division of Biochemical Toxicology, National Center for Toxicological Research, Jefferson, AR 72079

In 2007, a lethal renal failure syndrome was recognized among pets consuming products that were later found to contain melamine. This precipitated the recall of over 1,000 types of pet food products. Chemical analysis revealed co-contamination by cyanuric acid and related triazine compounds. Poisoned animals exhibited uremia, azotemia, and hyperphosphatemia in addition to renal tubule obstruction by spherical and polarizable crystals displaying a distinctive radial internal structure. Oral dosing studies combining melamine and cyanuric acid recreated the characteristic renal failure syndrome in laboratory animals. Detection of melamine and cyanuric acid in renal calculi collected from exposed animals confirmed that this renal failure syndrome is associated with the combination of these agents.

Numerous pet deaths caused by renal failure were associated with the national pet food recall of 2007. Initial chemical analyses of affected pet food formulations revealed that the wheat gluten used, a raw ingredient imported from the Peoples Republic of China (PRC), was contaminated with melamine, a nitrogen-rich industrial chemical commonly used in polymer manufacturing. The pet food crisis was compounded further by three unfortunate developments. First, melamine contamination was found in rice protein extract, another pet food ingredient imported from the PRC. Second, additional triazine compounds, including cyanuric acid, ammelide, and ammeline (Figure 1), were detected as co-contaminants with melamine in some products. Third, public anxiety was fomented by news that triazine-contaminated materials were unknowingly incorporated into feed formulations for poultry, swine, and farm-raised fish destined for human consumption. A food safety crisis was the overall result, undermining consumer confidence, causing public confusion, and inflicting significant adverse economic impact.

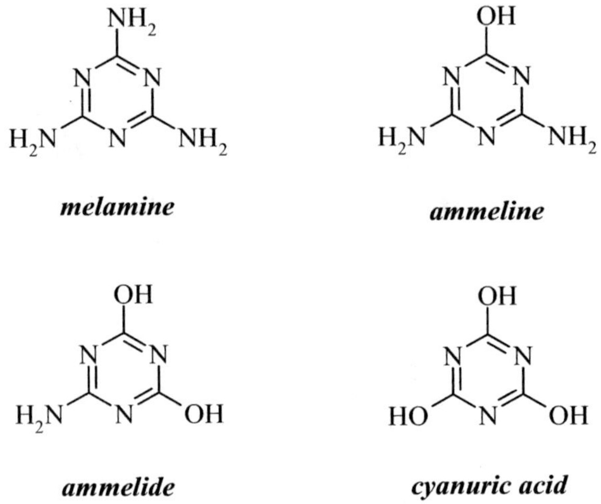

Figure 1. Triazines detected in contaminated pet food products and raw ingredients. *Melamine*, 2,4,6-triamino-1,3,5-triazine [108-78-1]; *ammeline,* 4,6-diamino-2-hydroxy-1,3,5-triazine [645-92-1]; *ammelide,* 6-amino-2,4-dihydroxy-1,3,5-triazine [645-93-2]; and *cyanuric acid,* 2,4,6-trihydroxy-1,3,5-triazine [108-80-5].

Past Examples of Melamine Contaminated Foods

Neither melamine nor cyanuric acid is included in the FDA EAFUS[1] database, is considered GRAS[2], or is approved by the FDA as an additive[3] in human or domestic pet foods. However, FDA regulations[4] permit the presence of up to 30% cyanuric acid in feed-grade biuret produced from the controlled pyrolysis of urea as an additive for ruminant feed products to provide a nonprotein source of nitrogen.

---

[1] EAFUS – Everything Added to Food in the United States database maintained by the FDA Center for Food Safety and Applied Nutrition (http://www.cfsan.fda.gov/~dms/eafus.html)

[2] GRAS - Generally Recognized As Safe; a designation created by Congress that includes components of foods recognized among scientific experts to be safe under the conditions of their intended uses (http://www.cfsan.fda.gov/~dms/opa-noti.html).

[3] FDA Food Additive website (http://www.cfsan.fda.gov/~dms/opa-addi.html)

[4] 21CFR573.220 – Feed-grade biuret; revised April 1, 2007.

Melamine-adulterated products responsible for the pet food recall of 2007 caused significant physical, psychological, or economic harm to pets, pet owners, or pet food producers and the ensuing crisis received intense domestic and international media coverage. Yet this was not the first reported occurrence of melamine contamination of food products associated with adverse effects; two previous episodes of melamine-contaminated food products in other countries received less media attention. Whether the current crisis could have been avoided if prior melamine contamination events received the consumer attention and scientific scrutiny precipitated by the 2007 recall is uncertain.

Melamine Contamination of Meat Meal Products in Italy, 1979-1987

An Italian survey of goods produced from 1979-1987 detected contamination of meat meal products with melamine at levels between 0.94% - 1.6% *(1)*. Following establishment of stricter guidelines, the incidence of melamine-contaminated meat meal decreased from 72% melamine-positive samples (28 of 39) to 5.4% positive (2 of 39). This pattern demonstrates that the deliberate adulteration of food products with melamine as a matter of routine commercial practice can be discouraged or eliminated through increased education, surveillance, and accountability by the industry. The prevalence of this practice during the era preceding the adoption of revised manufacturing procedures by producers is consistent with a common economic motivation for the artificial introduction of melamine into foods.

Total nitrogen analysis is often accepted as an indirect measure of protein content for food ingredients and products. Products meeting higher total nitrogen specifications achieve proportionally higher market prices. The nitrogen content (wt%) of the four triazine species shown in Figure 1 are: melamine (66.6%), ammeline (55.1%), ammelide (43.7%), and cyanuric acid (32.6%). For comparison, the generally accepted value for the average nitrogen composition of proteins is 16%. Therefore, the addition of 24% melamine to a low value material that contains neglible protein, such as wheat flour, will increase its apparent nitrogen content to that of a rich protein source that can be sold for a higher price *(2)*.

Melamine Contaminated Pet Foods in Asia, 2004

In March 2004 pets in several Asian countries, including Taiwan, South Korea, Cambodia, Vietnam, and Singapore, experienced an episode of illnesses and deaths due to acute renal failure *(3)*. Consumption of pet food products manufactured in Thailand at a single factory and sold under the US labels Pedigree and Mars were associated with the outbreak of renal failure in these animals and circumstantial data surrounding the ensuing pet food recall implicated potential mycotoxin contamination *(4, 5)*. This event precipitated a class action lawsuit involving 5,760 owners of injured dogs and a smaller number of cats, according to records of the Council of Agriculture of Taiwan reported in the media *(6)*.

Later, a group of alert American and Korean veterinary pathologists recognized similarities in the acute renal failure syndrome observed in domestic pets exposed to melamine-contaminated pet food products in 2007 with the episode of acute renal failures in 2004. Thorough chemical and histological analyses of tissue and urine samples from the 2004 outbreak together with specimens collected from animals involved in the 2007 pet food crisis were performed by Brown and coworkers *(7)* and by Thompson et al (*8*) which confirmed that acute renal failures in both episodes were associated with the presence of melamine and cyanuric acid.

2007 Pet Food Recall

The 2007 pet food recall crisis began in the US in February when, after altering pet food formulations to include wheat gluten supplied by ChemNutra, the Canadian-based pet food manufacturer Menu Foods received reports of adverse reactions involving pets that consumed its reformulated products. On March 15, 2007 Menu Foods notified the FDA of these developments and announced a voluntary pet food product recall the following day - the first of many to be made by North American pet food producers.

FDA public records[5] convey the scale of the 2007 pet food crisis; 1,154 pet food product types marketed under 139 different brand names and produced by 17 manufacturers were recalled. In addition to mobilizing hundreds of investigators, scientists, and technicians to address the expanding crisis, FDA handled over 18,000 calls from the public – more than it received for any issue in its history.

Because regulations do not exist for mandatory reporting of unexpected illnesses or deaths of domestic animals, official statistics for the number of pets affected by triazine-contaminated products are not available. However, clinical veterinarians estimate the number of pet deaths to be over 1,000 *(9)*. A *post hoc* survey of veterinarians performed by the American Association of Veterinary Laboratory Diagnosticians performed later in 2007 identified 348 verified cases of fatal acute renal failure associated with consumption of contaminated pet food products *(10, 11)*.

Exposed pets exhibited anorexia, vomiting, lethargy/depression, polyuria, and polydipsia together with additional clinical signs typical for acute renal failure, notably uremia, hyperphosphatemia, and increased anion gap (Table I) *(7, 8)*. The accumulation of stones (calculi) within various portions of the urinary tract was the histological feature noted most consistently by veterinary pathologists evaluating *post mortem* tissue specimens from affected pets, particularly renal calculi that partially or totally obstructed distal tubules and collecting ducts of the kidneys. Unique, spherical urinary calculi approximately 10 - 50 microns in diameter were observed using bright-field and polarized microscopy. These calculi exhibited radial internal striations and variable coloration described as yellow to greenish-brown. The presence of

---

[5] US Food and Drug Administration pet food recall website (http://www.accessdata.fda.gov/scripts/petfoodrecall/); website updated April 16, 2008, accessed June 9, 2008.

these characteristic uroliths is the distinguishing feature for renal failure associated with consumption of pet food contaminated with melamine and cyanuric acid; elucidating their origin and nature is under investigation.

Nephrotoxicity and Renal Failure in Animal Studies with Melamine and Cyanuric Acid

Past investigations of the acute and chronic toxicities of melamine or cyanuric acid in laboratory animals were performed by the National Toxicology Project (NTP) and others (Table II). Increased mortality due to acute renal failure was not observed in NTP studies with B6C3F$_1$ mice and F344/N rats exposed to 2,250 ppm or 4,500 ppm melamine in their diets for 103 weeks compared to untreated animals. However, chronic inflammation of the kidneys was observed in melamine-treated female rats and increased incidences of urinary calculi developed in melamine-treated rats and mice, typically detected in the bladder *(12)*.

Table I. Characteristics of pet food-induced acute renal failure

Behavioral changes
• Anorexia • Vomiting • Lethargy/depression • Polyuria • Polydipsia
Serum abnormalities
• Azotemia • Uremia • Hyperphosphatemia • Increased anion gap
Histopathology
• Yellow to greenish brown polarizable spherical crystals with radial striations located within distal tubules, collecting ducts, and urinary bladder • Tubular casts containing sloughed epithelial cell debris and polarizable material • Tubular necrosis, rupture, and degeneration; interstitial edema • Hemorrhage at corticomedullary junction • Chronic renal interstitial nephritis

NOTE: Compiled from references *(7, 8, 10, 11, 13, 14)*.

Table II. Studies with melamine or cyanuric acid administered separately

Species, strain (number)	*Agent (dose, route, duration)*	*Evidence of nephrotoxicity (reference)*
Mice, B6C3F$_1$ (50 per group)	Melamine (0, 2250, or 4500 ppm in feed; 103 weeks)	Increased incidence of bladder stones *(12)*.
Rats, F344/N (50 per group)	Melamine (0, 2250, or 4500 ppm in feed; 103 weeks)	Chronic renal inflammation *(12)*.
Rats, CD/Crj, (n.s.) [a]	Cyanuric acid (0 ,10, 40, 150, 600 mg/kg daily by gavage; 44 days)	Urinary calculi; hematuria; azotemia; renal tubule dilatation, necrosis, neutrophil infiltration, mineralization, and fibrosis *(15, 16)*.
Mice, B6C3F$_1$, (n.s.)	Sodium isocyanurate (0, 896, 1792, 5375 ppm in drinking water; 90 days)	No evidence of nephrotoxicity *(15, 17)*.
Guinea pigs and rats, (n.s.)	Cyanuric acid (0, 3.0, 30 mg/kg daily by gavage; 6 months)	Dystrophic changes in kidneys detected in the high dose group only *(18)*
Dogs, beagles (3 dogs)	Sodium isocyanurate (8.0% in diet, 2 years)	Renal fibrosis, 2/3 animals died at 16 and 21 months *(19)*

[a] n.s.; not specified

The acute and chronic toxicities of cyanuric acid or sodium cyanurate have been reviewed by Cannelli *(18)* and Hammond et al *(20)*. Early studies reported dystrophic changes in the kidneys of rats and guinea pigs receiving 30 mg/kg cyanuric acid orally for 6 months *(18)*. Increased urinary calculi were detected in CD rats receiving 500 - 700 mg/kg sodium cyanurate daily and in B6C3F$_1$ mice receiving 2,000 - 2,200 mg/kg sodium cyanurate daily in subchronic (14 week) and chronic (2 year) bioassays *(20)*.

Considering that published reports described sporadic cases of mild nephrotoxicity in animals receiving higher doses of melamine and cyanuric acid administered individually, new toxicological studies were necessary to determine whether co-administration of melamine and cyanuric acid would reproduce the severe renal failure syndrome associated with triazine-contaminated pet food products (Table III). Soon after the crisis developed in 2007, experiments of this type were initiated by an American pet food producer.

Rats were treated orally by gavage with a mixture of triazines at doses designed to approximate exposure to pets that consumed the most highly contaminated products, e.g. melamine (400 mg/kg), cyanuric acid (40 mg/kg), ammeline (40 mg/kg) and ammelide (40 mg/kg). Rats receiving triazine mixtures containing melamine and cyanuric acid exhibited clear evidence of renal failure (uremia, azotemia, hyperphosphatemia, increased anion gap), in contrast to no evidence of nephrotoxicity observed with either control rats or with rats exposed to the individual triazine species *(14)*.

Brown et al *(13)* reported that cats receiving melamine or cyanuric acid alone did not exhibit signs of nephrotoxicity but cats that consumed pet foods containing both melamine and cyanuric acid (0.2%, 0.5%, or 1% each melamine and cyanuric acid) once per day for two days each experienced acute renal failure (estimated total doses: 32 mg/kg, 121 mg/kg, or 181 mg/kg, respectively).

After learning that melamine-contaminated pet food scrap material was incorporated into feed pellets for farm-raised fish, FDA performed studies to evaluate the toxicities of daily oral doses of melamine and cyanuric acid individually or combined (400 mg/kg each) administered for three days in trout, salmon, tilapia, and catfish. In each of the four species of fish, only the combination of melamine and cyanuric acid produced renal failure symptoms, notably profuse accumulations of renal calculi obstructing the entire nephron; negligible signs of nephrotoxicity were observed in melamine-treated, cyanuric acid-treated, and control fish *(21)*.

Ensley and coworkers *(22)* treated, by daily gavage, groups of two male and two female pigs with melamine, cyanuric acid, or the combination of both for ten days. Evidence of renal failure developed in all pigs exposed to the combination of melamine and cyanuric acid (400 mg/kg each daily) and in one of four pigs in the highest melamine dose group (1,000 mg/kg daily). The investigators concluded that the combination of melamine plus cyanuric acid was a more potent nephrotoxin than either triazine alone.

Thus, in various vertebrate species, the nephrotoxicities of individual triazine species is typically moderate and occurs sporadically among animals receiving high doses (>400 mg/kg daily). In contrast, severe acute renal failure developed in all study animals when 400 mg/kg melamine and cyanuric acid were administered together. Of the animal studies described here, only Puschner et al *(13)* evaluated lower doses of the combination of triazines and demonstrated nephrotoxicity in a cat exposed to 32 mg/kg melamine and cyanuric acid. Additional NTP-supported studies are being planned to estimate the NOAEL[6] for the combination of melamine and cyanuric acid using rats and miniature pigs.

Although melamine cyanurate-induced acute renal failure has been observed in rats, cats, dogs, pigs, carnivorous trout and salmon, omnivorous catfish, and herbivorous tilapia, a careful comparison of interspecies differences in sensitivity to the combination of melamine and cyanuric acid is lacking. Thus, it is not known whether interspecies differences in diet, renal anatomy, or physiology influence their sensitivity to triazine-induced renal failure.

The anatomy and microanatomy of human and pig kidneys (see Figure 2) are similar in that both have comparable distributions of juxtamedullary nephrons that concentrate and dilute urine and cortical nephrons, responsible for regulatory and excretory functions (14% and 3% juxtamedullary nephrons in humans and pigs, respectively) *(23)*. Pigs and humans also have a multipapillary renal anatomy, unlike the unipapillary kidneys of rats, felines, and canines, which also have larger numbers of juxtamedullary nephrons. In fact, canines have no cortical nephrons and possess only juxtamedullary nephrons instead. Accordingly, the osmolarity of canine urine is about 86% higher than that of humans (2,610 mOsm/kg compared to 1,400 mOsm/kg) *(23)*. The terminal straight portion of the proximal tubule, termed the pars recta, is responsible for organic acid excretion and gives rise to a visible zone known as the outer stripe. The outer stripe is very obvious in rat kidneys, is present but thin in the kidneys of cats and humans, but it is virtually absent in dogs. Cats and dogs also exhibit abundant intrarenal fat deposits. Feline kidneys include sufficient fat to render them yellowish in color. Similarly, intracellular fat droplets present within epithelial cells lining the collecting ducts of canine kidneys color these structures. In summary, potentially important physiological differences associated with canines include higher osmotic gradients, reduced efficiency of organic anion excretion, and increased potential to accumulate or retain hydrophobic materials within intrarenal fat. It is not known whether any of these possible differences in renal physiology impact the severity of triazine-dependent nephrotoxicity.

---

[6] NOAEL – No observed adverse effect level.

Table III. Studies combining melamine and cyanuric acid

Species, strain, number	*Agent, dose, route, duration*	*Evidence of nephrotoxicity (reference)*
Rats (10 per group)	Melamine and cyanuric acid (400 mg/kg each by gavage; 2 days) or melamine (400 mg/kg by gavage) plus cyanuric acid, ammelide, and ammeline (40 mg/kg each by gavage); 2 days	Azotemia, uremia, hyperphosphatemia, increased anion gap, decreased plasma clearance, decreased urine osmolarity, abundant calculi obstructing renal tubules and collecting ducts *(14)*
Cats (1 per group)	Dosed feed containing 0, 0.2%, 0.5%, 1% melamine, cyanuric acid, or both; 11 exposure for individual triazines or 2 days exposure for combinations	Renal calculi obstructing tubules of distal nephron, severe renal interstitial edema, and hemorrhage at the corticomedullary junction *(13)*
Pigs (4 per group)	Melamine or cyanuric acid (0, 200, 400, 1000 mg/kg daily for 10 days) or melamine and cyanuric acid (400 mg/kg daily for 10 days)	Azotemia, uremia, anuria, anorexia, and lethargy *(22)*
Trout, salmon, tilapia, catfish (4 fish per group)	Melamine and cyanuric acid (single dose 400 mg/kg each by gavage; observed 14 days)	Very abundant calculi obstructing renal tubules and collecting ducts *(21)*

The variation in renal anatomy exhibited among fish species is even more profound than the distinctions described above among mammalian species *(24)*. For example, Henle's Loop is absent in fish. Some primitive species, such as monkfish and toadfish, lack glomeruli. Many saltwater species lack distal tubules. The nephrons of typical freshwater species include a glomerulus that is connected by a ciliated neck segment to two proximal tubule segments, followed by another ciliated intermediate segment that leads to the distal tubule segment which empties into the collecting duct system.

In spite of the diversity in renal anatomy and physiology exhibited among the species tested, a consistent acute renal failure syndrome developed among animals exposed to the combination of melamine and cyanuric acid that involved characteristic nephroliths obstructing renal tubules. These observations implicate common biochemical factors in the genesis of melamine cyanurate calculi but do not obviate the possibility that some species are less able to accommodate compromised renal function.

In addition to the evidence of renal obstruction among animals exposed to melamine cyanurate, some reports also described inflammatory changes that may have contributed to the severity of the disease *(7, 8, 11)*. Although inflammatory changes were not mentioned in an early case report by Jeong et al *(3)*, Thompson et al *(8)* described lymphoplasmacytic interstitial nephritis in two out of three affected dogs and Brown et al *(7)* confirmed this observation in five out of 11 animals. Furthermore, they described a pattern of interstitial inflammation and fibrosis involving moderate numbers of lymphocytes, plasma cells, macrophages, and (rarely) neutrophils, sometimes found surrounding the crystal-containing renal tubules. Inflammation and severe interstitial edema were also reported by Puschner et al *(13)* in cats treated with melamine and cyanuric acid. The presence of intracellular deposits of birefringent material engulfed by interstitial macrophages *(7)* provides further evidence that these renal calculi are recognized as inflammatory by the immune system.

Nephrotoxicity leading to renal failure may be classified as post-renal, meaning urinary obstruction downstream from the glomerulus interferes with renal output. Alternatively, renal failure may be considered pre-renal if afferent blood flow into glomeruli is restricted by decreased cardiac output, decreased arterial blood pressure, hypovolemia, thromboembolism, or arteriosclerosis. Also, nephrotoxicity may be associated with renal damage, which involves the death of kidney cells through direct cytotoxicity or via indiscriminant tissue effects including local ischemia or oxidative stress.

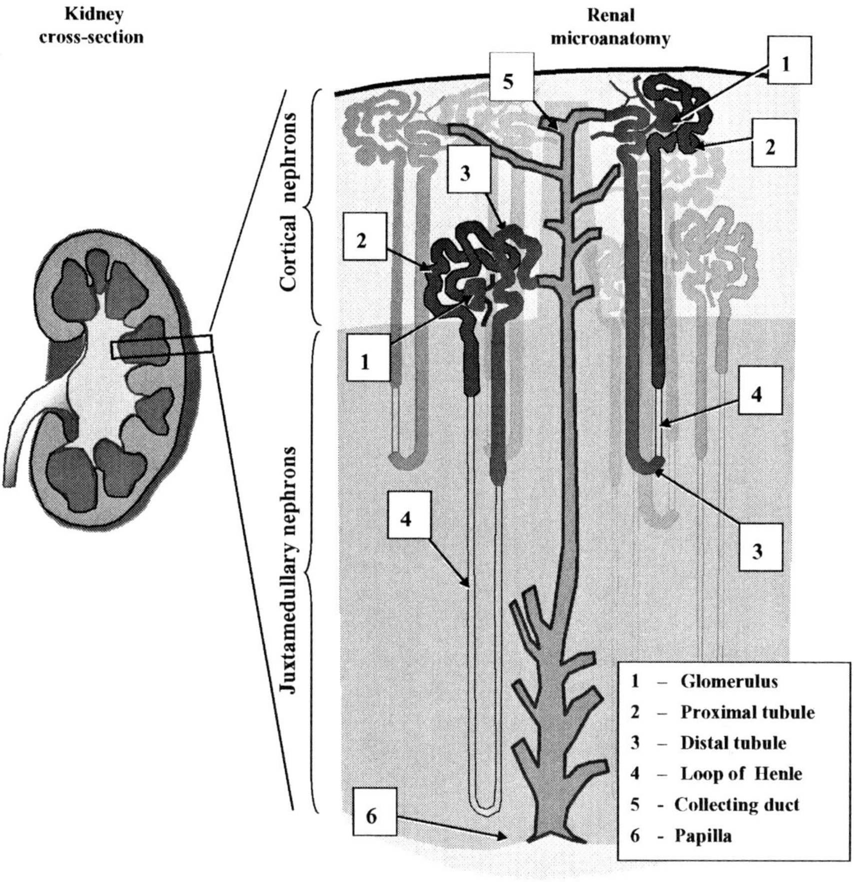

Figure 2. Renal anatomy. A coronal cross sectional diagram of a multipapillary mammalian kidney (left panel) and the microanatomies of cortical and juxtamedullary nephrons are presented (right panel). Renal calculi associated with consumption of melamine cyanurate were detected frequently within distal tubules **(3)** and collecting ducts **(5)** in most pets (references *3, 7, 8, 10,* and *11*). However, abundant renal calculi occluding the entire the nephron were detected in laboratory animals co-exposed to 400 mg/kg melamine and cyanuric acid (references *12, 13*, and *17*).

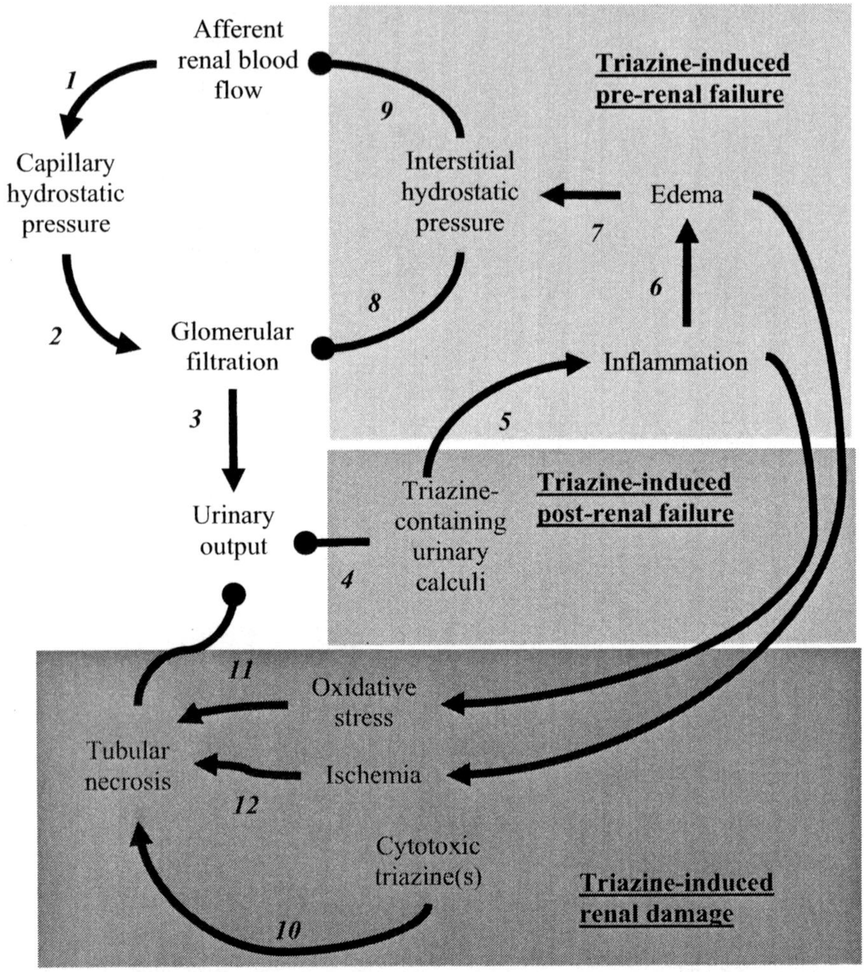

Figure 3. Renal failure mechanisms and triazine nephrotoxicity. Afferent blood flow ***(1)*** provides capillary hydrostatic pressure ***(2)*** that forces fluid through glomeruli to increase urinary output ***(3)***. Renal calculi restrict the flow of urine through nephrons ***(4)***, producing post-renal failure. Inflammation ***(5)*** caused by renal calculi stimulates edema ***(6)***, increasing interstitial hydrostatic pressure ***(7)*** that opposes filtration ***(8)*** and impedes the flow of blood into the kidneys ***(9)***, contributing to pre-renal failure. Renal damage may result when triazine compounds induce renal tubule cell death through direct cytotoxicity ***(10)*** or indirectly through oxidative stress ***(11)*** or ischemia ***(12)***.

Melamine and Cyanuric Acid Cytotoxicity

The cytotoxicities to tumor cells of mono-, di-, tri-, tetra-, penta-, and hexamethyl melamine derivatives and of pentamethylmonomethylolmelamine, hexamethylolmelamine, and trimethyltrimethylolmelamine have been known for decades *(25, 26)*. By contrast, unsubstituted melamine was up to 94-fold less potent than substituted melamines in assays with PC6 plasmacytoma cells and no cytotoxicity was detected for cyanuric acid using Chinese hamster lung cells *(15, 16)*. The unanticipated acute renal failure associated with triazine-contaminated pet foods led to a rapid re-evaluation of melamine and cyanuric acid cytotoxicities using three cell lines, COLO205 colon carcinoma cells, ACHN renal adrenocarcinoma cells, and RAW264.7 macrophage cells in our laboratory. COLO205 and ACHN cells were selected as models for triazine exposure in the colon and kidney. RAW264.7 macrophage cells were selected as an efficient phagocytic cell line to evaluate the cytotoxicity of engulfed triazine crystals. Interestingly, neither melamine nor cyanuric acid exhibited measurable cytotoxicity at ≤10 mM in colon or renal carcinoma cells, but melamine and the combination of melamine and cyanuric acid and a melamine-contaminated pet food sample were cytotoxic to RAW264.7 macrophage cells (Figure 4A-B).

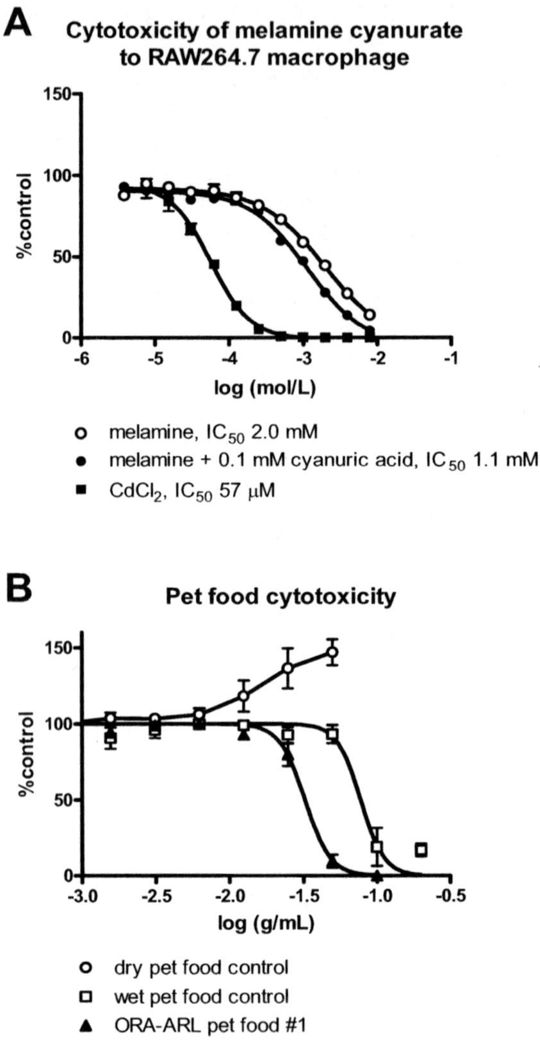

Figure 4. Melamine cyanurate cytotoxicity to RAW264.7 macrophage cells. **(A)** Melamine (open circles), melamine in the presence of 100 μM cyanuric acid (closed circles), or cadmium chloride positive control (closed boxes) were serially diluted in cell culture media and applied to RAW264.7 macrophage cells (20,000 cells/well) for 48 hrs in 96-well plates. Cell viability was measured using CellTiter Blue assays (Promega Biosciences) according to the manufacturer's instructions. **(B)** RAW264.7 macrophage cytotoxicity assays were applied to control pet foods (open symbols) and a pet food sample contaminated with melamine (filled triangles).

Physical and Chemical Properties of Melamine Cyanurate and Urinary Calculi

Chemical characterization of renal calculi in animals exposed to melamine and cyanuric acid is essential for elucidating a mechanistic scheme for the resulting renal failure syndrome. Fourier transform IR (FT-IR) analysis performed by investigators at the University of Guelph[7] and by the FDA Forensic Chemistry Center *(27)* confirmed that urinary calculi from exposed animals exhibited spectral features in common with laboratory crystals of melamine cyanurate, whether precipitated from aqueous solutions or prepared from melamine and cyanuric acid mixed with pooled feline urine. Thompson et al *(8)* reproduced the FT-IR results in their case study of three dogs and then utilized scanning electron microscopy with energy dispersive X-ray analysis (SEM-EDXA) to compare the elemental composition of melamine cyanurate-containing renal calculi in affected dogs with other common types of native kidney stones, such as calcium oxalate monohydrate (CaOx) and calcium phosphate (CaPi) nephroliths. As expected, SEM-EDXA analysis demonstrated clearly abundant carbon, nitrogen, and oxygen and the absence of calcium or phosphorus within triazine-containing crystals, compared with abundant calcium, carbon, and oxygen within CaOx crystals. The specific histological stains Oil Red O, Von Kossa Stain, and Alizarin Red S used to label nonpolar lipids or plastics, oxyanion-rich carbonates and phosphates, and metal cations, respectively, confirmed that triazine-containing crystals could be distinguished from CaOx and CaPi crystals by positive staining with Oil Red O but negative staining with Von Kossa and Alizarin Red S.

Furthermore, laboratory-formed triazine calculi produced from urine exhibited the distinctive microscopic coloration, size, and morphologic features of authentic triazine-containing renal calculi, unlike those from aqueous solutions. This observation is consistent with the possibility that constituents present in normal urine were sufficient to produce the spherical, colored, birefringent, polarizable crystals with radial internal structure instead of the smaller, white, simple needle-like crystals formed by precipitation of melamine cyanurate from water. An interesting serendipitous observation from our laboratory revealed that morphologically identical triazine crystals formed when melamine and cyanuric acid were mixed in cell culture media used for cytotoxicity assays (Minimal Essential Media supplemented with 5% fetal bovine serum). We also found that fetal bovine serum or crude preparations of bovine serum albumin are each capable of producing the characteristic crystals. GC-MS analysis of the crystals showed that equal amounts of melamine plus cyanuric acid accounted for 60% of the mass. Detergent extraction of the crystals followed by SDS-polyacrylamide gel electrophoresis exhibited major protein constituents consistent with the serum proteins albumin (68 kDa) and apolipoprotein (28 kDa). Serum proteins filtered into the urine, along with other

---

[7] Reported on-line (http://www.labservices.uoguelph.ca/urgent.cfm); updated May 1, 2007, accessed June 12, 2008.

proteins secreted by renal tubule cells, are believed to stimulate or interfere with the development of common calcium-containing urinary calculi *(28)*.

Melamine and cyanuric acid are known form a network of well-ordered intermolecular hydrogen bonds that self-assemble spontaneously *(29, 30)*. The keto-form of cyanuric acid (Figure 5) in equilibrium with the enol-form (Figure 1) is the form involved in the hydrogen bonded network with melamine. The nature and stability of melamine cyanurate complexes have been evaluated using x-ray crystallography, solid-state NMR, and calorimetry *(31-34)*.

Solvent pH affects the extent to which both melamine and cyanuric acid exist in their un-ionized forms that are available for hydrogen bonding. Although cyanuric acid is triprotic and melamine is tribasic, their first ionizations (pKa 4.74 and 5.34, respectively) are physiologically relevant within pH 5.0 – 7.3 found in the kidney. Below pH 6 melamine is converted from the uncharged free amine form to the melamine ammonium cation, destabilizing hydrogen bonding with the keto-form of cyanuric acid. Similarly, cyanuric acid dissociates above pH 4 to form its conjugate base, also destabilizing the melamine – cyanuric acid complex. Optimal concentrations of free acid and free base forms available to form hydrogen bonds can be calculated from the average of the two pKa values, e.g. pH 5.04 (Figure 6A). At conditions of constant ionic strength similar to plasma (I=0.15) and over a pH range spanning that of the nephron, a solubility minimum was revealed for melamine and cyanuric acid (Figure 6B) close to pH 5. Increased solubility was exhibited at pH 3.0, indicative that gastric acidity (pH 1-2) should facilitate dissolution of preformed melamine cyanurate complexes.

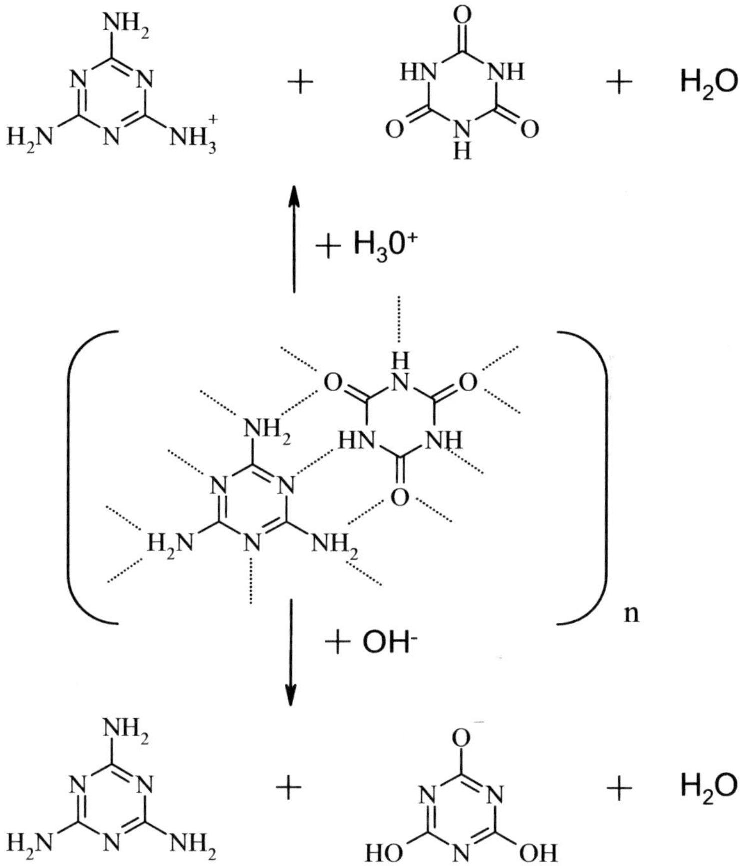

Figure 5. Melamine cyanurate acid-base ionization. Melamine protonation (upper reaction) and proton abstraction from cyanuric acid in water disrupt the stable self-assembling melamine cyanurate hydrogen bond network.

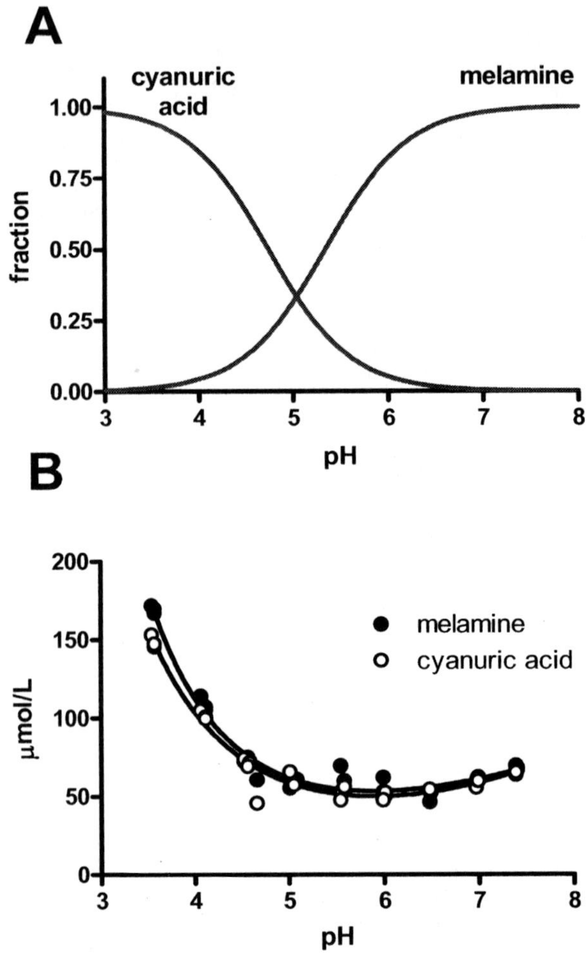

Figure 6. Solubility of melamine cyanurate. **(A)** Calculated influence of pH on ionization of melamine and cyanuric acid. **(B)** Measured solubilities of melamine and cyanuric acid over renal pH range (ionic strength 0.15).

Summary

Although neither melamine nor cyanuric acid were recognized as potent nephrotoxins, their presence together as major contaminants of pet food ingredients revealed that they comprise a toxic combination. The unusual stability of an intermolecular hydrogen bonding network associating these triazine species leads to the assembly of insoluble renal calculi with unique morphological, chemical, and biochemical properties. Enhanced solubility at the low pH of the gastric chamber enhances their migration from the gut into the circulation and delivery to the kidneys where they experience a solubility nadir in the pH range typical for the Loop of Henle. The characteristic morphologic properties of renal calculi formed from melamine and cyanuric acid can be reproduced by mixing the triazine species in urine or serum protein-containing solutions. Although melamine cyanurate-induced renal failure occurs primarily through a post-renal mechanism that occludes renal tubules and collecting ducts, the possibility exists that renal failure may be exacerbated through triazine-induced inflammation that provides a secondary pre-renal component. The presence of triazine calculi engulfed by renal institial macrophages and the direct cytotoxicity of melamine and melamine cyanurate to cultured macrophages is consistent with an immune component in triazine-induced renal failure. To enhance public safety and provide better information for science-based regulatory decision-making, further studies are necessary to determine the NOAEL for the combination of melamine and cyanuric acid in species that share similar renal physiology with humans.

Acknowledgement

The author wishes to thank Drs. Gonçalo Gamboa da Costa and Renate Reimschuessel for many helpful discussions and Russell Fairchild and Gregory Satterfield for triazine-contaminated and control pet food samples and for GC-MS triazine analyses. The contents of this report do not necessarily reflect the views and policies of the U.S. Food and Drug Administration, nor does mention of trade names or commercial products constitute endorsement or recommendation for use.

References

1. Cattaneo, P.; Ceriani, L. Situazione attuale della mellamina nelle farine di carne.*Tecnica Moliatoria* **1988,** 39, 28-32.
2. DeVries, J. W. *Economic drivers of intentional food and feed contamination*, 235th National Meeting of the American Chemical Society, New Orleans, LA, April 9, 2008.
3. Jeong, W. I.; Do, S. H.; Jeong da, H.; Chung, J. Y.; Yang, H. J.; Yuan, D. W.; Hong, I. H.; Park, J. K.; Goo, M. J.; Jeong, K. S. Canine renal failure syndrome in three dogs. *J Vet Sci* **2006,** *7,* 299-301.

4. Chen, K. Pedigree distributor recalls its pet food over mold concerns. *Taipei Times* March 12, 2004, p 10.
5. Ho, J. D. Pedigree makers say toxins to blame for dog food scare. *Taipei Times* March 24, 2004, p 11.
6. Chiu, Y. Pedigree sued by dog owners. *Taipei Times* July 2, 2004, p 2.
7. Brown, C. A.; Jeong, K. S.; Poppenga, R. H.; Puschner, B.; Miller, D. M.; Ellis, A. E.; Kang, K. I.; Sum, S.; Cistola, A. M.; Brown, S. A. Outbreaks of renal failure associated with melamine and cyanuric acid in dogs and cats in 2004 and 2007. *J Vet Diagn Invest* **2007,** *19*, 525-31.
8. Thompson, M. E.; Lewin-Smith, M. R.; Kalasinsky, V. F.; Pizzolato, K. M.; Fleetwood, M. L.; McElhaney, M. R.; Johnson, T. O. Characterization of melamine-containing and calcium oxalate crystals in three dogs with suspected pet food-induced nephrotoxicosis. *Vet Pathol* **2008,** *45*, 417-26.
9. Rovner, S. L. Anatomy of a pet food catastrophe. *Chemical and Engineering News* May 12, 2008, pp 41-43.
10. Rumbeiha, W.; Agnew, D.; Maxie, G.; Scott, M.; Hoff, B.; Powers, B. *AAVLD survey of pet food-induced nephrotoxicity in North America, April to June 2007*, American Association of Veterinary Laboratory Diagnosticians 50th Annual Conference, Reno, NV, October 20, 2007.
11. Agnew, D.; Maxie, G.; Scott, M.; Rumbeiha, W.; Hoff, B.; Powers, B. *Pet-food nephrotoxicity in cats and dogs: towards a case-definition*, American Association of Veterinary Laboratory Diagnosticians 50th Annual Conference, Reno, NV, October 21, 2007.
12. *Carcinogenesis Bioassay of Melamine (CAS No. 108-78-1) in F344/N Rats and B6C3F1 Mice (Feed Study)*; TR-245; National Toxicology Program: March, 1983; pp 110-198.
13. Puschner, B.; Poppenga, R. H.; Lowenstine, L. J.; Filigenzi, M. S.; Pesavento, P. A. Assessment of melamine and cyanuric acid toxicity in cats. *J Vet Diagn Invest* **2007,** *19*, 616-24.
14. Daston, G.; Dobson, R.; Motlagh, S.; Quijano, M.; Cambron, T.; Baker, T.; Pullen, A.; Regg, B.; Bigelow-Kern, A.; Vennard, T.; Fix, A.; Overmann, G.; Shan, Y. *Identification and Characterization of Toxicity of Contaminants in Pet Food Leading to an Outbreak of Renal Toxicity*, 2008 Annual Meeting of the Society of Toxicology, Seattle, Washington, March 17, 2008.
15. OECD SIDS: *Isocyanuric acid*; 1999; pp 219-279.
16. Ministry of Health and Welfare: Japan: *Toxicity testing reports of environmental chemicals*; Japan, 1997; pp 429-442.
17. Hazleton, U. S. *Thirteen week toxicity study in mice - monosodium cyanurate, Report 2169-100*; Hazleton Laboratories America: Vienna, VA, May 11, 1982; pp 506.
18. Canelli, E. Chemical, bacteriological, and toxicological properties of cyanuric acid and chlorinated isocyanurates as applied to swimming pool disinfection: a review. *Am J Public Health* **1974,** *64*, 155-62.
19. Hodge, H. C.; Panner, B. J.; Downs, W. L.; Maynard, E. A. Toxicity of sodium cyanurate. *Toxicol Appl Pharmacol* **1965,** *7,* 667-74.
20. Hammond, B. G.; Barbee, S. J.; Inoue, T.; Ishida, N.; Levinskas, G. J.; Stevens, M. W.; Wheeler, A. G.; Cascieri, T. A review of toxicology studies

on cyanurate and its chlorinated derivatives. *Environ Health Perspect* **1986,** 69, 287-92.

21. Reimschuessel, R. *Melamine related renal toxicity - renal crystal formation in fish using a combined dose of melamine and cyanuric acid*, 58th Meeting of the American College of Veterinary Pathologists, Savannah, GA, November 11, 2007.
22. Ensley, S.; Imerman, P.; Cooper, V.; Halbur, P.; Osweiler, G. *Determination of serum and tissue melamine and/or cyanuric acid concentrations in growing pigs*, American Association of Veterinary Laboratory Diagnosticians 50th Annual Conference, Reno, NV, Oct. 21, 2007.
23. Khan, N. M.; Alden, C. L. Kidney. In *Handbook of Toxicologic Pathology*, 2nd ed.; Hascheck, W. M.; Rousseaux, C. G.; Wallig, M. A., Eds. Academic Press: San Diego, CA, 2002; Vol. 2, pp 255-336.
24. Reimschussel, R. A fish model of renal regeneration and development. *Ilar J* **2001,** *42*, 285-91.
25. Rutty, C. J.; Abel, G. In vitro cytotoxicity of the methylmelamines. *Chem Biol Interact* **1980,** *29*, 235-46.
26. Cumber, A. J.; Ross, W. C. Analogues of hexamethylmelamine. The anti-neoplastic activity of derivatives with enhanced water solubility. *Chem Biol Interact* **1977,** *17*, 349-57.
27. Satzger, R. D. *Characterization of a pet food adulteration problem by utilizing a multidisciplinary analytical approach*, 235th National Meeting of the American Chemical Society, New Orleans, April 9, 2008.
28. Miller, N. L.; Evan, A. P.; Lingeman, J. E. Pathogenesis of renal calculi. *Urol Clin North Am* **2007,** *34*, 295-313.
29. Whitesides, G. M.; Mathias, J. P.; Seto, C. T. Molecular self-assembly and nanochemistry: a chemical strategy for the synthesis of nanostructures. *Science* **1991,** *254*, 1312-9.
30. Seto, C. T.; Whitesides, G. M. Self- Assembly Based on the Cyanuric Acid-Melamine Lattice. *J. Am. Chem. Soc.* **1990,** *112,* 6409-641 1.
31. Ranganathan, A.; Pedireddi, V. R.; RAo, C. N. R. Hydrothermal Synthesis of Organic Channel Structures: 1:1 Hydrogen-Bonded Adducts of Melamine with Cyanuric and Trithiocyanuric Acids. *J. Amer. Chem. Soc.* **1999,** *121*, 1752-1753.
32. Damodaran, K.; Sanjayan, G. J.; Rajamohanan, P. R.; Ganapathy, S.; Ganesh, K. N. Solid state NMR of a molecular self-assembly: multinuclear approach to the cyanuric acid-melamine system. *Org Lett* **2001,** *3*, 1921-4.
33. Bielejewska, A. G.; Marjo, C. E.; Prins, L. J.; Timmerman, P.; de Jong, F.; Reinhoudt, D. N. Thermodynamic stabilities of linear and crinkled tapes and cyclic rosettes in melamine--cyanurate assemblies: a model description. *J Am Chem Soc* **2001,** *123*, 7518-33.
34. ten Cate, M. G.; Huskens, J.; Crego-Calama, M.; Reinhoudt, D. N. Thermodynamic stability of hydrogen-bonded nanostructures: a calorimetric study. *Chemistry* **2004,** *10*, 3632-9.

Chapter 6

Mycotoxins of Concern in Imported Grains

Dojin Ryu[1] and Lloyd B. Bullerman[2]

**[1] Department of Nutrition & Food Sciences
Texas Woman's University
Denton, TX 76204
[2] Department of Food Science & Technology
University of Nebraska
Lincoln, NE 68583**

In the fiscal year 2007, the value of agricultural imports to the United States was expected to reach $70.5 billion with approximately $8.5 billion in grains, grain products and feed. Steadily rising volumes of imported foods suggest that a greater proportion of the U.S. food supply is produced and/or manufactured outside of direct food safety oversight by the public and private sectors. Cereal grains, including wheat, corn, barley, oats, rye and rice, may be imported from different countries and used as whole grains or ingredients in processed foods. These cereal grains may be contaminated with mycotoxins produced by certain filamentous fungi in the field, in storage and/or during transport. Among known mycotoxins, aflatoxins, ochratoxins, fumonisins, trichothecenes, including deoxynivalenol and T-2 toxin are major concerns. These mycotoxins are stable and may remain in finished grain based processed food products.

Introduction

With the growing economy and changing consumer demand, international trade has been increased and diversified in virtually all sectors of industry for the last few decades. International trade of agricultural products including foods and feeds has also significantly increased (*1*). The average share of imports in

U.S. food consumption increased from approximately 8% in the early 1980s to more than 11 percent in 2000 (*2*). Almost all groups of foods recorded increase of their shares including fruits, vegetables, fish, shellfish, tree nuts, pork and beef. Changes in import shares of grains such as wheat and rice were more significant than others, expanding from less than 1% to 11% by volume. These trends were also reflected in increased annual per capita food consumption from about 1,800 pounds to more than 2,000 pounds during the same period. Based on the volume of imported grains (Table 1), the import share of consumed grains and products in the U.S. reached 22.2% in 2007, up from 4.6% in the early 1980s (*3, 4*).

Table 1. U.S. Imports of Grains in Volume (metric tons)

	2003	2004	2005	2006	2007
Wheat	906,953	1,119,257	1,297,586	2,011,942	2,342,138
Oats	1,563,289	1,421,660	1,683,795	1,793,632	1,880,251
Rice	448,059	463,720	408,078	622,085	683,061
Barley	291,400	432,496	112,572	185,078	449,801
Corn	274,755	263,674	223,699	107,831	221,122

SOURCE: Data retrieved from the database FAS Online, U.S. Department of Agriculture. URL http://www.fas.usda.gov/USTrade/USTIMFATUS.asp?QI=

Mycotoxins occur in all major cereal grains worldwide. These toxic metabolites of certain fungi may cause significant economic loss and, more importantly, pose threats to animals and humans with their wide range of toxicities. This chapter provides a brief overview of mycotoxins considered significant in imported grains due to their occurrence and toxicities in animals and humans.

Mycotoxins of Concern

Mycotoxin Formation and Natural Occurrence

Mycotoxins are toxic compounds produced by filamentous fungi. They are considered to be secondary metabolites or compounds that are not utilized as components of the basic life processes of the organism. Not all fungal species produce mycotoxins, but certain toxigenic species and strains are known to produce mycotoxins with varying degrees of capabilities. Most of the mycotoxins considered important are produced mainly by the three fungal genera: *Aspergillus*, *Penicillium*, and *Fusarium*. These organisms may be found in soil and/or air in all agricultural areas worldwide, and thus contaminate many crops and commodities including cereal grains (Table 2).

Following invasion and growth of the fungi, the toxic metabolites tend to accumulate in the substrate or matrix as the organisms produce mycotoxins during their growth. Cereal grains are the most frequently affected commodity as the fungi may invade the plants in the field during the growing season as well as during the postharvest handling processes such as drying and storage. One of the most important factors affecting the growth of fungi and subsequent toxin production is moisture content. In general, cereal grains are dried to less than 12-13% of moisture content to prevent fungal growth and for safe storage. Temperature, drought stress, and insect infestation are also considered important factors in fungal contamination and mycotoxin production.

Table 2. Mycotoxins of Concern in Foods and the Producing Organisms

Mycotoxins	Major producing organisms	Commodities commonly affected
Aflatoxins	*Aspergillus flavus, A. parasiticus, A. nomius*	Corn, peanuts, tree nuts, rice, cottonseed
Ochratoxins	*A. ochraceus, A. carbonarius, Penicillium verrucosum*	Wheat, barley, oats, rye, sorghum, peanuts, peas, beans, green coffee beans, raisins, beer, wine, etc.
Fumonisins	*Fusarium verticillioides, F. proliferatum, F. subglutinans*	Corn, wheat, barley, rice
Deoxynivalenol (DON, Vomitoxin)	*F. graminearum, F. culmorum, F. crookwellense*	Wheat, barley, rye
T-2 Toxin	*F. sporotrichioides, F. poae*	Corn, wheat, barley, rice, rye, oats

The cosmopolitan nature of many toxigenic fungi and fluctuating environmental conditions result in mycotoxin contamination in a wide variety of commodities. As shown in Table 1, a mycotoxin can be produced by different strains/species of fungi. It is also common for a single strain of fungi to produce more than one toxin. These characteristics of fungi and their mycotoxin production lead to a prominent problem, i.e. existence of multiple mycotoxins in a single commodity. Thus, it is practically impossible to achieve zero concentration of all mycotoxins in agricultural crops (*5*).

Aflatoxins are produced mainly by *Aspergillus flavus* and *A. parasiticus*. *Aspergillus nomius* may also produce aflatoxins but this species has only been found in soils in the western United States. and to some extent in corn fields in

Thailand. Among the members of aflatoxins, B_1, B_2, G_1 and G_2 are considered major contaminants of cereal grains, cottonseed, tree nuts, peanuts, spices and a variety of other foods and feeds. These four aflatoxins have been found mostly in parts per billion (ppb; μg/kg) levels (*6*). Aflatoxins, particularly M_1 and M_2, may also be found in some animal products such as milk, eggs and meat, as metabolized products of ingested B_1 and B_2. Aflatoxin B_1 is the most toxic and most carcinogenic of the aflatoxins.

Ochratoxins are produced by *Aspergillus ochraceus* and related species, *A. carbonarius*, and some strains of *A. niger*. Ochratoxins are also produced by *Penicillium verrucosum* and certain other *Penicillium* species. Ochratoxin A is known to be the most toxic of the group and detected in greater amounts than other ochratoxins. Ochratoxin A has been reported in varying parts per billion levels as naturally occurring in a wide variety of agricultural crops and products including corn, wheat, sorghum, oats, rice, dried fruits, grapes, raisins, wine, beer, and green coffee beans (*6-11*). It should be noted that the occurrence of ochratoxin A is more widespread than any other mycotoxins, affecting a number of different field crops and their products. It has also been found in pork meat (especially kidney), cow's milk and human milk and serum (*6, 12*). Therefore, it is difficult to estimate the exposure of toxin from such variety of foods. This also implicates a significant food safety concern since insufficient or inaccurate exposure data may not warrant safe standards for regulation.

Mycotoxins produced by *Fusarium* species frequently contaminate many agricultural commodities worldwide. *Fusarium verticillioides* (formerly known as *F. moniliforme*) is the most common fungus found on corn (maize) and may produce several mycotoxins including fumonisins. Although many *Fusarium* species are considered as plant pathogens, *F. verticillioides* is an endophyte and may be present in corn without producing symptoms. Fumonisins can be found at varying levels even in corn that does not show visible signs of fungal infection. Other *Fusarium* species, including *F. proliferatum* and *F. subglutinans*, may also infect crops and produce fumonisins. Fumonisins have been reported frequently ranging from low levels (< 50 ppb) to over 5 ppm in many cereal grains and feed materials (*13*). In addition, surveys have shown that fumonisin B_1 occurs not only in corn grain, but also in finished corn-based processed food products worldwide (*14-16*). Among all fumonisins, fumonisin B_1 is produced in the greatest amount and is known to be the most toxic.

Other fungi in the genus *Fusarium* produce trichothecenes, another important group of mycotoxins including T-2 toxin, deoxynivalenol and nivalenol. With about 150 structurally related compounds, trichothecenes are the largest family of known mycotoxins (*17*). Genera other than *Fusarium*, i.e., *Cephalosporium*, *Myrothecium*, *Stachybotrys*, and *Trichoderma*, may also produce trichothecenes but they are not considered significant in food and feed. Similar to many other mycotoxins, the occurrence of trichothecenes in grains varies from year to year depending on climate conditions and the degree of fungal infestation. In addition, the occurrence of individual toxins may be affected by the geographical region, e.g. deoxynivalenol is commonly detected in North America, but T-2 toxins and nivalenol are more frequently observed in Europe and Asia (*18*). Deoxynivalenol (DON) is produced primarily by *Fusarium graminearum* (teleomorph: *Gibberella zea*), which causes Fusarium

Head Blight (or Scab) in wheat and barley and ear rot in corn. T-2 toxin is produced by *F. sporotrichioides* and *F. poae*.

Toxicity of Mycotoxins

Mycotoxins exert their biological effects in various target organs including liver, kidney, and skin. The adverse effect may extend to the reproductive, immune, and nervous system. In addition, some mycotoxins are considered as mutagenic, teratogenic, and/or carcinogenic. These effects are based on human epidemiological/clinical studies and numerous experiments with animal models (Table 3). The main route of exposure is ingestion while inhalation or direct dermal contact may also lead to toxic responses depending on the nature of mycotoxin. While acute disease or death may occur from high levels of exposure, an array of adverse effects from chronic exposure to low levels of mycotoxins are more commonly observed. In farm animals, chronic exposure to mycotoxins often result in reduced weight gain, decreased milk or egg production and decreased disease resistance and tumor formation. Among all known toxicities of mycotoxins, immunotoxic effects may cause the most significant health concerns and economic impact particularly in animal husbandry. As immunosuppression could lead to increased infections and secondary diseases in animals and humans alike, increased infections in food-producing animals may also implicate an increased risk of transmitting pathogens, such as *Listeria* and *Salmonella*, from animals to humans.

Table 3. Toxicity and target organs of major mycotoxins

	Aflatoxin	Ochratoxin	Fumonisin	DON	T-2
Liver	√	√	√		
Kidney		√	√	√	
Skin					√
Neurotoxic			√	√	
Immunotoxic	√	√	√	√	√
Teratogenic	√	√			√
Carcinogenic	√	√	√		

Synergistic effects of mycotoxins raise another significant concern. As discussed earlier, multiple mycotoxins could be produced by one or more fungal strains to contaminate one commodity. The probability of exposure from multiple toxins and subsequent toxic effects may be magnified when more than one grain is used to produce a food or feed (*5*). While additive or antagonistic effects may be expected in certain combinations, potentiative or synergistic effects have been observed with several mycotoxins (*19*).

Aflatoxins

Aflatoxins are potent hepatotoxins and known to cause or contribute to causing hepatocellular carcinoma. Acute aflatoxicosis, aflatoxin exposure at high doses, result in death in both animals and humans mainly via gross liver damage and intestinal hemorrhaging (*20*). Sublethal doses produce a chronic toxicity that may lead to weakened immune systems and greater susceptibility to infections. Exposure to very low doses over a period of time produces tumors in test animals and is known to contribute to liver cancer. Aflatoxin B_1 has been classified as a Class I human carcinogen by the International Agency for Research on Cancer (*21*).

The significance of aflatoxin may be found in the increased incidence of liver cancer in certain localities. As a number of epidemiological data suggest, the presence of aflatoxins in staple food has been considered a significant risk factor for human liver cancer (*22*). The mutagenic potential of aflatoxin B_1 has been shown by a specific DNA adduct found in urine from individuals in high liver cancer risk areas as well as laboratory animals (*23*). Biomarkers that may link carcinogenicity of aflatoxin to human liver cancer have also been developed and used to prove its involvement (*24*).

Another concern about aflatoxin is attributed to its metabolism in animals, particularly in dairy cattle. An average of 1-2% of aflatoxins in contaminated feed may be converted to the hydroxylated forms, i.e. from aflatoxin B_1 and B_2 to aflatoxin M_1 and M_2, and excreted in milk (*25*). Since aflatoxins are stable in most food processes, these aflatoxins in milk may contaminate products in the downstream processes.

Ochratoxin A

Ochratoxin A is a potent nephrotoxin and has been suggested as a causative agent of human nephropathy in Balkan countries known as Balkan Endemic Nephropathy and possibly elsewhere (*26, 27*). While ochratoxin A has been found in the blood of Europeans and Canadians, there is no direct evidence to blame this mycotoxin on human renal disease. Ochratoxin A also has shown to damage liver and kidney in animals including swine, dogs and rats. Furthermore, ochratoxin A is teratogenic, carcinogenic, and immunotoxic (*5*). In addition to ingestion, inhalation is another possible route of exposure since ochratoxin or grain dust containing the toxin may become airborne (*28, 29*).

As mentioned earlier, occurrence of ochratoxin in a wide variety of agricultural products and processed products is of major concern. Most frequently contaminated commodities include wheat, barley, oats, rye, sorghum, peanuts, peas, beans, green coffee beans, raisins, beer, and wine. Although ochratoxin can be reduced significantly during the roasting of coffee beans (> 50%), this water soluble toxin may persist in raisins, wine, beer and other products where heat processing is not as severe or is not required (*12*).

Fumonisins

Being the most potent and abundant, the toxicities of fumonisin B_1 have been demonstrated in domestic and laboratory animals, including equine leukoencephalomalacia (*30*), porcine pulmonary edema (*31*), and hepato- and nephrotoxicity in rodents (*32*). There also is evidence linking *F. verticillioides* infected corn to the high incidence of human esophageal cancer in South Africa and China (*33, 34*). More recently, fumonisin B_1 has been identified as a risk factor for neural tube defects in humans (*35-37*).

The importance of fumonisins is mainly attributed to their prevalence in cereal grains particularly in corn. Despite its lower toxicity, fumonisins may be found in virtually all corn samples regardless of the factors affecting fungal infection, growth, and toxin production. Stability of fumonisins is also of concern as the toxins are known to persist in most finished food products.

Trichothecenes

Trichothecenes in general are of concern due to their immunotoxicity and potency in inhibition of protein and DNA synthesis (*38*). Toxicities of trichothecenes may be observed as feed refusal and vomiting, weight loss, bloody diarrhea, hemorrhage, decreased egg production, abortion, and death. The two most potent and common trichothecenes are deoxynivalenol and T-2 toxin.

Deoxynivalenol is also known as vomitoxin since ingestion of this toxin causes emesis in animals including swine, dogs, cats and humans. Swine are particularly sensitive to DON as clinical signs may be observed with DON concentrations as low as 1.0 mg/kg (*39, 40*). While DON can be acutely toxic and lead death at high concentrations, altered immune function, reduced weight gain, and poor performance of domestic animals are more commonly observed at low to intermediate concentrations (*41*). Immunotoxicity of DON may be observed as immunosuppression and immunostimulation that may lead to increased susceptibility to infectious diseases and autoimmune disorders, respectively.

T-2 Toxin is another trichothecene that is of importance. Although the natural occurrence of T-2 toxin is lower, it is considered to be more potent than DON in inhibiting synthesis of protein, DNA and RNA. Poultry, particularly chickens and turkeys, are sensitive to T-2 toxin. The most notable adverse effect of T-2 toxin is dermal toxicity, causing irritation, reddening, and necrosis of skin. T-2 toxin is believed to be the primary cause of a human disease that occurred in the former Soviet Union during World War II known as Alimentary Toxic Aleukia. In comparison with T-2 toxin, DON has very low dermal toxicity (*42*).

Regulation of Mycotoxins

Many countries regulate the level of mycotoxins in foods and various agricultural commodities. Since mycotoxins are naturally occurring and not entirely controllable, certain levels of contamination may be expected in human foods and animal feeds. Efforts have been made to minimize their occurrence to avoid deleterious effect. These efforts include good agronomic and manufacturing practices, and employing regulatory limits. Countries including the U.S. and the European Union (EU) have established action levels and guidance levels to ensure the safety of foods and feeds based on the data from natural occurrence, toxicological, and epidemiological studies (Table 4). These action levels and guidelines in different countries do not always agree as they reflect diverse food intake, availability of commodities, and different geographical/climate conditions for mycotoxin contamination. Nonetheless, the presence of legal limits is important in international trade because they are applicable to the imported commodities as well as to the domestic counterparts.

Table 4. Action levels and guidelines in the U.S. and European Union

Mycotoxin	Commodity of Food	U.S.A.	E.U.
Aflatoxin	Human Food	20 ppb	2-15 ppb
Aflatoxin M_1	Milk	0.5 ppb	0.05 ppb
Aflatoxin	Dairy Feed	20 ppb	
Aflatoxin	Breeding Stock Feed	100 ppb	
	Mature Beef, Swine and Poultry Feed	300 ppb	
Ochratoxin A	Cereal Grains[2]		5.0 ppb
	Processed Cereal Products		3.0 ppb
	Dried Fruits, Wine		10.0 ppb
Deoxynivalenol	Finished Wheat Products	1.0 ppm	0.5 ppm
	Flour		0.75 ppm
	Raw Cereals		0.75 ppm
Fumonisins	Degermed Corn Meal	2.0 ppm	1.0 ppm
	Whole/Partly Degermed	4.0 ppm	
	Dry Milled Corn Bran	4.0 ppm	
	Cleaned Popcorn	3.0 ppm	
	Cleaned Corn for Masa	4.0 ppm	

NOTE: Where no levels are given, action levels or guidance levels have not yet been established.

Summary

Among hundreds of known mycotoxins, aflatoxins, ochratoxins, fumonisins and trichothecenes are considered most significant due to their occurrence, economic impact, and toxicity in animals and humans. Aflatoxins have been studied most extensively and have prompted investigations of other mycotoxins regarding their production, occurrence, and toxicities since its discovery in the early 1960s. In addition to the mycotoxins listed above, zearalenone should be noted with its unique estrogenic properties. This endocrine disruptor may be produced simultaneously with DON by the same organisms and can affect the reproductive system in animals. In recent years, ochratoxin A and many of the *Fusarium* toxins, including fumonisins and trichothecenes, have been given more attention. Although these mycotoxins are not as toxic as aflatoxins, they tend to occur more frequently in greater concentrations in a wide variety of commodities.

Despite the researches and technological advances, mycotoxins remain a threat to food safety. Globalization and increasing trends in trade of agricultural commodities also add to the importance of safeguarding our food supplies. This has prompted greater effort in surveillance and monitoring programs as well as in research to improve mycotoxin detection. It also becomes more plausible to adopt public policies to accommodate changing risks of exposure from the increasing numbers of commodities that may be contaminated by mycotoxins. Ochratoxin A is a good example of such a need since the U.S. has not yet established regulatory guidelines. Furthermore, development of uniform international standards and regulations for mycotoxins is necessary to ensure a safer supply of food and feed.

References

1. U.S. Census Bureau. *U.S. International Trade in Goods and Services.* CB08-42, BEA 08-08, FT-900 (08-01), March; U.S. Department of Commerce: Washington, DC, 2008.
2. Jerado, A. *Import share of U.S. food consumption stable at 11 percent.* FAU-79-01, July; U.S. Department of Agriculture, Economic Research Service: Washington, DC, 2003. http://www.ers.usda.gov/publications/fau/july03/fau7901/fau7901.pdf (accessed October 16, 2008).
3. Jerado, A. *What share of U.S. consumed food is imported?* Amber Waves, February; U.S. Department of Agriculture, Economic Research Service: Washington, DC, 2008. http://www.ers.usda.gov/AmberWaves/February08/DataFeature/ (accessed October 16, 2008).
4. Jerado, A. *Personal communication.* 2008.

5. *Mycotoxins: risks in plant, animal, and human systems;* Richard, J. L.; Payne, G. A. Eds.; Council for Agricultural Science and Technology Task Force (CAST) Report No. 139: Ames, IA., 2003.
6. Binder, E. M.; Tan, L. M.; Chin, L. J.; Handl, J.; Richard, J. Worldwide occurrence of mycotoxins in commodities, feeds and feed ingredients. *Ani. Feed Sci. Tech.* **2007**, 137, 265-282.
7. Jorgensen, K. Survey of pork, poultry, coffee, beer and pulses for ochratoxin A. *Food Addit. Contam.***1998**, *15*, 550-554.
8. MacDonald, S.; Wilson, P.; Barnes, K.; Damant, A.; Massey, R.; Mortby, E; Shepherd, M. J. Ochratoxin A in dried vine fruit: Method development and survey. *Food Addit. Contam.* **1999**, *16*, 253-260.
9. Otteneder, H.; Majerus, P. Occurrence of ochratoxin A (OTA) in wines: Influence of the type of wine and its geographical origin. *Food Addit. Contam.* **2000**, *17*, 793-798.
10. Park, J. W. Ochratoxin A in Korean food commodities: Occurrence and safety evaluation. *J. Agric. Food Chem.* **2005**, *53*, 4637-4642.
11. Patel, S.; Hazel, C. M.; Winerton, G. M.; Gleadle, A. E. Survey of ochratoxin A in UK retail coffees. *Food Addit. Contam.* **1997**, *14*, 217-222.
12. Bayman, P.; Baker, J. L. Ochratoxins: A global perspective. *Mycopathologia* **2006**, *162*, 215-223.
13. Shephard, G. S.; Thiel, P. G.; Stockenstrom, S.; Sydenham, E. W. Worldwide survey of fumonisin contamination of corn and corn-based products. *JAOAC Int.* **1996**, *79*, 671-687.
14. Sydenham, E. W.; Shephard, G. S.; Thiel, P. G.; Marasas, W. F. O.; Stockenstorm, S. Fumonisin contamination of commercial corn-based human foodstuffs. *J. Agric. Food Chem.* **1991**, *25*: 767-771.
15. Chu, F. S.; Li, G. Simultaneous occurrence of fumonisin B_1 and other mycotoxins in moldy corn collected from the People's Republic of China in regions with high incidence of esophageal cancer. *Appl. Environ. Microbiol.* **1994**, *60*, 847-852.
16. Gutema, T.; Munimbazi, C.; Bullerman, L. B. Occurrence of fumonisins and moniliformin in corn and corn-based food products of U. S. origin. *J. Food Prot.* **2000**, *63*, 1732-1737.
17. Grove, J. F. Non-macrocyclic trichothecenes. *Nat. Prod. Rep.* **1988**, *5*, 187-209.
18. McMullen, M.; Jones, R.; Gallenberg, D. Scab of wheat and barley: A re-emerging disease of devastating impact. *Plant Dis.* **1997**, *81*, 1340-1348.
19. Kubena, L. F.; Huff, W. E.; Harvey, R. B.; Corrier, D. E.; Phillips, T. D.; Creger, C. R. 1988. Influence of ochratoxin A and deoxynivalenol on growing broiler chicks. *Poultry Sci.* **1988**, *67*, 253-260.
20. Ngindu, A.; Kenya, P. R.; Ocheng, D. M.; Omonde, T. N.; Ngare, W.; Gatei, D.; Johnson, B. K.; Ngira, J. A.; Nandwa, H.; Jansen, A. J.; Kaviti, J. N.; Siongok, T. A. Outbreak of acute hepatitis caused by aflatoxin poisoning in Kenya. *Lancet* **1982**, *1*, 1346-1348.
21. *Some Traditional Herbal Medicines, Some Mycotoxins, Naphthalene and Styrene;* International Agency for Research on Cancer (IARC) Monographs on the Evaluation of Carcinogenic Risks to Humans February 12–19; 2002; Vol. 82, pp. 171-300.

22. Henry, S. H.; Bosch, G. X.; Troxell, T. C.; Bolger, P. M. Reducing liver cancer- global incidence of aflatoxin. *Science* **1999**, *286*, 2453-2454.
23. Essigmann, J. M.; Croy, R. G.; Nadzan, A. M.; Busby, Jr., W. F.;Reinhold, V. M.; Biechi, G.; Wogan, G. N. Structural identification of the major DNA adduct formed by aflatoxin B_1 in vitro. *Proc. Natl. Acad. Sci.* **1977**, *74*, 1870-1874.
24. Scholl, P.; Groopman, J. D. Epidemiology of human aflatoxin exposures and its relationship to liver cancer. In *Molecular Approaches to Food Safety: Issues Involving Toxic Microorganisms;* Eklund, M.; Richard, J. L.; Mise, K., Eds.; Alaken, Inc.: Fort Collins, CO, 1995; pp 169-182.
25. van Egmond, H. P. Aflatoxin M_1 occurrence, toxicity, regulation. In *Mycotoxins in Dairy Products;* van Egmond, H. P. Ed.; Elsevier Applied Science Publishers LTD: Barking, Essex, England, 1989; pp 11-55.
26. Krogh, P. Ochratoxins. In *Mycotoxins in Human and Animal Health;* Rodricks, J. V.; Hesseltine, C. W.; Mehlman, M. A., Eds.; Pathotox Publishers, Inc.: Park Forest South, IL, 1977; pp 489-498.
27. Wanigasuriya, K.P. Could Ochratoxin A in Food Commodities be the Cause of Chronic Kidney Disease in Sri Lanka? *Royal Soc. Trop. Med. Hyg.* **2008**, *102*, 726-728.
28. Richard, J. L.; Plattener, R. D.; May, J.; Liska, S. L. The occurrence of ochratoxin A in dust collected from a problem household. *Mycopathologia* **1999**, *146*, 99-103.
29. Skaug, M. A.; Wignand, E.; Stormer, F. C. Ochratoxin A in airborne dust and fungal conidia. *Mycopathologia* **2001**, *151*, 93-98.
30. Marasas, W. F. O.; Kellerman, T. S.; Gelderblom, W. C. A.; Coetzer, J. A. W.; Thiel, P. G.; van Lugt, J. J. Leukoencephalomalacia in a horse induced by fumonisin B_1 isolated from *Fusarium moniliforme*. *Onderstepoort J. Vet. Res.* **1988**, *55*, 197-203.
31. Harrison, L. R.; Colvin, B. M.; Greene, J. T.; Newman, L. E.; Cole, Jr., J. R. Pulmonary edema and hydrothorax in swine produced by fumonisin FB_1, a toxic metabolite of *Fusarium moniliforme*. *J. Vet. Diagn. Invest.* **1990**, *2*, 217-221.
32. Howard, P. C., Couch, L. H.; Patton, R. E.; Eppley, R. M.; Doerge, D. R.; Churchwell, M. I.; Marques, M. M.; Okerberg, C. V. Comparison of the toxicity of several fumonisin derivatives in a 28-day feeding study with female B6C3F1 mice. *Toxicol. Appl. Pharmacol.* **2002**, *185*, 153-165.
33. Gelderblom W. C. A.; Jaskiewicz, K.; Marasas, W. F. O.; Thiel, P. G.; Horak, R. M.; Vleggaar, R.; Kriek, N. P. J. Fumonisins-Novel mycotoxins with cancer promoting activity produced by *Fusarium moniliforme*. *Appl. Environ. Microbiol.* **1988**, *54*, 1806-1811.
34. Yoshizawa, T.; Yamashita, A.; Luo, Y. Fumonisin occurrence in corn from high- and low-risk areas for human esophageal cancer in China. *Appl. Environ. Microbiol.* **1994**, *60*, 1626-1629.
35. Marasas W. F. O.; Riley, R. T.; Hendricks, K. A.; Stevens, V. L.; Sadler, T. W.; Gelineau-van Waes, J.; Missmer, S. A.; Cabrera, J.; Torres, O. L.; Gelderblom, W.; Allegood, J.; Martínez de Figueroa, A. C.; Maddox, J.; Miller, J. D.; Starr, L.; Sullards, M. C.; Roman Trigo, A. V.; Voss, K. A.; Wang, E.; Merrill, Jr. A. H. Fumonisins disrupt sphingolipid metabolism,

folate transport and development of neural crest cells in embryo culture and in vivo: A risk factor for human neural tube defects among populations consuming fumonisin-contaminated maize. *J. Nutr.* **2004**, *134*, 711-716.

36. Gelineau-van Waes, J. B.; Starr, L.; Maddox, J. R.; Aleman, F.; Voss, K. A., Wilberding, J.; Riley, R. T. Maternal fumonisin exposure and risk for neural tube defects: mechanisms in an *in vivo* mouse model. Birth Defects Res. (Part A). *Clin. Mol. Teratol.* **2005**, *73*, 487-497.
37. Missmer, S. A.; Suarez, L.; Felkner, M.; Wang, E.; Merrill, Jr. A.E.; Rothman, K. J.; Hendricks, K. A. Exposure to fumonisins and the occurrence of neural tube defects along the Texas-Mexico border. *Envrion. Health Perspect.* **2006**, *114*, 237-241.
38. Pestka, J. J.; Smolinski, A. T. Deoxynivalenol: Toxicology and potential effects on humans. *J. Toxicol. Environ. Health* **2005**, *8* (Part B), 39-69.
39. Rotter, B. A.; Thompson, B. K.; Lessard, M.; Trenholm, H. L.; Tryphonas, H. Influence of low-level exposure to *Fusarium* mycotoxins on selected immunological and hematological parameters in young swine. *Fund. Appl. Toxicol.*, ***23***, 117-124.
40. Trenholm, H. L.; Hamilton, R. M. G.; Friend, D. W.; Thompson, B. K.; Hartin, K. E. Feeding trials with vomitoxin (deoxynivalenol)-contaminated wheat: Effects on swine, poultry, and dairy cattle. *J. Am. Vet. Med. Assoc.* **1984**, *185*, 527-531.
41. Pestka, J. J. Deoxynivalenol: Toxicity, mechanisms and animal health risks. *Ani. Feed Sci. Tech.* **2007**, *137*, 283-298.
42. Ueno, Y. Toxicological features of T-2 toxin and related trichothecenes. *Fund. Appl. Toxicol.* **1984**, *4*, s124-s132.

Chapter 7

Effect of Heat-Processed Foods on Acrylamide Formation

Fadwa Al-Taher

National Center for Food Safety and Technology, Illinois Institute of Technology, Summit-Argo, IL, 60501, USA

Acrylamide is formed via the Maillard reaction between reducing sugars and asparagine in carbohydrate-rich foods during heat treatment (>120 °C) processes such as frying, baking, roasting and extrusion. Acrylamide formation increases as product temperature increases (in the range of 120-180 °C). Although acrylamide is known to form during industrial processing of food, high levels of chemical have also been found in home-cooked foods, mainly potato- and grain- based products. There have been concerns about the potential health issues associated with the dietary intake of acrylamide. Much attention has been focused on finding ways to reduce or prevent the formation of acrylamide in foods without compromising food safety, organoleptic properties (e.g. taste, texture, and color) or adversely affecting nutritional quality. This chapter will review the studies on acrylamide that have been performed to date regarding its occurrence, formation, toxicity, prevention and mitigation, and analysis.

Introduction

Acrylamide (2-propenamide) is a colorless and odorless solid at room temperature in its pure form. It has a melting point of 84.5 °C, at which point it polymerizes to polyacrylamide *(1, 2)*. It is soluble in water, acetone, or alcohol. The main uses of polyacrylamide are as a flocculant in the treatment of municipal water supply and in paper and pulp processing. Other applications include soil conditioners to remove suspended solids from industrial waste water before discharge, reuse or disposal, cosmetic additives, and in the formulation of grouting agents. Polyacrylamide gels are also used in the research laboratory as a solid support for the separation of proteins by electrophoresis. Acrylamide is also present in tobacco smoke *(3)*. Acrylamide is known to be a neurotoxin and a carcinogen in animals, and a possible carcinogen in humans by the International Agency for Research on Cancer (IARC) *(4)*.

In 2002, researchers at the Swedish National Food Administration and Stockholm University reported finding acrylamide at levels up to 3 mg/kg in a wide range of potato- and cereal-based products such as potato chips, French fries, roasted and baked potatoes, bread, breakfast cereals, and biscuits *(5)*. This resulted in greatly heightened worldwide interest in understanding the formation and occurrence of acrylamide and its effects on humans. Acrylamide forms in carbohydrate-rich foods that are subjected to high-temperature (>120 °C) processes such as frying, baking and extrusion. Acrylamide can be found in processed foods as well as foods prepared by consumers. It is not present in uncooked food or in foods that are cooked at lower temperatures (e.g. boiled foods). Acrylamide is generated from certain food products during heat treatment as a result of the Maillard reaction between certain amino acids and reducing sugars *(6, 7, 8)*. This chapter will give a broad overview of the work that has been conducted so far on acrylamide.

Acrylamide Link to Cancer Risk

Acrylamide is recognized as a neurotoxin in humans and as a carcinogen in animals. It is classified as a "probable human carcinogen" by the IARC *(4)*. Acrylamide is carcinogenic to experimental mice and rats, causing tumors at many organ sites when given in drinking water or by other means *(9)*. In mice, acrylamide increased the incidence and collection of lung tumors and skin tumors *(10, 11)*. In two bioassays in rats, acrylamide administered in drinking water consistently produced mesotheliomas of the testes, thyroid tumors, and mammary gland tumors *(12)*. Also, brain tumors increased. In one of the rat bioassays, pituitary tumors, pheochromocytomas, and uterine tumors were noticed *(13)*.

More than one third of the calories that is consumed by the U.S. and European populations are derived from food that contains acrylamide *(14)*. Generally, the foods that contribute the most to acrylamide intake are potato products, breads and coffee. Estimated dietary acrylamide intake in populations has been determined by national food administrations for several countries (France, Germany, The Netherlands, Sweden, United States). The mean dietary intake of acrylamide averages 0.5 μg/kg of body weight per day, whereas intake is higher (1.0 μg/kg) among children. Studies have been done to determine whether the amount of acrylamide in the human diet is an important cancer risk factor *(15)*.

Several epidemiological studies examined the relationship between dietary intake of acrylamide and cancers of the colon, rectum, kidney, bladder, and breast *(15)*. There was no evidence that intake of specific foods containing acrylamide were associated with the risk of these cancers. Furthermore, there was no relationship between estimated acrylamide intake in the diet and cancer risk. Therefore, the epidemiological studies conducted so far do not support acrylamide intake in the diet as an important public health concern *(15)*.

Occurrence

High levels of acrylamide are found primarily in widely consumed processed foods such as potato products (French fries, potato chips, baked potatoes), bakery and cereal products (bread, cereal, crackers, cookies, cakes), and coffee. An LC-MS/MS analytical method was developed by the U.S. Food and Drug Administration (FDA) for the determination of acrylamide in food products *(16)*. The method was used by the U.S. FDA to determine the levels of acrylamide in a wide variety of food products obtained in the U.S. *(17)*. Initially, food groups were chosen for analysis if they were previously reported to contain acrylamide or if they contributed significantly to the diet of infants or young adults. The products included baby foods, bagels, breads, crackers, crisps, coffee, doughnuts, French fries, infant formula, gravy, jelly, meats, nuts, potato chips, pastries, pies, pretzels, pudding, seasonings, soups, tortillas, canned and frozen vegetables, and canned fruits. Most products were analyzed as received, while others were examined before and after cooking. The exploratory data on acrylamide levels in the foods that the U. S. FDA tested are available on the internet (http://www.cfsan.fda.gov) and are summarized in Table 1 *(18)*.

Although cereals are a major dietary source of acrylamide, the percentage of total acrylamide that cereals contribute varies for different populations, ranging from ~24% in the diet of Swedish adults to ~44% in the diet of Belgian adolescents, with the American diet at 40% *(17)*. The U.S. FDA reported that cereals contain 52-1057 ng/g acrylamide, and acrylamide levels in breads and bakery products range from non-detectable to 364 ng/g in dark, rye or toasted bread *(18)*. The contribution of potato products to total acrylamide intake ranges from ~29% for adult Norwegian women to ~69% for Dutch children/adolescents with ~38% for the U.S. population *(17)*. The U.S. FDA reported that potato products (French fries and potato chips) contain 20-2762 ng/g acrylamide *(18)*.

The contribution of coffee to the dietary intake of acrylamide varies widely demographically and can be high in countries with a high coffee consumption. It ranges from ~8% for the U.S., to ~13% for The Netherlands, to ~28% for Norwegian adults, and to ~39% for Swedish adults *(17, 19)*. This indicates that in many cases the amount of acrylamide that coffee contributes to the diet may be important in some populations. The U.S. FDA reported that ground coffee contains 27-609 ng/g acrylamide *(18)*. Table 2 summarizes acrylamide data for coffee from a few studies and provides the range of acrylamide concentrations reported for various coffee types.

Wide variations in levels of acrylamide have been observed in different food categories as well as in different brands of the same food category. This seems to result from different levels of acrylamide precursors present as well as from variations in processing or cooking conditions such as the temperature and time. Foods rich in reducing sugars and amino acids are derived primarily from plant sources such as potatoes and cereals (barley, rice, wheat) *(6, 15, 20)*.

Formation

Mechanism of formation of acrylamide

Table 1. Summary of acrylamide concentrations reported for various food categories by the U.S. FDA as purchased.

Food Category	**Concentration (ng/g)*	*Food Product with highest level of acrylamide*
Baby food	ND-130	Teething biscuits
French fries	20-1325	Baked fries
Potato chips	117-2762	Sweet potato chips
Infant formula	ND	-
Protein foods	ND-116	Grilled veggie burgers
Breads and bakery products	ND-364	Dark, rye bread, toasted
Cereals	52-1057	Wheat cereal
Snack food (other than potato chips)	12-1340	Veggie crisps
Gravies and seasonings	ND-151	Pecan liquid smoke
Nuts and nut butters	ND-457	Smokehouse almonds
Crackers	26-1540	Graham crackers
Chocolate products	ND-909	Hershey's cocoa
Canned fruits and vegetables	ND-83	Oven baked beans
Cookies	ND-955	Ginger snap
Coffee	27-609	Blend coffee and chicory; ground, not brewed
Frozen vegetables	<10	-
Dried foods	11-1184	Onion soup and dip mix
Dairy	ND-43	Evaporated milk
Fruits and vegetables	ND-1925	Ripe olives
Hot beverages	93-5399	Instant hot beverage; powdered, not brewed
Juice	267	Prune
Taco, tostada, and tortilla products	29-794	Original tostadas
Miscellaneous	ND-804	Toasted corn

*Results are summarized from the U.S. FDA. Survey data on acrylamide in food: individual food products. United States Food and Drug Administration, Center for Food Safety and Applied Nutrition, 2006, http://www.cfsan.fda.gov/~dms/acrydata.html.

ND – none detected or less than LOQ, which is 10 ppb (ng/g)

Miscellaneous – jelly, jam, cheddar and bacon potato skins, chicken quesadilla rolls, cheese pizza rolls, organic breakfast, flax, soy, blueberry

Research suggests that acrylamide forms in foods mainly through the Maillard reaction between reducing sugars and certain amino acids. Studies with model systems demonstrated that asparagine is the major amino acid precursor *(6, 21, 22, 23)*. This explains the occurrence of acrylamide in potato- and grain-

based foods, which are particularly rich in free asparagine *(6)*. Acrylamide in food is derived mainly from heat-induced reactions (temp. >120 °C) between the amino group of free asparagine and the carbonyl group of reducing sugars such as glucose during baking, frying and other thermal treatments. The products of the Maillard reaction also generate flavor and color during these processes. An important reaction is the Strecker degradation of amino acids by these intermediates in which the amino acid is decarboxylated and deaminated to form an aldehyde. Significant quantities of acrylamide were found when equal

Table 2. Acrylamide concentration in coffee products.

Coffee Type	*Number of samples*	*Range*	*Reference*
Roasted and ground coffee beans	31	45-374 ng/g	*(3)*
Instant coffee (powder)	12	169-539 ng/g	*(3)*
Brewed coffee	8	6-16 ng/ml	*(3)*
Brewed Espresso	18	11.4-36.2 μg/L	*(50)*
Coffee blends with cereals	2	200.8-229.4 μg/L	*(50)*
Brewed Cappuccino	1	6.4 μg/L	*(50)*
Soluble (instant) coffee	5	47.4-95.2 μg/L	*(50)*
Turkish coffee	5	29-75 ng/g	*(47)*
Roasted, ground coffee beans from different origins	11	12-29 ng/g	*(47)*
Instant coffee	3	42-338 ng/g	*(47)*
Filtered coffee	1	50 ng/g	*(47)*
Classic roast,coffee mixes,decaffeinated coffee	20	62-385 μg/kg	*(48)*

amounts of asparagine and glucose were allowed to react in a buffer system *(6)*. The formation of acrylamide from asparagine through the Maillard reaction involves several reactions with different intermediates.

The Schiff base is one proposed intermediate formed early in the Maillard reaction due to elimination of water from the conjugate of glucose and asparagine (21, 24). The sugar - asparagine adduct, N-glycosylasparagine, formed by reaction of reducing sugars with asparagine, when heated, resulted in significant amounts of acrylamide, while the N-glycosides formed by the reaction of glutamine and methionine only formed minor amounts of acrylamide (22). Studies have proven that the major mechanistic pathway in the formation of acrylamide in foods involves a Schiff base where decarboxylation is necessary, followed by further degradation of the decarboxylated Schiff base. This degradation step involves cleavage of a nitrogen-carbon bond, which can occur by two different mechanisms. One involves direct degradation of the decarboxylated Schiff base to form acrylamide via elimination of an imine and the other involves hydrolysis of the decarboxylated Schiff base to yield aminopropionamide and a carbonyl compound *(21, 24)*.

A model system using pyrolysis-gas chromatography/mass spectrometry and Fourier transform infrared (FTIR) showed that the ability of the open-chain form of *N*-glycosylasparagine (the Schiff base) to undergo intramolecular cyclization and formation of oxazolidin-5-one is the key step that allows decarboxylation of asparagine and subsequent formation of acrylamide *(25)*.

Temperature-time processing conditions

Potato products

Cooking temperature and time had the greatest impact on the level of acrylamide formation in potato products *(6, 7, 20, 22, 26, 27)*. The manner in which heat was transferred to foods (e.g. frying, baking, roasting, microwave-heating) did not impact the rate of acrylamide formation *(24)*. Acrylamide content increases in the temperature range of 120-175 °C, then decreases when the food is heated at higher temperatures *(6, 7, 8)*. However, acrylamide levels for home-cooked deep-fried French fries ranged from 265 μg/kg for potatoes fried at 150 °C for 6 min to 2130 μg/kg for French fries prepared at 190 °C for 5 min. Similarly, acrylamide levels in frozen French fries baked at 232 °C increased with baking time (16-24 min) and ranged from 198-725 μg/kg *(27)*. However, in an asparagine/glucose (molar ratio 1:1) model system heated for 10-30 min, there was a reduction in acrylamide levels with time except for the lowest temperature (155 °C) where the amount of acrylamide went through a maximum at 20 min *(8)*. In potato shapes with low surface to volume ratios (SVRs) such as potato chips, acrylamide levels increased with increasing frying temperatures as well as with frying time, reaching maximum levels of 2500 μg/kg. However, in samples with higher SVRs, acrylamide levels were the greatest at 160-180 °C with maximal acrylamide formation of 18000 μg/kg, then decreased with higher frying temperatures and more prolonged frying times *(26)*. At higher frying temperatures (180-190 °C), acrylamide levels in French fries increased exponentially at the end of the frying process *(27)*. Similar results were noticed with the baking of French fries *(27)*. This phenomenon may be due to the higher rate of acrylamide formation at the surface of French fries during cooking. As the French fry surface becomes dry, the temperature rises >120 °C, allowing acrylamide to form in the dry crust. It was determined that the most important measure for minimizing the formation of acrylamide in French fries was to avoid overcooking by preventing surface browning *(27)*.

Bakery products

Processing temperatures and times also influenced acrylamide formation in bakery products. The amount of acrylamide in bread crusts increased with time and temperature with a maximum in the level of acrylamide found at approximately 190-210 °C *(28, 29)*. Both baking temperature (>200 °C) and time increased acrylamide levels in the crust of wheat bread from less than 10 to 1900 μg/kg *(28)*. Studies performed on the effects of baking times and temperature (180 °C and 200 °C) on acrylamide formation in gingerbead showed that acrylamide formation occurred linearly over the 20 min baking period *(30)*. This is in contrast to acrylamide formation in some potato products.

Coffee

Coffee is a complex matrix in terms of acrylamide formation and reduction. It has been shown that roasting time and temperature, coffee type, and the amount of precursors in the raw material had an impact on acrylamide formation in coffee beans. The amount of acrylamide in the roasted coffee bean and the type of brewing method affect the amount of acrylamide detected in the brewed beverage. Although coffee beans are roasted at very high temperatures (240-300 °C), significant amounts of acrylamide are formed during the first few minutes of roasting then levels decrease exponentially toward the end of the roasting cycle *(31)*. Maximum acrylamide levels in Robusta (3800 μg/kg) and Arabica (500 μg/kg) coffee beans were found during the first minutes of the roasting process *(19)*. Kinetic models and spiking experiments with isotope-labeled acrylamide have shown that >95% of acrylamide formed during the roasting is degraded during the process *(32)*. These findings may explain why light roasted coffees contain higher amounts of acrylamide than very dark roasted coffees *(33)*. Robusta coffee contained significantly greater amounts of acrylamide (708 ng/g) than Arabica coffee (374 ng/g) when roasted at 240 °C for 7.5 min *(19)*. During the brewing of coffee, almost all acrylamide present in the ground coffee is transferred to the liquid phase of the coffee beverage, due to its high solubility in water *(34)*. Unlike with filtered coffee, the acrylamide is not transferred completely to the beverage when espresso coffee is prepared *(19)*.

Agricultural Practices

Potato products

A strong correlation was reported between the sugar content of the raw potato tuber and the potential for acrylamide formation *(23, 35)*. Acrylamide levels were highly correlated with both glucose and fructose concentrations of potato tubers, whereas asparagine levels did not predict acrylamide formation in cooked potato products *(23)*. Acrylamide contents of thermally-processed potato products varied depending on potato cultivar when prepared under identical conditions *(20)*. Studies showed that reducing sugar concentrations varied by a factor of 32 among potato cultivars while levels of free asparagine contents were higher than those of reducing sugars but varied slightly *(23, 35)*. The amount of reducing sugars in potato products increased while crude protein and free amino acids decreased when less nitrogen fertilization was applied to potatoes during growth *(36)*. Lowering the amount of nitrogen fertilization resulted in a 30-65% increase in acrylamide levels in fried potato products *(36)*. In addition to fertilizer application rate, dry and hot weather seems to increase acrylamide formation by increasing the content of reducing sugars in potato cultivars *(36)*. These studies indicate that acrylamide content can be reduced through agricultural practices and by carefully selecting potato cultivars with low levels of reducing sugars.

The high variability in reducing sugar content among potatoes of the same cultivar suggest that storage conditions may have a stronger influence on sugar content of potato tubers than cultivar *(37)*. Short-term storage of potatoes at 4 °C (e.g. in the refrigerator of a supermarket) significantly increase the potential for acrylamide formation *(37, 38)*. Cooling potatoes to temperatures less than 10 °C causes the reducing sugars to increase thus, increasing the potential for acrylamide formation *(27, 37, 38)*. A study was conducted whereby potatoes of the cultivar Ernestolz were stored for 15 days at 4 °C. This resulted in an increase in the level of reducing sugars and the acrylamide forming potential *(38)*. The reducing sugars may participate in both nonenzymatic Maillard browning reactions and acrylamide formation during the processing of the potatoes *(17)*. The phenomenon whereby potatoes accumulate sugars during storage at cold temperatures is referred to as "cold sweetening" and is believed to be the way the potato plant protects the tuber from freezing *(35)*. The reconditioning of potatoes at room temperature following cold storage results in significant reductions in the content of reducing sugars and may reduce acrylamide formation potential *(17)*. In addition to increased acrylamide formation, the increase in reducing sugar content can also result in enhanced browning during thermal processing of potato products.

Bakery products

In starch-based and cereal systems, acrylamide levels are more highly correlated with levels of asparagine rather than reducing sugars *(29)*. In a bread model system, asparagine dramatically increased the amount of acrylamide, but not with the levels of reducing sugars, such as glucose *(29)*. The acrylamide content of bread depends on the wheat cultivar used to prepare the dough. Reducing sugars in wheat flour are also cultivar dependent. Acrylamide levels are related to free asparagine and crude protein content *(39)*. Crop cultivar was reported to strongly affect the acrylamide concentration of breads, mainly due to varying asparagine contents *(39)*. No correlation was reported between reducing sugar and acrylamide contents of heated flour or breads. In contrast, asparagine content of the flour and asparagine levels in dough significantly affected acrylamide contents *(39)*. Nitrogen fertilization caused an increase in amino acid and protein contents, thus increasing acrylamide levels in bread *(39)*. When wheat was grown with sulfate depletion, asparagine levels were greater compared to levels in wheat grown in soils with proper amounts of sulfate fertilizer *(17)*. This was also shown in acrylamide levels of baked products. The asparagine levels in wheat grown in sulfur-depleted soil ranged from 2600 to 5200 μg/kg and those from wheat grown under normal conditions, from 600 to 900 μg/kg. This suggests that wheat should be grown in soil with adequate amounts of sulfur *(17)*. Harvest year, climate, and sprouting of the grain also affect acrylamide levels in breads. Warmer temperatures and increased sunshine during crop growth result in increased protein and amino acid contents in grain and flour. Heavy rainfall before harvest can result in grain sprouting. Sprouting of the grain causes significant increases in acrylamide levels, which is due to high enzyme activities and the formation of precursors from protein and starch *(39)*. More research is needed to understand the seasonal variations and the influence of environmental factors on asparagine levels.

Coffee

A relationship between acrylamide formation in coffee during roasting and with sucrose and asparagine levels in green coffee was observed *(19)*. Higher sucrose content led to a reduced acrylamide formation and an increased content of asparagine resulted in a higher formation of acrylamide. Asparagine is a limiting factor in acrylamide formation during the roasting process *(19)*.

Acrylamide was stable in brewed coffee that was held at room temperature for 5 h and in a sealed can of ground coffee stored at -40 °C over 8 months *(34)*. However, other data showed that acrylamide is not stable in commercial roast and ground coffee stored in a sealed container *(34, 41, 42)*. There was a 40-65% decrease in opened ground coffee stored at room temperature over 6 months *(41)*. Acrylamide levels decreased less substantially when soluble coffee or coffee substitutes were stored in the dark for 3-6 months at 10-12 °C *(41)*. Acrylamide reduction in vacuum-packed roasted and ground coffee was studied over a year at four different temperatures (-18, +4, ambient and +37 °C). The rate of decrease in acrylamide levels was correlated to the storage temperature with the highest reduction rates at 37 °C (>7-fold reduction in acrylamide concentration after 6 months storage time) *(42)*. Reductions in acrylamide levels during storage of coffee may be because of reactions of acrylamide with SH- and NH_2-containing amino acids, peptides, and proteins in these foods in the solid state. Other possibilities include hydrolysis, degradation, and polymerization of acrylamide during storage *(19)*.

Frying oil

Several researchers *(20, 27, 43, 44, 45)* tried to determine whether there was a correlation between the type of frying oil used and the age of oil on the formation of acrylamide. There is conflicting information from this research. One study showed that potato samples fried in olive oil had 60% more acrylamide than those fried in corn oil and an increase in frying time caused higher amounts of acrylamide *(20)*. Another study demonstrated that virgin olive oil (VOO) phenol compounds which are not degraded during frying affect acrylamide formation. Acrylamide was generated more rapidly during frying in the oil having the lowest concentration of phenolic compounds. Furthermore, the VOO with the highest concentration of ortho-diphenolic compounds inhibited acrylamide formation in potato crisps *(43)*.

In contrast to the above studies, other investigators determined that the type of frying oil did not affect acrylamide levels *(27, 44, 45)*. One study found that commonly used frying oils (peanut, canola, corn, safflower, olive and hydrogenated soybean) had no significant effect on acrylamide levels in French fries deep-fried at 180 °C for 4 min *(27)*. Similarly, experiments with model systems and with fried potato products determined that acrylamide formation during frying (175 °C, 2-5 min) was not affected by the type of frying oils (rapseed, olive, sunflower, soybean, corn, grapeseed, and palm fat). Overall, the body of evidence suggests that the type of vegetable oil does not seem to impact the acrylamide formation in potatoes during frying *(44)*.

Greater amounts of acrylamide were found in potato chips fried in thermally aged cooking oil than those fried in fresh cooking oil *(16)*. This may be due to the formation of carbonyls (Maillard browning precursors) in oils that are less thermally stable. However, another study indicated that normal degradation of cooking oils, which was monitored by measuring peroxide values, did not affect acrylamide levels *(45)*. Addition of oil oxidation products, such as pentanal, hexanal, octanal, and decanal to an asparagine-containing silica gel model system heated at 170 °C for 5 min was found not to increase the formation of acrylamide compared to the control (without oil oxidation products) *(46)*. Several oil hydrolysis compounds were subsequently evaluated, including diacylglycerol, monoacylglycerol, glycerol, acrolein, and acrylic acid. Only the heated model system containing acrolein and asparagine showed a significantly higher acrylamide content compared to the control to which only asparagine was added. However, the contribution of acrolein to the overall formation of acrylamide appeared to be negligible in the presence of a reducing sugar. This indicates that the importance of acrolein and other oil degradation components for formation of acrylamide is insignificant *(46)*.

Browning

Potato products

Acrylamide concentration and the brown color of thermally processed products were reported to show a high degree of correlation *(27, 35)*. Since acrylamide and the brown color of cooked foods are formed during the Maillard reaction, it is likely that acrylamide is formed parallel with browning *(35)*. A study by Jackson et al. *(27)* demonstrated that as fries were fried for long periods of time and at higher temperatures, the "L" component of color (a measure of the white/black component of color) decreased while the "a" values (degree of redness) increased. Statistical analysis (regression analysis) indicated that the "a" and "L" components of color correlated highly (r^2 = 0.8858 and r^2 = 0.8551, respectively) with the log of acrylamide levels in the French fries. In contrast, the "b" color values (a measure of the yellow/blue component of color) for the samples correlated poorly with acrylamide levels (r^2 = 0.089). This study showed that French fries fried to a golden color with light browning at the edges of fries had lower acrylamide levels than those with a brown surface *(27)*.

Surface browning as a function of acrylamide levels was also studied in baked French fries. The degree of surface browning between replicate baked samples was much more variable that between replicate fried potato samples. Similarly, colorimetric measurements of surface color for baked sample were more variable than for fried samples and indicate a weak, but positive correlation for the "L," "a" and "b" color values with acrylamide *(27)*. This greater variability in the results for baked samples than fried samples is expected because baking produces less even heating of the food surface than frying.

A linear relationship between browning levels and acrylamide concentrations was observed for fried potato slices that had lower-surface-to-volume ratios *(26)*. However, the surface color had a low degree of correlation with acrylamide levels in potatoes with a high surface area (shredded potatoes). This may be due to the degradation of acrylamide at the end of the frying process *(26)*. Color measurements indicated that the degree of surface browning is the major determinant of the final acrylamide content of French fries irregardless of frying conditions. Since acrylamide formation increases exponentially towards the end of the frying or baking process, an important factor for minimizing acrylamide formation is to determine the proper cooking end-point. The degree of surface browning could be used as a visual indicator of acrylamide formation during cooking.

Bakery products

The degree of browning in gingerbread was reported to be an excellent predictor of acrylamide levels as browner gingerbread products had higher acrylamide levels (30). A strong correlation was found between the brown color of the bread crust and acrylamide content. Acrylamide appears to form in the crust at elevated temperatures and time indicating that color could be used as a gauge of acrylamide formation during bread making (28, 39). Jackson and Al-Taher (unpublished data) showed that as cookies were baked for longer periods of time, the "L" component of color decreased while the "a" values increased and that both components of color correlated highly (r2 = 0.8079 and r2 = 0.8633, respectively) with the log of acrylamide levels in the cookies. In contrast, the "b" values for the samples correlated poorly with acrylamide levels (r2 = 0.1567). Overall, the available information suggests that prolonged baking or excessive browning should be avoided to minimize acrylamide formation in baked foods. Since acrylamide formation increases linearly in the baking process, an important factor for minimizing acrylamide formation is to determine the proper cooking end-point. This indicates that the degree of surface browning could be used as a visual indicator of acrylamide formation during cooking.

Coffee

There was only one study that examined the correlation of browning with the formation of acrylamide in roasted coffee beans *(47)*. A headspace vial with a small amount of green coffee beans was heated in a laboratory convection oven at three different temperatures (150, 200, 225 °C) for different times (5-30 min). Acrylamide levels and the bean color were measured. The amount of acrylamide increased rapidly at the beginning of roasting at 200 and 225 °C, reaching a maximum, and then decreased exponentially. Although "L" and "b" values decreased exponentially with time, "a" values increased quickly at the beginning of roasting at 200 and 225 °C, reaching a maximum, and then decreased exponentially. The amount of acrylamide in coffee beans increased continuously during roasting at 150 °C. The "a" component of surface color followed the same trend. Values for "a" increased continuously during roasting at 150 °C. A high degree of correlation was reported between acrylamide and the "a" value (r^2 = 0.9286) of roasted coffee indicating that acrylamide levels in coffee can be estimated from the "a" value *(47)*.

Analysis

Analytical method development and modification to identify and quantify acrylamide in food products have been carried out since its discovery in 2002. Gas chromatography with mass spectrometric detection (GC-MS) and high-performance liquid chromatography (HPLC) with tandem mass spectrometric detection (LC-MS/MS) are the most widely used analytical tools for the detection of acrylamide in most food products. There is no significant difference between results obtained by GC-MS with derivatization and LC-MS/MS. The advantage of LC-MS/MS over GC-MS is that analysis time is reduced since acrylamide can be analyzed without prior derivatization *(48)*.

GC-MS methods

Two types of GC-MS methods are typically used for the determination of acrylamide in foods: (1) with bromination and (2) without derivatization. Bromination of acrylamide to 2-bromopropenamide is usually carried out overnight at temperatures slightly above or at the freezing point of water. The use of isotopically labeled internal standards (e.g. methacrylamide) reduces the reaction time from overnight to 1 h *(49)*. Excess bromine is removed by titration with thiosulfate solution until the endpoint is reached (i.e light yellow color). The non-polar brominated acrylamide is then extracted in non-polar organic solvents (e.g. ethyl acetate or cyclohexane) to remove the analyte from the aqueous phase. Several cleanup steps can then be used including centrifugation, fractionation on silica-gel cartridges or florisil cartridges and gel permeation chromatography. Typical internal standards include D_3-acrylamide, $^{13}C_3$-acrylamide or both. Quantitation is achieved by the standard addition method. A standard medium to high polarity column is used for the GC/MS analysis *(49)*.

A few methods have been developed that omit the derivatization step and measure acrylamide directly after extraction and clean-up. The sample preparation and measurement procedures are significantly different from those used for the GC method with bromination. The extraction solvents used are mainly aqueous or a combination of aqueous and organic (i.e. *n*-propanol or 2-butanone). The homogenized sample is mixed with the extraction solvent and an internal standard at a prespecified temperature for 10-20 min. This swells the matrix so the extraction solvent can better access potentially adsorbed or enclosed acrylamide, thus, providing some time for the development of matrix/internal standard interactions *(49)*. Subsequently, several cleanup and defatting steps are then employed (e.g. hexane, fractionation of the aqueous phase on graphitized carbon cartridges, centrifugation followed by removal of water fraction by azeotropic distillation, liquid/liquid extraction with *n*-hexane/acetonitrile mixture, or Soxhlet extraction). Because of the high polarity of non-derivatized acrylamide, sample extracts can be injected on-column into the GC. For analyte separation, columns with polar phases (e.g. polyethyleneglycol) can be used. Quantification is carried out by addition of different kinds of internal standards, ranging from propionamide to isotopically labeled acrylamide *(49)*.

A few researchers *(47, 48, 50)* developed improved sample extraction methods for the determination of acrylamide in coffee and coffee products. A method developed by Soares et al. *(50)* uses some methods to purify acrylamide prior to GC/MS analysis. A GC/MS method for purifying acrylamide from sample matrices involves two main purification steps: the first with ethanol and Carrez solutions in order to precipitate polysaccharides and proteins, respectively; and the second with a layered solid-phase extraction column that is effective in getting rid of the main chromatographic interferences *(50)*.

LC-MS/MS methods

Most LC methods use water to extract acrylamide and added internal standard (e.g. $^{13}C_3$-acrylamide, D_3-acrylamide or $^{13}C_1$-acrylamide) from the sample matrix. A defatting step with hexane, toluene or cyclohexane is used before or combined with the extraction step. Subsequently, the aqueous phase is centrifuged. Several solid phase extraction (SPE) cartridges can be used for clean-up, but filtration through a 0.22 μm nylon filter has been shown to be an effective method for sample clean-up before HPLC analysis *(49)*.

Reversed-phase chromatography is widely used for the separation of acrylamide with Hypercarb (5 μm) being the most frequently used column *(17, 20)*. An alternative to reversed-phase chromatography is ion-exchange chromatography. Ion-exchange chromatography provides good separation of acrylamide from food matrix compounds even in absence of sample clean-up steps. Tandem mass spectrometry is the method of choice for the detection of acrylamide after chromatographic separation. UV detection and single quadruple mass spectrometry showed a lack of selectivity when they were used to measure acrylamide levels in complex food matrices *(49)*. On the other hand, LC-MS/MS, working in multiple reaction-monitoring mode (MRM), in which the transition from a precursor ion to a product ion is monitored, has a high selectivity *(49)*.

Few analytical methods perform well for measuring acrylamide levels in complex matrices, such as coffee and chocolate products Multiple responses are observed at retention times close to that of acrylamide, which may cause interference *(15, 40)*. During a validation of a LC-MS/MS for acrylamide in coffee, it was shown that coffee is a difficult matrix to analyze because it requires frequent cleaning of the HPLC system *(15)*.

The LC-MS/MS method developed by Senyuva and Gokmen *(47)* involves extraction of acrylamide with methanol, purification with the Carrez solutions, evaporation and solvent change to water, and clean-up with an Oasis HLB solid-phase extraction cartridge. The chromatographic conditions allow separation of acrylamide with accurate and precise quantification of acrylamide during MS detection in SIM mode *(47)*. Pardo et al. *(48)* developed a selective and sensitive method using pressurized fluid extraction (PFE) and isotope dilution LC-MS/MS for acrylamide quantitation in coffee and chocolate samples. This method includes PFE with acetonitrile, florisil clean-up purification inside the PFE extraction cell and detection by liquid chromatography coupled to atmospheric pressure ionization in positive mode tandem mass spectrometry (APCI-MS/MS). Clear extracts and higher signal responses are obtained due to less ion suppression effects. The LC-MS/MS method using atmospheric pressure ionization and florisil inside the PFE extraction cell minimizes interferences in coffee and chocolate samples *(48)*. Currently, there is no method for the determination of acrylamide that is applicable to all food matrices.

Prevention and Mitigation

Potato products

Acrylamide levels can be decreased by reducing cooking times and temperatures, lowering the pH, and/or choosing potato tubers with low reducing sugar content. However, since the Maillard reaction which is responsible for acrylamide formation also guarantees desirable flavor and color compounds in heated food, reducing the cooking time and temperature or lowering the pH may compromise color and flavor.

To minimize acrylamide formation in potatoes cooked at high temperatures (fried, baked, roasted), it is imperative that raw potatoes not be stored at temperatures <10 °C *(27)*. Potato tubers are often chilled to prevent sprouting at 4 °C, but this causes starch to break down and reducing sugars to accumulate *(51, 52)*. Conditions that minimize acrylamide in French fries involve frying or baking potato pieces as long as necessary to get the surface golden in color and the texture crispy *(27, 51)*. Using raw potatoes with low amounts of reducing sugars, a 15 min soak to remove the precursors, frying temperatures of 165-170 °C, and an oil:potato ratio of 1L:100g, French fries were shown to consistently generate 40-70 µg acrylamide/kg *(52)*. It was also suggested to use fresh potatoes with reduced amounts of reducing sugars and to use cooking temperatures less than 250 °C to limit acrylamide formation in baked or roasted potatoes *(52)*.

Rinsing and soaking treatments are effective at reducing acrylamide formation in French fries prepared from fresh-cut potatoes. Studies in our laboratory found that soaking potato slices in room temperature water for at least 15 min before frying resulted in 63% reduction in acrylamide *(27)*. This work was in agreement with another study that found that soaking potato pieces for 10 min in cold or warm water resulted in desirable flavor and texture when fried, yet had only half the acrylamide content of the comparable untreated slices *(52)*. Blanching for long times (50 °C for 80 min or 70 °C for 45 min) resulted in the lowest levels of acrylamide formation. Blanching removed more glucose and asparagine from the potatoes than did water immersion *(53)*. Soaking and blanching treatments reduce acrylamide formation by leaching out sugars and asparagine from the surface of the potato slice.

Using acid solutions (i.e. citric acid or acetic acid) instead of water to soak potatoes resulted in greater reductions in acrylamide levels in fried potato products. This decreased the pH of potato juice and increased the extraction of amino acids and sugars. Soaking potatoes in 1:3 vinegar:water rinse reduced the acrylamide forming potential by 75% when compared to the untreated control *(27)*. Further reductions in acrylamide formation were not observed when the soaking solution contained higher vinegar levels (1:1 vinegar:water) *(27)*. Dipping potato cuts in 1% and 2% citric acid solutions for one hour before frying inhibited acrylamide formation in French fries by 73.1% and 79.7% *(54)*. Major reductions (> 90%) in acrylamide formation were observed when ascorbic acid was added to homogenized potato and then microwave heated (3 min, 750

W). The internal pH of the potatoes was reduced from 5.72 to 2.96, which resulted in a 70% decrease acrylamide formation *(8)*.

Acid treatments are effective at preventing acrylamide formation since they lower pH into the range where acrylamide formation is minimized (< pH 5) *(54)*. Increased acidity below pH 6 may be used to decrease acrylamide formation in potatoes and possibly other foods. The benefits of a low pH can result from protonation of the reactive free α-NH_2 group of asparagine to the nonreactive α-NH_3^+ form and from partial acid-catalyzed hydrolysis of asparagine to aspartic acid and of acrylamide to acrylic acid. However, low pH may cause undesirable taste *(55)*.

Treating potato with asparaginase, an enzyme that hydrolyzes asparagine into aspartic acid and ammonia, reduced acrylamide levels by 99% *(21)*. Furthermore, soaking blanched potato strips (75 °C, 10 min) in an asparaginase solution at 40 °C for 20 min effectively reduced acrylamide formation during frying by reducing the amount of asparagine *(53)*.

Several additives influenced acrylamide formation in a model system. In a potato powder model system, sodium acid pyrophosphate, citric, acetic, and L-lactic acid significantly reduced the final acrylamide content by lowering the pH *(55)*. Free glycine, L-lysine, and L-cysteine also lowered acrylamide, while keeping the pH at its original level. L-cysteine seemed to reduce the acrylamide content in the most effective way (92%), followed by L-lysine (39%), and glycine (24%) *(55)*. These amino acids might reduce acrylamide formation by competing with asparagines to react with reducing sugars in the Maillard reaction.

Bakery products

Several studies *(27, 28, 30, 39, 57)* have shown that the formation of acrylamide in cereal or bakery products can be reduced by doing the following: avoiding high temperatures during baking; extending fermentation times where feasible; substituting ammonium bicarbonate with sodium bicarbonate; avoiding or minimizing use of reducing sugars where possible; and maintaining consistent control of surface browning during baking. Prolonged baking or excessive browning should be avoided in order to minimize the acrylamide content. Agronomic conditions also influence the formation of acrylamide.

Sprouting is a major factor influencing acrylamide formation in cereal grains. It leads to the degradation of starch and proteins, and the release of sugars and asparagine. A 500% increase in acrylamide formation was reported, from 54.5 to 273.3 ppb, in bread prepared from sprouted grain compared with non-sprouted grain *(39.)* These results suggest that sprouted grain should not be used for bakery products *(39)*.

More than 99% of acrylamide was observed to be formed in the crust rather than the crumb during baking of bread. The acrylamide level of the crusts was related to temperature and time of baking and the surface color of the crust correlated strongly with acrylamide content *(28)*. Potato bread that was toasted until it was dark brown in color had a high amount of acrylamide (>600 μg/kg) (27). Potato bread formed more acrylamide during toasting than other types of bread (white, whole wheat, rye) because of the higher concentrations of asparagine in breads containing potato flour than those without. To minimize acrylamide content, consumers should avoid toasting bread to a "dark" color and scrape the surface to remove the dark parts of toasts and other baked goods *(27)*.

Several ingredients may increase acrylamide formation during baking. A study showed that the baking agent, ammonium hydrogen carbonate, increased acrylamide formation in bakery products. The promoting effect of ammonium hydrogen carbonate on the formation of acrylamide might be indirect by providing more reactive carbonyls originating from the reaction of ammonia with reducing sugars. Use of sodium hydrogen carbonate as an alternative baking agent reduced the acrylamide concentration by more than 60% *(30)*. Also, using sucrose instead of honey or inverted sugar syrup in gingerbread reduced acrylamide content by a factor of 20 *(30)*. Acrylamide levels were reduced by 50% or more by using sucrose instead of reducing sugars (i.e. glucose) in cookie dough *(57)*. Both honey and inverted sugar syrup contain reducing sugars, precursors of acrylamide. Furthermore, *(56)* found that adding the disaccharide trehalose, which is used in many commercial food applications to glucose/asparagine or ascorbic acid/asparagine mixtures prevented acrylamide formation. This may be due to the suppression in the generation of the intermediate carbonyl compounds such as pyruvaldehyde.

Extensive fermentation with yeast is one possible way to reduce acrylamide content in bread by reducing free asparagine, a precursor for acrylamide in cereal products *(58)*. Prolonged fermentation (2 h) of whole-wheat dough caused an 87% reduction in acrylamide concentrations of breads as compared to those subjected to a shorter fermentation time. The corresponding acrylamide reduction in rye bread was 77%. Sourdough fermentation was less effective than yeast fermentation in reducing the asparagine content of the dough *(58)*.

A study showed that a significant reduction of acrylamide could be achieved by choosing ingredients with a low content of free asparagine or by applying an asparaginase, an enzyme, to the dough. Addition of asparaginase preparation of gingerbread dough caused 75% decrease in free asparagine and a 55% decrease in acrylamide level *(30)*.

The ability of different amino acids to reduce acrylamide formation in gingerbread was measured. The amino acids, glycine and L-glutamine, did not affect acrylamide levels, but addition of L-cysteine was effective at reducing the acrylamide content of gingerbread. However, cysteine addition is not a practical method for reducing acrylamide formation since the amino acid imparts an unpleasant odor and flavor to food *(30)*. Citric acid added at levels of 0.5 and 1.0 g per 100 g of gingerbread reduced acrylamide formation by factors of 4 and 40, respectively. A moderate addition of citric acid ($\leq$ 5000 mg/kg) reduced the acrylamide content in gingerbread but it also affected browning, leavening and taste *(30)*. Acrylamide should be reduced without compromising the quality and nutritional aspects of cereal and bakery products.

Coffee

To date, efforts to reduce acrylamide content in coffee have been unsuccessful. Because of the rigorous quality attributes of coffee, there are not many options available for mitigation of acrylamide during the roasting process. Since the aroma is a result of the roasting process and related to the chemical composition of the raw material, changes in the raw material of roasting process leads to totally different product characteristics. It may be best to select coffee varieties with low asparagine contents to reduce the acrylamide content *(32)*.

Conclusion

Wide variations of acrylamide levels in different food categories and even in different brands of the same food categories have been observed. This is due to the different levels of acrylamide precursors, reducing sugars and asparagine, present as well as from variations in processing or cooking conditions, such as the temperature and time. Several researchers discovered that the sugar content in potato tubers varies and that the asparagine content in cereal grains differs between varieties. Consequently, there is a wide range in the potential for acrylamide formation among food products made from these commodities.

It is important that methods be found that can reduce acrylamide levels in foods, wherever possible, without compromising food safety, organoleptic properties (e.g. taste, texture, and color) or adversely affecting nutritional quality. Studies have shown that acrylamide formation in potato products can be reduced substantially by selecting potato cultivars with low concentrations of reducing sugars and by cooking until the foods are fully cooked, but not overly browned on the surface. A reduction in acrylamide in wheat products could be achieved by growing wheat cultivars that do not accumulate very high levels of asparagine in the grain. Research has indicated that the use of asparaginase and the application of glycine in baked wheat-based products (dough systems) can result in a reduction in acrylamide levels. Acrylamide levels can also be decreased by reducing cooking times and temperatures or lowering the pH in potato and bakery products.

Acrylamide concentration and brown color of thermally processed products were reported to show a high degree of correlation. The degree of browning could be used as a visual indicator of acrylamide formation during cooking.

Coffee is a complex matrix in terms of acrylamide formation and reduction. Acrylamide content in coffee beans decreases during storage. It is also reduced during the roasting process before optimal roast is achieved. Attempts to reduce acrylamide content in the coffee brew have been unsuccessful.

Although acrylamide is known to be a neurotoxin and a carcinogen in animals, there is no evidence yet to support the fact that it may also be a carcinogen in humans. To date, no epidemiological study has found a link between dietary exposure to acrylamide and cancer risk.

References

1. Smith, E. A.; Pruen, S. L.; Oehme, F. W. Environmental degradation of polyacrylamides. 1. Effects of artificial environmental conditions: temperature, light and pH. *Ecotoxicol. Environ. Saf.* **1996,** *35*, 121-135.
2. Smith, E.A.; Pruen, S. L.; Oehme, F. W. Environmental degradation of acrylamide. II. Effects of environmental (outdoor) exposure. *Ecotoxicol. Environ. Saf.* **1997,** *37,* 76-91.
3. Friedman, M. Chemistry, biochemistry, and safety of acrylamide. A review. *J. Agric. Food Chem.* **2003,** *51*, 4504-4526.
4. IARC. Acrylamide. *Some Industrial Chemicals;* Monographs on the Evaluation of Carcinogenic Risks to Humans; International Agency for Research on Cancer: Lyon, France, 1994.
5. Acrylamide in food, Swedish National Food Administration, 2002.
6. Mottram, D. S.; Wedzicha, B. L.; Dodson, A. T. Acrylamide is formed in the Maillard reaction. *Nature* **2002**, *419*, 448-449.
7. Tareke, E.; Rydberg, P.; Karlsson, P.; Eriksson, S., Törnqvist, M. Analysis of acrylamide, a carcinogen formed in heated foodstuffs. *J. Agric. Food Chem.* **2002,** *50,* 4998-5006.
8. Rydberg, P.; Erikson, S.; Tareke, E.; Karlsson, P.; Ehrenberg, L.; Törnqvist, M. Investigations of factors that influence the acrylamide content of heated foodstuffs. *J. Agric. Food Chem.* **2003**, *51*, 7012-7018.
9. Friedman, M. A.; Dulak, L. H.; Stedham, M. A. Lifetime oncogenicity study in rats with acrylamide. *Fund. Appl. Toxicol.* **1995,** *27,* 95-105.
10. Bull, R. J.; Robinson, M.; Laurie, R. D.; Stoner, G. D.; Greisiger, E.; Meier, J. R.; Stober, J. Carcinogenic effects of acrylamide in SENCAR and A/J mice. *Cancer Res.* **1984,** *44,* 107-111.
11. Bull, R. J.; Robinson, M.; Stober, J. A. Carcinogenic activity of acrylamide in the skin and lung of Swiss-ICR mice. *Cancer Lett.* **1984,** *24,* 209-212.
12. Johnson, K. A.; Gorzinski, S. J.; Bodner, K. M.; Campbell, R. A.; Wolf, C. H.; Friedman, M. A.; Mast, R. W. Chronic toxicity and oncogenicity study on acrylamide incorporated in the drinking water of Fisher 344 rats. *Toxicol. Appl. Pharmacol.* **1986,** *85,* 154-168.
13. Klaunig, J. E. Acrylamide carcinogenicity. *J. Agric. Food Chem.* **2008,** *56,* 5984-5988.
14. Petersen, B. J.; Tran, N. Exposure to acrylamide: Placing exposure in context. In *chemistry and Safety of Acrylamide in Food;* Friedman, M., Mottram, D., Eds.; Springer Press: New York, 2005; pp 63-76.

15. Mucci, L. A., Wilson, K. M. Acrylamide intake through diet and human cancer risk. *J. Agric. Food Chem.* **2008,** *56,* 6013-6019.
16. Roach, J. A. G.; Andrzejewski, D.; Gay, M. L.; Northrup, D.; Musser, S. M. Rugged LC-MS/MS survey analysis for acrylamide in foods. *J. Agric. Food Chem.* **2003**, *51,* 7547-7554.
17. Friedman, M.; Levin, C. E. Review of methods for the reduction of dietary content and toxicity of acrylamide. *J. Agric. Food Chem.* **2008,** *56,* 6113-6140.
18. U. S. FDA. Survey data on acrylamide in food: individual food products. United States Food and Drug Administration, Center for Food Safety and Applied Nutrition, 2006, http://www.cfsan.fda.gov/~dms/acrydata.html.
19. Bagdonaite, K.; Derler, K.; Murkovi, M. Determination of acrylamide during roasting of coffee. *J. Agric. Food Chem.* **2008,** *56*, 6081-6086.
20. Becalski, A.; Lau, B. P. Y.; Lewis, D.; Seaman, S. W. Acrylamide in foods: occurrence, sources, and modeling. *J. Agric. Food Chem.* **2003,** *51*, 802-808.
21. Zyzak, D. V.; Sanders, R. A.; Stojanovic, M.; Tallmadge, D. H.; Eberhart, B. L.; Ewald, D. K.; Gruber, D. C.; Morsch, T. R.; Strothers, M. A.; Rizzi, G. P.; Villagran, M. D. Acrylamide formation mechanism in heated foods. *J. Agric. Food Chem.* **2003,** *51,* 4782-4787.
22. Stadler, R. H.; Blank, I.; Varga, N.; Robert, F.; Hau, J.; Guy, A. P.; Robert, M.-C.; Rieliker, S. Acrylamide formed in the Maillard reaction. *Nature* **2002**, *419,* 449.
23. Becalski, A.; Lau, B. P.-Y.; Lewis, D.; Seaman, S. W.; Hayward, S.; Sahagian, M.; Ramesh, M.; Leclerc, Y. Acrylamide in French Fries: Influence of free amino acids and sugars. *J. Agric. Food Chem.* **2004,** *52*, 3801-3806.
24. Stadler, R. H.; Roberts, F.; Rideiker, S.; Varga, N.; Davideck, T.; Devaud, S. In-depth mechanistic study on the formation of acrylamide and other vinylogous compounds by the Maillard reaction. *J. Agric. Food Chem.* **2004**, *52*, 5550-5558.
25. Yaylayan, V. A.; Wnorowski, A.; Locas, C. P. Why asparagine needs carbohydrates to generate acrylamide. *J. Agric. Food Chem.* **2003,** *51,* 1753-1757.
26. Taubert, D.; Harlfinger, S.; Henkes, L.; Berkels, R.; Schömig, E. Influence of processing parameters on acrylamide formation during frying of potatoes. *J. Agric. Food Chem.* **2004,** *52*, 2735-2739.
27. Jackson, L.; Al-Taher, F. Effects of consumer food preparation on acrylamide formation. In *Chemistry and Safety of Acrylamide in Food;*

Friedman, M., Mottram, D., Eds.; Springer; New York, 2005; Vol. 561, pp 447-465.

28. Surdyk, N.; Rosén, J.; Andersson, R.; Åman, P. Effects of asparagine, fructose, and baking conditions on acrylamide content in yeast-leavened wheat bread. *J. Agric. Food Chem.* **2004,** *52,* 2047-2051.
29. Bråthen, E.; Knutsen, S. Effect of temperature and time on the formation of acrylamide in starch-based and cereal model systems, flat breads and bread. *Food Chem.* **2005,** *92,* 693-700.
30. Amrein, T. M.; Schonbachler, B.; Escher, F.; Amado, R. Acrylamide in gingerbread: Critical factors for formation and possible ways for reduction. *J. Agric. Food Chem.* **2004,** *52,* 4282-4288.
31. Cirilo, M. P. G.; Coelho, S.; Araujo, C. M.; Gonçalves, F. R. B; Nogueira, F. D.; Gloria, M. B. A. Profile and levels of bioactive amines in green and roasted coffee. *Food Chem.* **2003,** *82,* 397-402.
32. Skog, K.; Alexander, J. The formation of acrylamide in cereal products and coffee. In *Acrylamide and Other Hazardous Compounds in Heat-Treated Foods;* Woodhead Publishing: Cambridge, U. K., 2006; pp 36.
33. Guenther, H.; Anklam, E.: Wenzl, T.; Stadler, R.H. Acrylamide in coffee: Review of progress in analysis, formation and level reduction. *Food Addit. Cont.* **2007,** *24,* 60-70.
34. Andrzejewski, D.; Roach, J. A. G.; Gay, M. L.; Musser, S. M. Analysis of coffee for the presence of acrylamide by LC-MS/MS. *J. Agric. Food Chem.* **2004**, *52*, 1996-2002.
35. Amrein, T. M.; Bachman, S.; Noti, A.; Biedermann, M.; Barboa, M. F.; Biedermann-Brem, S.; Grob, K.; Keiser, A.; Realini, P.; Escher, F.; Amado, R. Potential of acrylamide formation, sugars, and free asparagine in potatoes: A comparison of cultivars and farming systems. *J. Agric. Food Chem.* **2003,** *51,* 5556-5560.
36. De Wilde, T.; De Meulenaer, B.; Mestdagh, F.; Govaert, Y.; Ooghe, W.; Fraselle, S.; Demeulemeester, K.; Van Peteghem, C.; Calus, A.; Degroodt, J.-M.; Verhé, R. Influence of fertilization on acrylamide formation during frying of potatoes harvested in 2003. *J. Agric. Food Chem.* **2006,** *54,* 404-408.
37. Noti, A.; Biedermann-Brem, S.; Biedermann, M.; Grob, K.; Albisser, P.; Realini, P. Storage of potatoes at low temperatures should be avoided to prevent increased acrylamide formation during frying or roasting. *Mitt. Lebensm. Hyg.* **2003,** *94,* 167-180.
38. Biedermann, M.; Noti, A.; Biedermann-Brem, S.; Mozzarti, V.; Grob, K. Experiments on acrylamide formation and possibilities to decrease the potential of acrylamide formation in potatoes. *Mitt. Lebensm. Hyg.* **2002**, *93*, 668-687.

39. Claus, A.; Schreiter, P.; Weber, A.; Graeff, S.; Herrmann, W.; Claupein, W.; Schieber, A.; Carle, R. Influence of agronomic factors and extraction rate on the acrylamide contents in yeast-leavened breads. *J. Agric. Food Chem.* **2006,** *54, 8969-8976.*
40. Delatour, T.; Périsset, A.; Goldmann, T.; Riediker, S.; Stadler, R. H. Improved sample preparation to determine Acrylamide in difficult matrixes such as cocoa powder, cocoa, and coffee by liquid chromatography tandem mass spectroscopy. *J. Agric. Food Chem.* **2004,** *52,* 4625-4631.
41. Hoenicke, K.; Gatermann, R. Studies on the stability of acrylamide in food during storage. *J. AOAC Int.* **2005,** *88,* 268-273.
42. Lantz, I.; Ternité, R.; Wilkens, J.; Hoenicke, K.; Guenther, H.; van der Stegen, G. H.D. Studies on acrylamide levels in roasting, storage and brewing of coffee. *Mol. Nutr. Food Res.* **2006**, *50*, 1039-1046.
43. Napolitano, A.; Morales, F.; Sacchi, R.; Fogliano, V. Relationship between virgin olive oil phenolic compounds and acrylamide formation in fried crisps. *J. Agric. Food Chem.* **2008,** *56,* 2034-2040.
44. Mestdagh, F. J.; De Meulenaer, B.; Van Poucke, C.; Detavernier, C.; Cromphout, C.; Van Peteghem, C. Influence of oil type on the amounts of acrylamide generated in a model system and in French fries. *J. Agric. Food Chem.* **2005,** *53,* 6170-6174.
45. Williams, J. S. E. Influence of variety and processing conditions on acrylamide levels in fried potato crisps. *Food Chem.* **2005,** *90,* 875-881.
46. Mestdagh, F.; Castelein, P.; Van Peteghem, C.; De Meulenaer, B. Importante of oil degradation components in the formation of acrylamide in fried foodstuffs. *J. Agric. Food Chem.* **2008,** *56,* 6141-6144.
47. Senyuva, H.; Gökmen, V. Study of acrylamide in coffee using an improved liquid chromatography mass spectrometry method: Investigation of colour changes and acrylamide formation in coffee during roasting. *Food Addit. Cont.* **2005,** *22,* 214-220.
48. Pardo, O.; Yusa, V.; Coscolla, C.; León, N.; Pastor, A. Determination of acrylamide in coffee and chocolate by pressurized fluid extraction and liquid chromatography-tandem mass spectrometry. *Food Addit. Cont.* **2007,** *24,* 663-672.
49. Wenzl, T.; Calle, B.; Anklam, E. Analytical methods for the determination of acrylamide in food products: a review. *Food Addit. Cont.* **2003,** *20,* 885-902.
50. Soares, C.; Cunha, S.; Fernandes, J. Determination of acrylamide in coffee and coffee products by GC-MS using an improved SPE clean-up. *Food Addit. Cont.* **2006,** *23*, 1276-1282.

51. Blenkinsop, R.W.; Copp., L. J.; Yada, R.Y.; Marangoni, A.G. Changes in compositional parameters of tubers of potato *(Solanum tuberosum)* during low-temperature storage and their relationship to chip processing quality. *J. Agric. Food Chem.* **2002**, *50,* 4545-4553.
52. Grob, K.; Biederman, M.; Biedermann-Brem, S.; Noti, A.; Inhof, D.; Amrein, T.; Pfefferle, A.; Bazzocco, D. French fries with less than 100 µg/kg acrylamide. A collaboration between cooks and analysts. *Eur Food Res Technol.* **2003,** *217*, 185-194.
53. Pedreschi, F.; Kaack, K.; Granby, K. The effect of asparaginase on acrylamide formation in French fries. *Food Chem.* **2008,** *109*, 386-392.
54. Jung, M.Y.; Choi, D.S.; Ju, J.W. A novel technique for limitation of acrylamide formation in fried and baked corn chips and in French fries. *J. Food Sci.* **68,** 1287-1290.
55. Frédéric, M.; Maertens, J.; Cucu, T.; Delporte, K.; Van Peteghem, C.; De Meulenaer, B. Impact of additives to coger the formation of acrylamide in a potato model system through pH reduction and other mechanisms. *Food Chem.* **2008,** *107,* 26-31.
56. Oku, K.; Kurose, M.; Ogawa, T.; Kubota, M.; Chaen, H.; Fukuda, S.; Tsujisaka, Y. Suppressive effect of trehalose on acrylamide formation from asparagine and reducing saccharides. *Biosci., Biotechnol., Biochem.* **2005**, *69,* 1520-1526.
57. Gökmen, V.; Açar, O.; Köksel, H.; Acar, J. Effects of dough formula and baking conditions on acrylamide and hydroxymethylfurfural formation in cookies. *Food Chem.* **2007,** *104,* 1136-1142.
58. Fredriksson, H.; Tallving, J.; Rosén, J.; Åman, P. Fermentation reduces free asparagine in dough and acrylamide content in bread. *Cereal Chem.* **2004,** *81,* 650-653.

Chapter 8

Furan in Thermally Processed Foods

Patricia J. Nyman

Center for Food Safety and Applied Nutrition, U.S. Food and Drug Administration, HFS-706, 5100 Paint Branch Parkway, College Park, Maryland, 20740

Furan is a volatile organic compound that has been classified as a potential human carcinogen. In 2004, furan was unexpectedly found in a broad range of thermally processed foods. Thermal decomposition and rearrangement of organic compounds was proposed for its formation. This chapter will present an overview on furan in thermally processed food and will discuss analysis, occurrence, formation, and exposure.

Furan (C_4H_40) is a colorless volatile organic compound with a boiling point close to room temperature (31.4°C). It is used in the manufacture of agricultural and pharmaceutical products and other organic compounds such as thiophene and tetrahydrofuran and is naturally occurring in certain woods. Furan is an animal carcinogen at high doses and has been identified as a possible or anticipated human carcinogen by the International Agency for Research on Cancer and the U.S. Department of Health and Human Services *(1, 2)*.

Until recently, the occurrence of furan in food was not believed to be widespread. Furan had been reported in a few foods such as cooked canned meat and poultry, roasted coffee, roasted filberts, beer, heated soy and rapeseed proteins, fish and milk proteins, wheat bread, and caramel *(3, 4)*. In a review by Maga, carbohydrate thermal decomposition and rearrangement was proposed as the principal formation pathway of furan and furan derivatives in food.

In 2004, renewed interest in the occurrence of furan in food was stimulated by a U.S. Food and Drug Administration (U.S. FDA) investigation on furan formation in foods subjected to non-thermal ionizing radiation. During that study, furan was unexpectedly found in a wide range of thermally processed canned and jarred foods. Lower detection limits resulting from improved analytical instrumentation and improved chromatographic techniques were considered important factors contributing to this discovery. In view of the analytical improvements, it is not unreasonable to speculate that low parts per billion (ppb, ng/g) or trace levels of furan (ng/g) have occurred in thermally processed and roasted foods for decades if not longer.

More data on the occurrence of furan in food was needed to evaluate the public health impact from long-term exposure to low ng/g furan levels. The U.S. FDA and European Food Safety Authority (EFSA) began to investigate the occurrence of furan in food, especially thermally processed canned and jarred foods. Requests for additional data from both the U.S. FDA and EFSA resulted in several studies on furan in food *(5, 6)*. These studies were the subject of a review on the analysis, occurrence, and formation of furan in heat-processed foods by Crews and Castle *(7)*. In May 2006, a workshop on 'Furan in Food' organized by the European Commission, Director General, Health and Consumer Protection, was held in Brussels. The focus of the workshop was to identify additional data needed to determine a reliable exposure assessment *(8)*. The outcome of the workshop resulted in several publications on the analysis, occurrence, and formation of furan in food *(9-13)*.

Analysis

Furan is a gas at room temperature and is ideally suited for headspace analysis (HS). A number of HS methods were developed in response to renewed interest in the occurrence of furan in food. Several of these methods *(9, 14-17)* are a modification of a U.S. FDA method that used static HS gas chromatography/mass spectrometry (GC/MS) to quantitatively determine furan in canned and jarred foods *(18)*. Other methods used solid phase micro extraction (SPME) with a carboxen/polydimethylsiloxane fiber (75-80 µm film) followed by GC/MS *(12, 19-23)*. These methods generally analyzed between 1 to 10 g test portions. Limits of quantitation (LOQs) were matrix dependent and ranged from 0.5 to 13 ng/g for static HS and from 0.02 to 0.8 ng/g for SPME. All of these methods used deuterated furan (furan-d_4) as an internal standard. Most GC separations were conducted using a PLOT (porous layer open tubular) capillary column with a poly-styrene divinyl benzene stationary phase, which retains apolar compounds without cryofocusing. Two studies of model systems with furan precursors were conducted using pyrolysis GC/MS and proton transfer reaction MS *(24, 25)*.

Sample Preparation

The high volatility of furan at room temperature required certain precautions when samples were handled and prepared for analyses. Most studies followed U.S. FDA specifications for sample preparation with a few modifications *(18, 26, 27)*. In general, samples were chilled (ca. 4°C) for several hours prior to handling and were held on ice during sample preparation. Foods with a viscosity similar to water were transferred directly to HS vials. Non-homogeneous semi-solid and solid foods were homogenized using a food processor, transferred to HS vials, and then diluted to slurry consistency. Test portions were fortified with furan-d_4 and immediately sealed. For standard addition analysis, test portions were also fortified with furan.

A number of techniques were used to improve furan sensitivity. Hasnip *et al.* added a few 2-4 mm glass beads to the HS vials to improve mixing *(17)*. Zoller *et al.* added 0.2 g of amylase to HS vials containing foods that can form a starchy gel *(9)*. Some researchers used sodium chloride or sodium sulfate to reduce the solubility of furan in the aqueous phase thereby increasing the concentration of furan in the vapor phase *(9, 14, 15, 20)*.

HS Thermal Equilibration

Headspace sampling often uses elevated temperatures to increase the amount of the analyte in the vapor phase. For furan analysis, excessive temperatures and long thermal equilibration times can lead to furan formation. As a result, most researchers used a lower HS oven temperature (50°C) than specified in the original U.S. FDA method (80°C). The original validation of the U.S. FDA method showed that it performed reliably for canned and jarred foods containing relatively high levels of furan (52 to 118 ng/g) *(27)*. However, additional HS studies using the original conditions (30 min equilibration in a 80°C HS oven) showed that furan formed at low levels ($\leq$ 3 ng/g) in some fatty foods containing relatively low levels of incurred furan (1 to 6 ng/g) *(28)*. As a result, the HS oven temperature was lowered. Additional oven temperature and thermal equilibration time studies showed that the method performed reliably using the modified conditions (30 min equilibration in a 60°C oven) *(26)*. The U.S. FDA survey data obtained prior to the temperature change was shown to be valid by conducting analyses at both the original (80°C) and modified (60°C) temperatures for several canned and jarred foods previously found to contain low levels.

Quantitation by using External Standards and the Method of Standard Additions

External standards and the method of standard additions were used to quantify furan in foods. The U.S. FDA used the method of standard additions to avoid matrix effects *(18, 26, 27)*. Matrix effects were characterized by a decrease in the integrated peak areas for furan and furan-d_4 and a change in the slope determined from linear regression analyses of the furan/furan-d_4 response ratio versus concentration for calibration standards prepared in water and the same curve prepared in the food matrix (data not reported). Altaki *et al.* compared SPME HS results determined with external standards and standard additions for apple juice, honey, powdered instant coffee, and rice/potato with chicken baby food *(19)*. Comparable results were obtained for all the foods by both methods of quantitation. However, the data are limited in comparison with the hundreds of samples analyzed by various organizations conducting furan analysis. An interlaboratory trial comparing data obtained by static and SPME HS using both external standards and standard additions would be useful to alleviate any uncertainty with respect to the matrix effects and quantitation of furan.

Occurrence

Table I summarizes the data from a number of studies and provides the range of furan concentrations reported for various food categories and the corresponding literature citation. Most of the foods analyzed by the U.S. FDA, EFSA, and other studies were found to contain measurable amounts of furan and, in general, comparable furan concentrations were found. Table II reports the number of samples, median furan concentrations, and number of samples reported as none detected or as less than the limit of detection (LOD) for some of the food categories including baby foods, infant formula, and adult foods in which higher furan concentrations were reported. Most of the data in Table I were reported as individual values and were used to compile Table II. Data originally reported as a range, average, or median were not included. For example, Hoenicke *et al.* found <5 to 100 ng/g furan in 9 minced meat samples (median of 20 ng/g) *(29)*. These data are not represented in Table II.

Table I. Summary of furan concentrations reported for various food categories and corresponding references

Food Category	*ng/g*	*Reference*
Baby[1] food	<0.8 – 153	(9, 14, 15, 17, 19, 30, 31)
Baby fruit and fruit juice	1.0 – 31.7	(9, 14, 15, 19, 30, 31)
Baby food (homemade)	0.1 – 1.0	(21)
Baked cookies, pancakes, waffles, french toast	ND[2] – 35	(29, 31)
Beer	ND – 4.4	(9, 30)
Bread, toast	ND – 39	(9, 15, 17, 30)
Bread, crust	24.0 – 193	(9, 15)
Breakfast cereal	9.2 – 47.5	(31)
Candy	0.8 – 5.5	(31)
Chocolate drinks/mixes, cocoa, chocolate syrup	0.4 – 10.3	(15, 31)
Coffee (brewed)	2.0 – 199	(9, 14-16, 19, 20, 29, 30)
Coffee, instant (prepared)	<2 – 51.3	(9, 19, 20, 30)
Crackers, low moisture snack foods	<3.2 – 143	(9, 26, 29, 31)
Dairy, eggs	ND – 2.9	(15, 31)
Dessert (puddings, gelatin)	<0.8 – 27	(17, 30, 31)
Evaporated milk	10.9 – 15.3	(30)
Fats, oils	ND – 5.4	(31)
Fish	1.5 – 8.1	(30, 31)
Fruit, fruit juice	ND – 30.5	(15, 19, 30, 31)
Gravies	13.3 – 174	(31)
Infant formula	ND – 27	(30, 31)
Jams, jellies, preserves	ND – 37.4	(30, 31)
Meals (homemade)	<2.0 – 16	(33)
Meals (ready-to-eat)	<2.0 – 27	(33)
Meat substitute	ND – 4.4	(31)
Meats	ND – 100	(14, 15, 29-31)
Miscellaneous[3]	ND – 91.0	(9, 15, 19, 30, 31)
Nutritional drinks	ND – 174	(30)
Nuts, nut butter, seeds	ND – 7.5	(9, 13, 30, 31)
Pet food	7.8 – 9.1	(22)
Soups, sauces, stews, chili, broths	3.3 – 240	(9, 14, 15, 19, 30, 31, 33)
Sweets and pastries	ND – 169	(9, 31)
Vegetables, fresh	ND – <2.0	(9, 15)
Vegetables, vegetable juice	ND – 122	(9, 15, 30, 31, 33)

[1]Includes infant and toddler foods

[2]ND – none detected or less than LOD

[3]Miscellaneous – soy sauce, syrup, caramel sauce, ketchup, mayonnaise, honey, sweetened & condensed milk, pie filling, marshmallows

With the exception of two samples, all of the jarred baby foods contained quantifiable levels of furan. Vegetable and vegetable-meat baby foods were found to contain higher concentrations; the highest level found was 153 ng/g in jarred vegetable/veal baby food *(9, 15)*. In comparison, low concentrations of furan ($\leq$ 1 ng/g) were found in a relatively small number of homemade baby foods. Higher concentrations of furan were also found in adult vegetables and vegetable juices. The highest level found was 122 ng/g in baked beans *(30)* and, in a related category, 240 ng/g in chili *(14)*.

Table II. Furan concentrations found in select food categories, number of samples, median concentration, and number of samples reported as none detected or less than LOD

Food Category	*ng/g*	*Samples*	*Median*	*ND*[1]
Baby[2] food	<0.8 – 153	154	42	0
Baby fruit and fruit juice	0.1 – 31.7	80	3.2	0
Baked cookies, pancakes, waffles, french toast	ND – 35	15	ND	10
Bread, toast	ND – 39	37	5	14
Bread, crust	24.0 – 193	14	58	0
Breakfast cereal	9.2 – 47.5	6	36.5	0
Coffee (brewed)	1 – 199	38	39.2	0
Coffee, instant (prepared)	<2 – 51.3	23	9.8	0
Crackers, low moisture snacks	<3.2 – 143	62	20.2	0
Fruit, fruit juice (prune juice)	ND – 30.5	65	2	6
Gravies (roasted turkey gravy)	13.3 – 174	8	29.8	0
Infant Formula	ND – 27	42	8.2	13
Jams, jellies, preserves	ND – 37.4	47	4.6	2
Meats	ND – 39.2	31	1.0	8
Miscellaneous[3]	ND – 91.0	38	6.6	11
Nutritional drinks	ND – 174	22	12.2	2
Soups, sauces, stews, chili, broths	3.3 – 240	114	18.2	0
Sweets and pastries	ND – 169	31	5.1	2
Vegetables, vegetable juice	ND – 122	87	7	2

[1]ND – none detected or less than LOD

[2]Includes infant and toddler foods

[3]Miscellaneous – soy sauce, syrup, caramel sauce, ketchup, mayonnaise, honey, sweetened & condensed milk, pie filling, marshmallows

With the exception of prune juice (30 ng/g) and carrot juice (40 ng/g), adult and baby fruit and fruit juices in cans or jars contained lower concentrations of furan (1–16 ng/g) *(9, 15, 19, 30, 31)*. This is interesting in light of the fact that model systems of ascorbic acid (AsA, pH = 4) were found to produce the highest levels of furan compared with model systems of other precursors (see

Formation) *(12, 24, 32)*. This illustrates that food ingredients may not reliably predict furan levels in a particular food due to complex competing reactions that may favor other end-products.

In general, higher levels of furan were found in toasted bread (39 ng/g) and especially the crust of some bread (193 ng/g) *(9, 17)*. The loaf shape and surface-to-volume ratio were suggested as important factors contributing to the presence of higher furan levels in some breads; smaller loafs might be expected to contain higher amounts of furan. Similar to crusty products, low moisture snack foods and breakfast cereals were also found to contain higher levels of furan; the highest level found was 143 ng/g in a long salted stick and 47.5 ng/g in corn flakes *(9, 26, 31)*. Furan was observed to be stable during the shelf-life of these products *(9)*.

Coffee was shown to be a significant source of exposure to furan for coffee drinkers. The amount of furan in the roasted coffee bean and the brewing method were important factors affecting the amount of furan found in brewed coffee. The highest level found (199 ng/g) was from an espresso-type coffee brewed with ground roasted coffee containing about 6000 ng/g furan *(9)*. Compared to brewed coffee (median of 39.2 ng/g), prepared instant coffee reportedly contained lower furan concentrations (median of 9.8 ng/g).

Higher furan concentrations were found in soy sauce (≤ 91 ng/g) and gravy (≤ 174 ng/g). A few of the nutritional/diet drinks, syrups, and caramel sweets were found to have high furan levels (174 ng/g, strawberry shake nutritional drink; 88 ng/g, maple syrup; 169 ng/g, sweet caramel biscuit); however, no consistent trend was identified from the limited sampling of these foods *(9, 30, 31)*.

Formation

Complex competing reactions involving oxidation and/or reduction of precursors have been proposed for the formation of furan in food *(24, 25, 32)*. Furan formation in food has been shown to be associated with thermal degradation and rearrangement of ascorbic acid (AsA), sugars, amino acids, amino acids with sugars, and lipids. Figures 1 and 2 represent simplified mechanisms of furan formation from these precursors and show some of the more important intermediates. The presence of these precursors may not necessarily be a strong indicator of high furan levels in a particular food. Processing conditions and competing reactions between precursors and other food ingredients may promote or inhibit furan formation. Recent studies of model systems have shown that AsA followed by unsaturated fatty acids or triglycerides formed the highest amounts of furan *(25, 32)*. However, direct comparisons between model systems of precursors and foods may not be meaningful. Mark *et al.* showed that furan formation decreased by 46 percent in binary mixtures of AsA and linoleic acid and by as much as 95 percent in binary mixtures of AsA with glycine, serine, or erythrose *(25)*. In a similar study by Limacher *et al.*, furan formation was shown to decrease by as much as 80 percent in binary mixtures of AsA with erythrose, glucose, or phenylalanine *(12)*.

Perez Locas and Yaylayan *(24)* conducted studies using pyrolysis gas chromatography/mass spectrometry (GC/MS) analysis and model systems of ascorbic acid, amino acids, sugars, amino acid/sugar mixtures (model Maillard systems), and ^{13}C-labeled serine and glucose to propose mechanisms for furan formation. AsA and dehydroascorbic acid (DHA) showed the highest efficiency for furan formation. Other model systems also formed furan as shown in decreasing order of efficiency: glycolaldehyde/alanine > erythose > ribose/serine > sucrose/serine > fructose/serine > glucose/cysteine. With the exception of erythrose, model systems of sugars or amino acids subjected to pyrolysis GC/MS did not form significant amounts of furan. On the other hand, model systems containing a combination of amino acids and sugars produced higher amounts of furan, which can be attributed to thermal degradation and rearrangement of carbohydrates and proteins associated with the Maillard reaction.

Furan Formation from AsA and DHA

Perez Locas and Yaylayan proposed oxidative and non-oxidative mechanisms for furan formation from AsA *(24)*. DHA is a known oxidative product of AsA that forms 2,3-diketogulonic (DKG) acid in food. In the simplified pathway shown in Figure 1, DKG undergoes α-carbonyl cleavage and decarboxylation to form aldotetrose. Aldotetrose can undergo cyclization and dehydration to form 3-furanone, which undergoes dehydration and reduction to form furan. In the mechanism proposed for non-oxidative pyrolytic formation of furan from AsA (not shown), 3-deoxy-pentosulose and then 2-deoxy-aldotetrose are formed; the latter can undergo cyclization and dehydration to form furan.

Studies conducted by Becalski and Seaman showed that DHA formed the highest amount of furan (381 ng/g) when subjected to thermal treatment under aqueous conditions *(32)*. In contrast, studies conducted by Mark *et al.* under dry thermal conditions showed that AsA formed higher furan levels (9950 µmol/mol AsA) than DHA (270 µmol/mol DHA) *(25)*. Similar results were observed by Limacher *et al.* using AsA and DHA model systems simulating roasting and pressure cooking conditions *(12)*.

Limacher *et al.* conducted mechanistic studies using ^{13}C-labeled AsA and showed that furan formed via various routes from the loss of CO_2 or formic acid (see Figure 1). The furan generated contained an intact C_4 unit (C-3 to C-6) from AsA. Quantitative results support the assertion that pathways that include 2-deoxyaldotetroses and 2-furfural as intermediates are major routes in dry or aqueous model systems. The ^{13}C-labeled AsA studies were extended to pumpkin vegetable puree, carrot juice and orange juice. Samples were fortified with unlabeled and [6-^{13}C]-labeled AsA and heated to simulate sterilization (123°C, 22 min). The highest increase in furan concentration (124% over background (48.6 ng/g)) was observed in the pumpkin vegetable puree. However, the labeled furan was not detected and was barely detected in the other juices. These results showed that AsA was a minor precursor and other compounds such as lipids may be more important furan precursors in complex foods *(12)*.

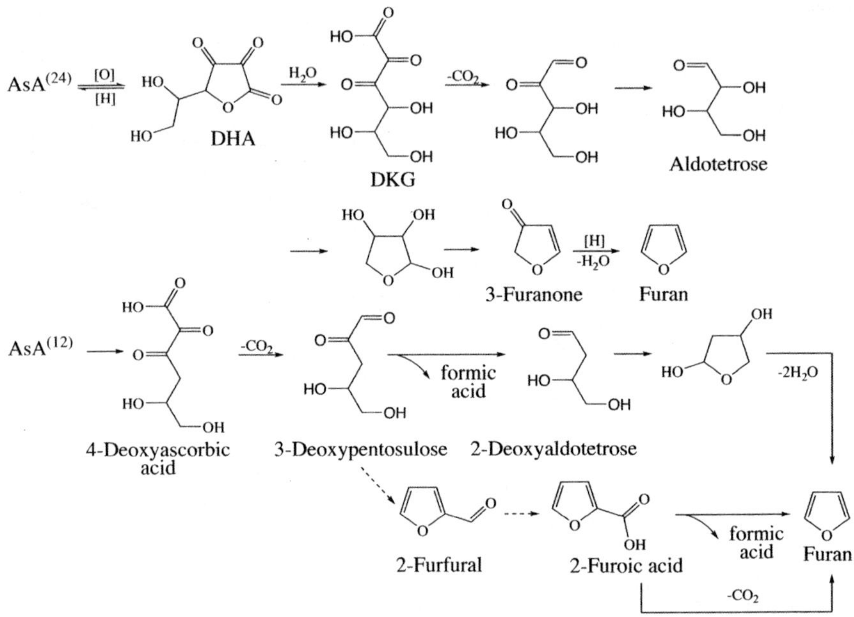

Figure 1. Proposed pathways for furan formation from AsA modified from references 12 and 24; [O] – oxidation; [H] – reduction.

In another study, apple and orange juices were subjected to immersion in a 100°C water bath for 5 min or an autoclave at 125°C for 25 min *(23)*. The amount of furan formed was correlated with the loss of AsA. Prior to heating, apple juice was found to contain only 3.6 µg/mL AsA compared with 373 µg/mL in orange juice. The autoclaved apple and orange juices formed the highest amount of furan (apple juice, 14 ng/g; orange juice, 7 ng/g), and all or most of the AsA was destroyed. No explanation was presented for the higher levels of furan found in the autoclaved apple juice, but it was speculated that antioxidants present in orange juice may inhibit furan formation. Negligible amounts of furan formed in the apple juice heated for 5 min at 100°C. No significant change in AsA was observed. As a result, the furan formed was attributed to carbohydrates rather than AsA. For orange juice, the furan formed was attributed to AsA, because furan formation coincided with a decrease in AsA for both methods of heating. Lower amounts of furan (1.4 ng/g) formed in the orange juice heated at 100°C for 5 min.

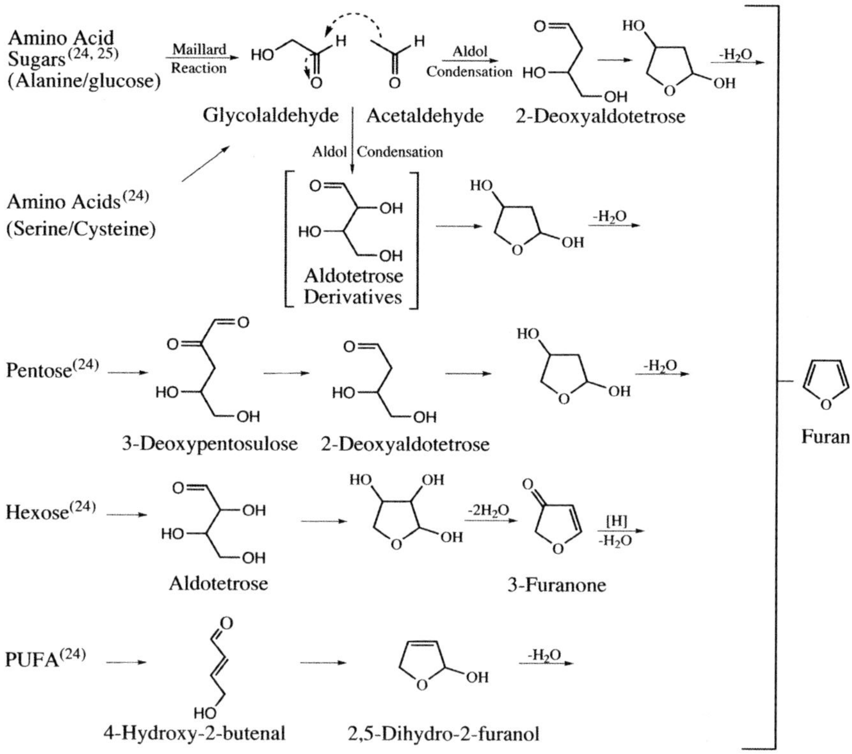

Figure 2. Proposed pathways for furan formation from amino acids with sugars, amino acids, sugars, and PUFA modified from references 24 and 25; [O] – oxidation; [H] – reduction

Furan Formation from Lipids

Perez Locas and Yaylayan *(24)* proposed a mechanism for furan formation from polyunsaturated fatty acids (PUFA). The mechanism includes the formation of lipid hydroperoxides from PUFA by reactive oxygen species. Transition metal ions can promote oxidation and homolytic cleavage of PUFA hydroperoxides to form 4-hydroxy-2-butenal. Subsequent cyclization occurs to form 2,5-dihydro-2-furfuranol, which undergoes dehydration to form furan. Studies of a model system of linoleic, linolenic, trilinolein, or trilinolenin showed that furan formed following thermal treatment. Higher furan amounts were generated from linolenic (3270 μmol/mol) and trilinolenin (4747 μmol/mol) as compared with linoleic (681 μmol/mol) and trilinolein (1727 μmol/mol). This was attributed to the ease of oxidation of the three double bonds in linolenic acid (C18:3) as compared with the two double bonds in linoleic acid (C18:2). Model systems of linoleic acid containing Fe^{3+} generated higher amounts of furan but gave conflicting results in studies of linolenic acid and the triglycerides of linoleic and linolenic acids *(25, 32)*.

Furan Formation from Amino Acids

Model systems of ^{13}C-serine were studied to evaluate the formation of furan from amino acids. Pyrolysis GC/MS analysis indicated that serine can degrade to acetaldehyde and glycolaldehyde and undergo aldol condensation to form aldotetrose derivatives. The aldotetrose derivatives such as erythrose can undergo cyclization and dehydration to form furan. The same mechanism was proposed for furan formation from cysteine. α-Alanine needed a source of glycolaldehyde (reducing sugars) to generate furan. Similarly, aspartic acid and threonine may also need a source of glycolaldehyde to undergo aldol condensation and generate furan.

Furan Formation from Maillard Type Reactions and Reducing Sugars

The formation of 749 μmol furan/mol from a reaction mixture containing glucose, alanine, and threonine during proton transfer reaction GC/MS was attributed to the Maillard reaction *(25)*. The proposed mechanism included the formation of a Strecker aldehyde and glycolaldehyde. Strecker aldehydes, such as acetaldehyde, form from the degradation of amino acids in the presence of sugars; glycolaldehyde forms from the degradation of sugars. These intermediates undergo aldol condensation to form aldotetrose derivatives, which can undergo cyclization and dehydration to form furan.

Serine/^{13}C-glucose model systems subjected to pyrolysis GC/MS showed increased efficiency for furan formation *(24)*. The increased efficiency was attributed to the ability of hexose and ribose sugars to catalyze the formation of deoxyosone derivatives in the presence of amino acids. Four pathways were proposed by Perez Locas and Yaylayan for furan formation from hexose. Two major pathways (50%) incorporate C3 to C6 carbon atoms from glucose and were initiated by the formation of 1-deoxyosone in the presence of amino acids or through a retro-aldol cleavage. Both pathways can form aldotetrose followed by 3-furanone, which subsequently undergoes dehydration to form furan. The remaining pathways (not shown) incorporate 1) C1 to C4 carbon atoms to form furan from retro-aldol cleavage and dehydration of a 2-deoxy-3-ketoaldotetrose intermediate or 2) incorporate C2 to C5 carbon atoms to form a 3-deoxyosone intermediate, a precursor of furan. An amino acid assisted pathway for furan formation from pentose was proposed. In this pathway furan formation was initiated by the formation of 3-deoxyosone derivatives, which through α-carbonyl cleavage form 2-deoxyaldotetrose – a direct precursor of furan.

Other Factors Affecting Furan Formation and Occurrence in Food

Hasnip *et al.* investigated the effects of domestic cooking on foods and the migration and/or formation of furan from can coatings, sealing gaskets, and epoxidized oils used to manufacture gaskets *(17)*. Select foods were heated according to manufacturers' instructions using a domestic electric toaster, microwave oven, and gas and electric cookers. With the exception of bread, heating did not have a significant effect on the amount of furan found in the foods. Furan increased from less than 2 ng/g to an average of 39 ng/g in breads that were toasted. In general, the results of the study indicated that any furan formed during heating was offset by a concurrent loss due to evaporation.

Select foods, epoxidized linseed oil (ELO), and epoxidized soybean oil (ESBO) were heated in the presence of lacquer coated coupons of can material or gaskets commonly used in food packaging. Furan levels did not increase in comparison to foods heated in the absence of these materials. Higher furan levels were found in heated ELO and ESBO compared with unepoxidized oils. Heated ELO formed higher furan levels than ESBO, which was attributed to the higher degree of unsaturation associated with linolenic acid (60%; C18:3) in linseed oil compared with soybean oil (10%). It was concluded that the impact on food from migration or formation was negligible due to the low level of epoxidized oils used in food packaging.

Roberts *et al.* compared the effects of heating on ready-to-eat convenience foods vs. their homemade equivalent, foods heated in a saucepan vs. in a microwave oven, and foods that were heated and then allowed to cool for up to 10 min with and without stirring *(33)*. Foods were cooked according to the manufacturers' instructions. No consistent trends were observed between furan levels found in ready-to-eat vs. homemade foods or foods heated in a saucepan vs. a microwave oven. Furan levels actually increased in some foods subjected to microwave heating and decreased in the same food after heating in a saucepan. Furan levels found in heated foods did not change significantly when the foods were allowed to cool for up to 10 min. In comparison, heated foods that were cooled and stirred for up to 10 min reportedly showed a near linear decrease in the amount of furan found. With the exception of baked beans, a similar trend was observed when foods were tested from 0.2 to 10 min and manipulated to simulate consumption from a dinner plate. On the basis of these results, it was recommended that consumers could reduce their exposure to furan by stirring foods before eating.

Exposure

The U.S. FDA and EFSA have conducted dietary exposure assessments using recent survey data on furan in foods. The U.S. FDA dietary exposure was derived using food intake values and the U.S. FDA survey data on furan in foods *(30, 31, 34)*. Estimated mean exposures were determined on the basis of heat processed adult foods, baby foods, and infant formula. The mean exposures to furan from adult foods (2 years and older) and infant foods (0-1 years, excluding infant formula) were determined to be 0.26 μg/Kg-bw/day and 0.41 μg/Kg-bw/day, respectively. Brewed coffee was found to be the major source of furan exposure for the average adult and represented 0.15 μg/Kg-bw/day. The exposure from infant formula was determined to be 0.9 μg/Kg-bw/day.

The EFSA reported a range of exposure estimates on the basis of 11 food categories using consumption data and the survey data from the U.S. FDA and the Swiss Federal Office of Public Health (11). Assuming an average body weight equal to 60 Kg, brewed coffee represented the highest dietary exposure for the average adult (0.04 to 1.9 μg/Kg-bw/day) accounting for more than twice the dietary exposure from other sources. Assuming an average body weight of 7.5 Kg, the estimated exposure from baby food was determined to be <0.03 to 3.5 μg/Kg-bw/day and 0.2 to 1.5 μg/Kg-bw/day from infant formula.

In some cases, the exposure estimates conducted by the U.S. FDA and EFSA may be exaggerated. Studies have shown that consumer handling of foods prior to consumption can reduce furan exposure *(33)*. Jarred baby food that was warmed in a microwave and subsequently stirred showed furan losses of up to 55 percent, and coffee analyzed 1 h after brewing showed furan losses of about 50 percent *(9)*.

Conclusion

In several independent surveys, furan was found in a wide range of foods. Static and SPME HS followed by GC/MS analysis can reliably quantify furan with LOQs ranging from 0.02 to 13 ng/g depending on the food matrix. To avoid furan formation during analysis, the HS oven temperature should not exceed 60°C.

Thermal processing during production is considered a major cause for the occurrence of furan in food, and multiple pathways can lead to furan formation. Ascorbic acid, amino acids, amino acids with sugars, and lipids were identified as precursors for furan formation in model systems. The occurrence of one or more of these precursors in a particular food may not be a reliable indicator of the presence of furan; competing reactions may favor other end-products. For adults, the major source of furan exposure is brewed coffee.

References

1. IARC Monographs on the Evaluation of the Carcinogenic Risk to Humans. International Agency for Research on Cancer: **1995**; pp 393-407.
2. *Toxicology and carcinogenesis studies of furan in F344/N rats and B6C3F1 mice*; NTP Technical Report No. 402; NTP (National Toxicology Program), U.S. Department of Health and Human Services, Public Health Service, National Institutes of Health: Research Triangle Park, N.C., **1993**.
3. Maga, J.A. Furans in Foods. In *CRC Critical Reviews in Food Science and Nutrition*, CRC Press: Boca Raton, FL, **1979**; pp 355-400.
4. Maarse, H.; Visscher, C. A. *Volatile Compounds in Foods, Volume II;* 6th ed.; TNO-CIVO:Zeist: The Netherlands, **1989**.
5. Furan in Food; Thermal Treatment; Request for Data and Information. *Fed. Regist.,* **May 10, 2004,** pp 25911-25913.
6. Invitation to submit data on furan in food and beverages, 2006. *European Food Safety Authority* **2008,** http://www.efsa.europa.eu/en/science/data_collection/furan.html

7. Crews, C.; Castle, L. A review of the occurrence, formation and analysis of furan in heat-processed foods. *Trends in Food Sci. and Tech.* **2007,** *18*, 365-372.
8. Stadler, R. H. Update on the progress in acrylamide and furan research. *Food Addit. Contam.* **2007,** *24*, 1-2.
9. Zoller, O.; Sager, F.; Reinhard, H. Furan in food: Headspace method and product survey. *Food Addit. Contam.* **2008,** *24*, 91-107.
10. Crews, C.; Hasnip, S.; Roberts, D. P. T.; Castle, L. Factors affecting the analysis of furan in heated foods. *Food Addit. Contam.* **2008,** *24*, 108-113.
11. Heppner, C. W.; Schlatter, J. R. Data requirements for risk assessment of furan in food. *Food Addit. Contam.* **2008,** *24*, 114-121.
12. Limacher, A.; Kerler, J.; Conde-Petit, B.; Blank, I. Formation of furan and methylfuran from ascorbic acid in model systems and food. *Food Addit. Contam.* **2007,** *24*, 122-135.
13. Senyuva, H. Z.; Gokmen, V. Potential of furan formation in hazelnuts during heat treatment. *Food Addit. Contam.* **2008,** *24*, 136-142.
14. Becalski, A.; Forsyth, D.; Casey, V.; Lau, B. P. Y.; Pepper, K.; Seaman, S. Development and validation of a headspace method for determination of furan in food. *Food Addit. Contam.* **2005,** *22*, 535-540.
15. Reinhard, H.; Sager, F.; Zimmerman, H.; Zoller, O. Furan in foods on the Swiss market: Method and results. *Mitt. Lebensm. Hyg.* **2004,** *95*, 532-535.
16. Senyuva, H. Z.; Gokmen, V. Analysis of furan in foods. Is headspace sampling a fit-for-purpose technique? *Food Addit. Contam.* **2005,** *22*, 1198-1202.
17. Hasnip, S.; Crews, C.; Castle, L. Some factors affecting the formation of furan in heated foods. *Food Addit. Contam.* **2006,** *23*, 219-227.
18. Determination of furan in foods. *U. S. Food & Drug Administration* **2006,** http://www.cfsan.fda.gov/~dms/furan.html
19. Altaki, M. S.; Santos, F. J.; Galceran, M. T. Analysis of furan in foods by headspace solid-phase microextraction-gas chromatography-ion trap mass spectrometry. *J. Chromatogr. A.* **2007,** *1146*, 103-109.
20. Ho, I.-P.; Yoo, S.; Tefera, S. Determination of furan levels in coffee using automated solid-phase microextraction and gas chromatography/mass spectrometry. *J. AOAC Int.* **2005,** *88*, 574-576.
21. Bianchi, F.; Careri, M.; Mangia, A.; Musci, M. Development and validation of a solid phase micro-extraction-gas chromatography-mass spectrometry method for the determination of furan in baby-food. *J. Chromatogr. A.* **2006,** *1102*, 268-272.
22. Goldmann, T.; Perisset, A.; Scanlan, F.; Stadler, R. H. Rapid determination of furan in heated foodstuffs by isotope dilution solid phase micro-extraction-gas chromatography–mass spectrometry (SPME-GC-MS). *Analyst.* **2005,** *130*, 878-883.
23. Fan, X. Impact of ionizing radiation and thermal treatments on furan levels in fruit juice. *J. Food Sci.* **2005,** *70*, E409-E414.
24. Perez Locas, C.; Yaylayan, V. A. Origin and mechanistic pathways of formation of the parent furan – A food toxicant. *J. Agric. Food Chem.* **2004,** *52*, 6830-6836.

25. Mark, J.; Pollien, P.; Lindinger, C.; Blank, I.; Mark, T. Quantitation of furan and methylfuran formed in different precursor systems by proton transfer reaction mass spectrometry. *J. Agric. Food Chem.* **2006,** *54*, 2786-2793.
26. Nyman, P. J.; Morehouse, K. M.; Perfetti, G. A.; Diachenko, G. W.; Holcomb, J. R. Single-laboratory validation of a method for the determination of furan in foods by headspace gas chromatography/mass spectrometry, Part 2: Low-moisture snack foods. *J. AOAC Int.* **2008,** *91*, 414-421.
27. Nyman, P. J.; Morehouse, K. M.; McNeal, T. P.; Perfetti, G. A.; Diachenko, G. W. Single-laboratory validation of a method for the determination of furan in foods by using static headspace sampling and gas chromatography/mass spectrometry. *J. AOAC Int.* **2006,** *89*, 1417-1424.
28. Method Ruggedness Testing, Determination of Furan in Food. *U. S. Food & Drug Administration* **2006,** http://www.cfsan.fda.gov/~dms/furan2.html
29. Hoenicke, K.; Fritz, H.; Gatermann, R.; Weidemann, S. Analysis of furan in different foodstuffs using gas chromatography mass spectrometry. *Czech J. Food Sci.* **2004,** *22*, 357-358.
30. Morehouse, K. M.; Nyman, P. J.; McNeal, T. P.; Dinovi, M. J.; Perfetti, G. A. Survey of furan in heat processed foods by headspace GC/MS and an estimate of adult exposure. *Food Addit. Contam.* **2008,** *25*, 259-264.
31. Exploratory Data on Furan in Food: Individual Food Products. *U. S. Food and Drug Administration* , **2007,** http://www.cfsan.fda.gov/~dms/furandat.html
32. Becalski, A.; Seaman, S. Furan precursors in food: A model study and development of a simple headspace method for determination of furan. *J. AOAC Int.* **2005,** *88*, 102-106.
33. Roberts, D.; Crews, C.; Grundy, H.; Mills, C.; Matthews, W. Effect of consumer cooking on furan in convenience foods. *Food Addit. Contam.* **2008,** *25*, 25-31.
34. DiNovi, M.; Mihalov, J. An updated exposure assessment for furan from the consumption of adult and baby foods. *U. S. Food and Drug Administration,* **2007,** http://www.cfsan.fda.gov/~dms/furanexp.html

Chapter 9

What chemists need to know about very low levels of chemicals in food

Richard W. Lane

Research & Development, Unilever, Englewood Cliffs, NJ 07632

Advances in analytical techniques over the last few decades allow detections of chemicals at very low levels – parts per billion or less. Any detection, no matter how infrequent or how low, and especially in food, inevitably raises a question of safety. This causes concern in consumers, which poses problems for industry and regulators. Improved detection capabilities will lead to more findings, so a process is needed to determine whether unexpected chemicals occurring at very low levels are of true concern or not. The level of concern can be assessed relatively easily when there are regulatory limits or sufficient toxicology data, but most chemicals have little or no data associated with them. This paper proposes an approach that could help rank concern by using conservative methods developed over the past forty years. The proposed approach uses established generic safe levels for chemicals, with little or no safety data, and expected length of exposure. This approach could help manage risk by prioritizing detections according to the level of concern they raise, allowing resources to be directed at important issues while those determined to be of little or no public health importance to be addressed as resources permit. It also provides a common scientific basis that can be used by all interested parties when deciding what to do when a detection occurs. This will ensure a safe food supply, increase consumer confidence, save resources, and allow manufacturers to work with the knowledge that negligible levels of risk will be handled in a consistent manner.

Analytical chemistry has made extraordinary advances in recent years. Chemicals that were once undetectable can now be routinely found at the parts per billion (ppb) level and even lower in some cases, in complex matrices. Composite foods can now be analyzed routinely, which means that chemicals can be found when just a few years ago they could not. While not expected to be common, very low level detections will happen more and more frequently as chemistry continues to advance. Therefore, it is important that chemists and all others involved in food safety will know what to do when an unexpected situation occurs.

Some chemicals found at very low levels will have a regulatory level already in place, making assessment of the suitability of the food straightforward. Other chemicals will have sufficient toxicological data in the scientific literature to make a safety determination based on accepted risk assessment practices. This too will allow a relatively easy decision to be made about whether there is a health risk or not. But still other chemicals will inevitably be found that have few or no toxicology data associated with them. What can be done in these cases?

It turns out that there are ways to handle these occurrences based on toxicology data gathered for chemicals unrelated to the one found. This paper will review work already done in this area and propose a way to handle very low level detections in food based on expected length of exposure.

Chemicals of concern

"Chemicals," for the purpose of this paper, are substances with covalent bonds and molecular weights generally less than 1500. It excludes proteins, nanomaterials, metals and their salts, and intentionally biologically active substances such as pharmaceuticals, pesticides, vitamins and other nutrients. Many of these will have adequate toxicology data anyway.

"Concern" simply means to be of importance to. For health professionals, concern arises when there is a reasonable likelihood of an adverse effect in a significant portion of the population. Most of the chemicals of concern for public health have already been identified and handled by regulatory processes such as bans, denying approvals, and the setting of tolerances, action levels, and specifications. For these chemicals, levels that cause adverse effects are reasonably well established and by applying a suitable safety factor, levels that are safe in food have been promulgated. Making decisions about detections of these chemicals is easy because there is confidence in the data, the process, and the result. When a detection occurs at a very low level and exposure is within that established to be safe, then consumption of the food is considered to be safe. Public health is conserved. When a new chemical of true health concern is found (e.g., acrylamide), governments and industry work to understand its toxicity and do what is possible to control exposure. From this it can be seen that (1) there are mechanisms already in place that deal with chemicals of true concern in food, and (2) low exposures to unwanted chemicals do occur and can be safe.

To the general public, though, almost every chemical found in food is unwanted and thus is a "chemical of concern." It does not usually matter whether it poses a significant health risk or not. This is especially true in the Western world where chemophobia is common, and is unlikely to change any time soon with the media and some public interest groups actually encouraging this thinking. The general public, thus, can react very strongly to the presence of something that has a chemical-sounding name, is portrayed as a contaminant or adulterant, is found in food for the first time, suggests poor management practices, or is a substance that cause particularly fearful adverse effects. Any detection, no matter how low, has the ability to cause concern to the public. This concern, based on fear rather than risk, becomes an issue for regulatory officials and food manufacturers and thus, needs to be handled in a forthright manner based on sound scientific principles.

The Problems

Chemists now have the ability to detect less and less of more and more, meaning there is an increasing likelihood of chemicals of unknown hazard being discovered. While the analysis is usually being done to understand off-flavors or some other issue in a particular food, the detection of the unexpected chemical still occurs and the consequences need to be addressed.

A few decades ago there was a rough correlation between a "low level" detection – parts per million – and the possibility of a health concern. Levels tested in toxicology studies were not too much greater than those what could be detected and any effects that might have been observed in animals could be reasonably anticipated to occur in humans. Today, levels detected may be many orders of magnitude below doses used in toxicology studies, and it is not always certain that the effects seen in toxicology studies predict what might be seen in humans at very low levels of exposure. So if the hazard may not be relevant at the very low level of exposure, certainly the risk is not either. Chemistry keeps moving ahead, but toxicology testing and risk assessment cannot keep up.

Laws and regulations cannot keep up either. Many that were written in the 1950s and 1960s mean or are interpreted to mean that "none" means absolutely no exposure. This was fine when the level of detection was approximately the level of concern. But today, the level of detection may be orders of magnitude below a level of concern. So a very low level detection can be interpreted to mean an unacceptable presence even when there is negligible risk to public health. Amending general laws and regulations to reflect the new power of chemistry takes time and effort and would not necessarily result in greater flexibility. A regulation cannot be written in a timely manner for every chemical that is found, especially for those for which there are no toxicology data and a robust risk assessment is not possible. Other than good manufacturing practices, which are already in place, regulations cannot be written in anticipation of detections of unknown chemicals. Thus, an approach is needed to adapt to the new reality of very low level detections occurring, but allow those that are truly safe to be considered acceptable.

People's reactions to detections of very low levels have not kept up with the chemistry. They still react to hazards determined in toxicology studies, not to the risks based on the very low-level of exposures. They react to the presence of the chemical, not the amount that could be a health concern.

The food supply continues to globalize because it means more food for more people, greater convenience, better nutrition, and new experiences. But with globalization comes vulnerability. Regulatory oversight will be harder to maintain as we continue to search for newer, cheaper foods from more remote areas. Foods will be co-mingled and shipped great distances, further diminishing control. Pollution will increase, meaning more chemicals will be found more frequently. Populations will increase, changing where food moves. It seems reasonably foreseeable that with such a food supply it will become increasingly more common for very low-level detections to occur. If such detections become roadblocks to trade, where will we be in 20 years?

Since chemists will always be better at detecting chemicals than toxicologists will be at testing them and explaining their risks, than laws and regulations will be at controlling them, and than people will be at responding to them, as a society we need to change our approach to these detections. We need to understand that low-level exposures happen, that there will not be a full answer for each one, and that reacting to every detection will have both intended and unintended consequences. We might be able to remove a low-level exposure, but that could impact nutrition and other health issues, adversely affect the economics of food production, and reduce confidence in the food supply. None of this can be seen as very beneficial to society – or the world as a whole – if there is no meaningful health hazard involved.

In short, society has not evolved to handle the issue of very low level detections. We need to be clear what we want. If the only acceptable exposure is below the level of detection, then we can stop doing toxicology and risk assessments and just react when chemists find molecules in places they do not belong.

If that is not to happen, then we need to find a way forward with very low-level detections. Such an approach need to allow health professionals and regulatory officials to transparently demonstrate to a scientifically illiterate public that very low exposures are not a health concern. The approach needs to acknowledge the emotional concern, not dismiss it, and be able to properly characterize the potential risk. It needs to fit within existing laws. While this seems to be asking a lot, there is a way to handle these situations. All that is needed is an estimate of likely intake (the approximate level in food and the amount of food that is eaten in a day), the likely length of time that consumption might occur, and generic safe levels to which the exposure can be compared.

Role of the Chemist

When a detection of a chemical occurs, chemists can provide crucial information to assist in determining the true risk of the chemical. Most important is an estimate of concentration, followed by an approximate structure. From the structure, certain estimates of the inherent toxicity of the compound can be made. If there are serious structural alerts, then immediate action may be required to make sure the food supply remains safe. Whatever can be done to make both of these as specific as possible will help immensely, especially in the crucial first hours after the detection occurs.

Next, the likely source of the exposure needs to be determined. The food in which the chemical was detected will, of course, be known, but it is also important to determine the ingredient or process from which it might have originated. The chemist can help greatly with that process too. From that comes estimates of the extent of the unexpected chemical's presence in the food supply and the length of time it was present.

Dealing with Very Low Levels Detections: Duration of Exposure and Food Consumption

Long-term exposure

When long-term exposure is suspected, it is assumed that it occurs for most of a person's lifetime. Further, unless it is known with certainty that exposure to the chemical is for a shorter duration and limited to certain foods, the default position is to treat a new exposure as chronic and pervasive in the food supply.

If the chemical is thought to be limited to one type or a few types of food, then one would estimate the intake for each at the 75th percentile of consumption. Using consumption at the 75th percentile would ensure that people with high, consistent intakes are not at risk for adverse effects, but recognizes that the chemical's long-term presence in food makes it unlikely that people would be able to consume extremely high levels throughout the entire period. If the chemical is thought to be in more than a few types of food it will be necessary to estimate the intake as a proportion of the food supply depending on how widespread the exposure may be or assume that it could be in much of the food supply.

One simple approach would be to assume that the chemical is in all solid foods, which is conservatively estimated to result in a consumption of 1500 grams/day (g/d) *(1)*. If the chemical is present at 10 ppb, intake would be 150 micrograms/day (µg/d). If its presence in the food supply can be narrowed down, then daily intake would be lower.

Intermediate-term exposure

Some examples of intermediate-term exposures are when a particular food or food group has an inadvertent contact with a non-food material or an inappropriate use of a chemical is not found right away, or there could be an ongoing issue with a type of food produced or the region from which it originates. Exposures that occur for intermediate periods would cover a crop season or maybe up to a few years. Exposure is more likely to be limited to one or a few types of food, but still could be spread over much of the food supply.

If the unexpected chemical is in one or a few types of food, estimate intake using 90th percentile consumption to ensure that people who consume a food at a high level will not have adverse effects. If the chemical is found in more than a few types of food, it would be necessary to estimate intake as a proportion of the food supply or even as much of the food supply.

As an example, assuming a detection occurs at 10 ppb found in one type of food, the consistently high consumer might ingest 250 g in a day. The intake of the chemical would be 2.5 μg/d.

Short-term exposure

When there is inadvertent contact between a food and a non-food material or an inappropriate use of a chemical, and the contact is discovered and stopped right away, a short-term exposure would result. It will generally occur in one particular food and maybe even in one particular lot, so exposure is limited in both portion of diet and duration.

Few chemicals found at very low levels in foods for a short period of time will be of true concern because experience tells us that very low levels of substances rarely cause acute or short-term adverse effects in humans. Those that do are *very* well known! Even with little expectation of an acute effect, residual concern can be reduced by using a high percentile of intake of the food in question. This will ensure that people who might have a large consumption during the short period the food is in the supply chain will not be at risk. Intake at the 95th percentile may be appropriate.

A hypothetical short-term, low-level intake would be estimated as follows. Assume consumption of the food in question at the 95th percentile is 500 g/d (which is a lot of food, about a third of all consumed in a day) and the chemical is found at 10 ppb. This high, short-term intake of the chemical would be 5 μg/d.

Now that we have some insight into possible intake levels and durations of exposure, there needs to be something against which to compare them. In order to do that, some thought experiments will be used to frame the issue.

Approach to Very Low, Safe Levels

In order to understand exposure at very low levels, it helps to first understand something about high levels. Consider the absolute upper limit of exposure to any compound. This would be about three kilograms, a high estimate of what a person eats and drinks in a day *(1)*. All three kilograms would be the chemical. This is unlikely to ever happen and makes little practical sense. A better concept to consider is a *practical* upper limit of exposure. That is, at what level is a chemical's acute toxicity so low that it is not a concern because such high exposures are unlikely to occur? Convention has arbitrarily set the upper level at 2 g/kg in animals, or approximately 120 g in humans, at one time. Levels above that are not considered a concern to health because people are extremely unlikely to consume that much at one time. While they theoretically could, it has been decided that dealing with such remote possibilities is not worth using scare time and resources.

At the other end of exposure, the absolute lower limit is zero molecules, but this is not sensible either – no matter what some laws say. A better concept to consider is a *practical* lower limit of exposure. That is, at what level is a chemical's likelihood for causing harm of no concern because exposures are so low? This practical lower limit used to be the limit of detection, but as pointed out above, that does not work any more. So instead of working with detections, we need a way to deal with the issue based on the chance of having an effect, just like what was done for high exposures. The same convention that accepts that the likelihood of an astoundingly high exposure being so remote as to be trivial and not worth bothering about can be used to accept that an extremely low exposure leads to the same trivial risk and is not worth bothering about either. If society can make a decision about the insignificance of high exposures, why can it not reach the same decision about low ones? In both cases the possibility of an adverse effect occurring is so remote that the issue is not worth applying resources. With such a convention in place, a chemical with little or no toxicology data can still be determined to a low risk if exposure is below an assigned no-significant-risk level. Above that level there would be need for further assessment and possibly product action if the concern cannot be sufficiently addressed.

Levels of exposure below which there is no practical concern can be determined from the toxicology data of all compounds tested. Since exposure through food is oral, only toxicology studies done by oral exposure need to be considered. Taking the set of all the tests ever done – all compounds, all doses, all durations, all end points, all species – one could then split them into subsets of short-, intermediate- and long-term studies, which would align with the exposure scenarios described above.

For acute and subchronic studies, no-effect levels in animals would be plotted and a low percentile of all no-effect levels would be selected as a benchmark. The benchmark would have to be low enough to cover almost all scenarios, but not so low as to include the very unusual chemicals that need to be considered individually. A suitable safety factor, perhaps 100, would be applied to the benchmark to produce a level that would be expected to not cause adverse effects in humans. A different method would be used for cancer because society has determined that a different approach is needed for protection *(1)*, but the approach is essentially the same.

Three points can be gleaned from this exercise. First, protective values can be set that are not compound specific. They are derived from all compounds ever tested. So no matter what compound is detected, exposure below the protective level for the duration of exposure will not cause a meaningful risk. In other words, there are levels of exposure for any chemical below which there is no *practical* risk of an adverse effect.

Second, while it would be very protective of health to base safe values on the most potent compound tested in the most sensitive species, we would be back where we started: everything would be a concern because the level set would be so low that all exposures would be above it and thus be of concern and will not lead to a practical solution to the problem. We know from experience that not everything is like botulinum toxin or dioxin. Agreement is needed that we are not trying to attain zero risk, but zero *practical* risk. Practical risk is a negligible additional risk compared to the risks that people already incur as part of their daily lives. We need to accept that there is a very small chance that some incredibly potent chemical might slip through the process, but the likelihood is so remote that it is acceptable. Exposures need to be set that appropriately protect public health without trying to attain the unattainable, zero risk.

Third, the power of using all data from all studies allows us to know with great certainty that very low level exposures to compounds, even those without toxicology data, have no meaningful risk of causing adverse effects. The levels are based on the totality of our knowledge of all chemicals tested, giving us great confidence that there is little likelihood of increased risk of an adverse effect, no matter what the chemical. We do not claim 100% certainty, but we are sufficiently confident because of the size of the data set. We will be as certain of this position as the data from any single study.

Dealing with Very Low Levels of Detection: From theory to practice

Finding a level below which there is minimal likelihood of an adverse effect occurring has been called the "threshold of toxicological concern," (TTC) although a better term might be "toxicologically insignificant exposure." The concept was first mentioned 500 years ago by Paracelsus (c. 1508) who wrote, "All substances are poisons; there is none that is not a poison. The right dose differentiates a poison from a remedy." It is frequently simplified to, "The dose makes the poison." Paracelsus knew that an adverse effect depends on the exposure. Unfortunately, he was not able to offer a way to find the level of exposure that makes something not a poison.

More recently, Ben Oser applied this concept to foods as part of the National Academy of Science's Food Protection Committee *(2)*. He clearly stated that exposure to chemicals in food is unavoidable and specifically noted that removing every trace of chemicals is a major undertaking even for something as small and simple as a toxicology study. Oser noted that there are safe levels of every chemical, a "zone of toxicological inconsequence," but did not state what it actually is. Further, he believed that the zone would be difficult to determine and that every chemical would need to be expertly evaluated on its own. He was clear that when the intake of a chemical is within the zone of inconsequence, regulatory action was neither helpful nor necessary.

Long-term exposure

Most concerns about chemicals in food revolve around their potential chronic toxicity and carcinogenicity, which are associated with long-term, very low level exposures. Jack Frawley wrote the seminal paper on this topic proposing that exposure below a specified level for *any* food packaging component could be considered safe regardless of its degree of toxicity *(3)*. He examined the no observed effect levels (NOELs) from 220 chronic studies based on a survey of the toxicological literature and company files. Heavy metals and pesticides were excluded since they are not used in food packaging. Of the remaining 132 chemicals, only one had a NOEL below 100 ppm and none below 10 ppm. Frawley applied a safety factor of 100 to the 100 ppm level, giving a presumably safe dietary concentration of 1 ppm. He added another 10-fold safety factor because his database was admittedly incomplete and concluded that a substance present below 0.1 ppm of the total diet was safe for chronic exposure. This was the first quantification of a safe level of exposure to any chemical. The paper also included experimental work of the migration of packaging ingredients into food to show how the concept could be applied.

In the 1980s, Alan Rulis, Dave Hattan, Gary Flamm, and others at FDA re-examined low-level exposure to packaging materials *(4-8)*. Instead of chronic studies, they used carcinogenicity studies performed by the National Cancer Institute and the National Toxicology Program that had been conducted since Frawley's work. Using the lowest TD_{50} values (exposures that give tumors in 50% of animals) for 477 substances and a probabilistic method, they obtained a theoretical 10^{-6} risk of cancer in humans. It was also assumed, very conservatively, that packaging ingredients could migrate into the entire food supply so that intake would be based on consuming 3000 g of food per day. They concluded that 0.5 ppb in the total diet was a safe level for any chemical for any adverse effect. That resulted in a safe daily exposure below 1.5 μg. This work became the cornerstone of FDA's regulation for substances migrating out of packaging, the Threshold of Regulation *(1)*. It has simplicity at its core since it is based on one protective value and one level of intake.

Because this approach is a rigid, worst-case scenario, though, others have used data about chemicals – structural alerts, genotoxicity, and short-term toxicity – to determine higher safe levels. Cheeseman et al. *(9)* found that when these toxicology data were available for a compound the protective level could be increased from 0.5 to 5 to 15 ppb (1.5 to 15 to 45 μg/d), depending on the specific data available.

ILSI Europe has been using international teams to further polish the approach to chronic, very low level exposures. A series of publications *(10-15)* gives new avenues to handling any compound that might be encountered. By following decision trees and employing various information, thresholds of toxicological concern were established ranging from 0.15 – 1,800 μg/d. This represents the most advanced thinking to date on the threshold of toxicological concern. The greater power and flexibility means that the simplicity of earlier work was lost, so the full utility of the work might be difficult to apply when time is short or expertise is limited.

Intermediate-term exposure

Few chemicals found at very low levels in foods for an intermediate period of time will be of true concern because experience shows that these situations rarely cause toxicity in humans. Those that do are well known and regulated, but still this cannot be ignored. Work was done in the 1990s by Ian Munro and coworkers *(16)* to determine exposures to ingredients in foods that are so low that toxicology testing was not necessary to establish safety. This is a slightly different emphasis than previous work: rather than establishing a safe level *per se*, the idea was to avoid doing (unnecessary) toxicology testing because the results would be meaningless with the low exposure that was known to occur. But the outcome is the same. Munro et al. reviewed subchronic toxicology tests of 613 organic chemicals, with 2900 NOELs. They divided the chemicals into three groups based on Cramer et al.'s classification of chemicals' structural concern *(17)*. They used the 5th percentile of NOELs, applied a safety factor, and determined that exposures below 90, 540, and 1800 µg/d, for the highest to lowest structural concern levels, did not require testing. This approach is now part of the Joint FAO/WHO Expert Committee on Food Additives' decision-making process for flavoring substances *(18)*. A key element is knowing the structure of the chemical, but if it is not known then the lowest level would be used as the most protective.

Short-term exposure

Few chemicals found at very low levels in foods for a short period of time will be of true health concern because experience tells us that very low levels of substances rarely cause acute or short-term adverse effects in humans. Those that do are very well known. And any short-term effect would be seen immediately and recalls or other product actions started. For this reason there is no need for a generic threshold of toxicological concern for short-term exposures from food There is the lingering possibility that a short-term exposure to a very low level of a genotoxic compound could be a health concern, which has stimulated some work on genotoxic impurities in pharmaceuticals *(19)*. This work has established that intakes of even potential carcinogens are safe when less than 120 µg/d is consumed for up to a month.

Summary

As the power of analytical chemistry finds more chemicals with more frequency, we enter the paradoxical situation where increasing detections of decreasing importance leading to increasing concern about the safety of the food supply. We risk becoming victims of our fear and ignorance when we assume that presence equals hazard. This cycle can only be broken if there is a fair way to deal with the presence of very low levels of chemicals in food. It needs to allow the food supply to operate, while maintaining public health and trust, as the situation is corrected. That will prevent trade, the food supply, and business from being unnecessarily disrupted. It will allow consumers' expectations of convenience and proper nutrition to be met.

Fortunately, we now have an approach to the risk of very low levels of chemicals in food. Animal cancer data have been used to establish a level below which chronic exposures to food contact materials have no meaningful impact on public health. An intake of less than 1.5 µg/d is accepted as the threshold of toxicological concern for long-term exposure when there are no data about the chemical. This level can be increased or decreased as more is known about the chemical. For intermediate-term exposures, levels below 90, 540, and 1800 µg/d are accepted as being safe for flavoring chemicals of differing structural classifications, and these values have now been accepted for chemicals of any use. If the structure is not known, then the lowest value would be the default. Short-term exposures have not been studied in this regard because very low detections do not usually translate into any health concerns. If they do, the effect is immediate and recalls would be started. Only the lingering possibility that a short-term exposure to a very low level of a genotoxic compound could be a health concern has stimulated some work, which indicates that intakes of less than 120 μg/d are safe for these materials.

The TTC concept has been extended to ingredients in personal and household care products *(20)* and pharmaceuticals *(19, 21, 22)*. The US Environmental Protection Agency is using a TTC approach for pesticide active ingredients *(23)*. All this work has been necessitated by the collision of well intentioned, but overly conservative safety principles with analytical techniques that can now find very low levels of almost anything. The use of the TTC to new areas demonstrates the need to resolve the conflict between detection and risk on a broader scale.

Despite the scientific and regulatory advances that have been made, there is still more to be done. This paper has outlined an approach to handling unexpected chemicals in foods. But this cannot be taken lightly. As indicated, society generally is not ready for this. Consumer trust can be lost easily and the present situation will be reinstituted to nobody's advantage. The only way to expand the use of the TTC and have it widely accepted is to set some rules.

Next Steps

In order for the TTC to be widely accepted by the general public, some basic points need to be stressed.

First, the TTC concept is widely published and has been well received by the scientific community internationally. The publications are transparent and have laid out all the thinking behind the TTC. They demonstrate that this is a rigorous, conservative approach for establishing negligible concern. In fact, there is so much built-in conservatism that the possibility of a chemical slipping through the process and actually causing adverse effects is extraordinarily remote. Thus, the proposed approach is completely defensible.

Second, the TTC is already being applied to food-contact materials and flavoring substances. Just like any other chemical that could be found in food, they constitute a broad spectrum of structures and toxicities. If the TTC model works for these groups of chemicals it can work for all (with obviously unacceptable chemicals excluded).

Third, there are societal benefits to using the TTC when applied correctly. By allowing very low exposures to be categorized as trivial risks, regulatory officials can give attention to truly important matters, those that can actually affect public health. Reducing alarms about non-issues will reduce anxiety which will also improve public health *(21)*. Allowing products with insignificant levels of chemicals to remain in commerce reduces food costs, prevents shortages, and maintains consumer nutrition. Manufacturers do not suffer unnecessary losses, which means more money can be applied to paying employees and compensating shareholders. This results in better living standards and it is well known that higher incomes are directly related to healthier, longer lives *(22, 23)*. So overall, a prudent approach to very low level exposures will lead to better public health.

But this will only work if certain conditions are met. First, the TTC cannot be used frequently. Overuse will lead to questions and will been seen as a backdoor route save money and increase profits while causing harm to the consumer. While, in fact, it is just the opposite, that will not be easy for consumers to accept if this is used frequently. Second, it needs to be used correctly. "The dose makes the poison" is a basic tenet of toxicology and is being employed by toxicologists all the time, in some form, everywhere. Promoting an accepted TTC to be used openly, consistently, and correctly would benefit society. Otherwise, toxicologists and nontoxicologists will be applying it on their own without a proper methods to follow, which could actually be detrimental. It needs to be monitored so that it is not abused. If using the TTC is seen as a "back room" process with no oversight, then broad acceptance will be hard to find. Perhaps instituting a notification process or something similar would show everyone that it works. This would also allow for learning and improvements, but it gets tricky with the risk of recalls, so certain safeguards need to be in place. Finally, the TTC is not an escape mechanism for sloppy manufacturing practices. Whatever caused the detection needs to be traced and corrected to the extent possible. If there are data to support taking action at very low levels, then action must be taken. As with any work in progress, there is always more to do. While the concept is clear, enacting it will be difficult.

This paper was not intended to give any sort of depth, but simply to help the chemist understand the general issue and what he can do to help with the process. The toxicological basis for the TTC has been exhaustively examined. The next aspect that needs to be addressed is how exposures, both in duration and amount, are determined. A way to work mixtures into the TTC concept is also needed. And although already very conservative, a way to assure ourselves that sensitive subpopulations are properly covered would help.

With these pieces in place, society should be able to agree that exposures below the TTC do not result in any meaningful additional risk. It should be assured that safety is maintained as well as any other process being used. With acceptance, society will be able to move from decisions about unexpected findings made on presence and hazard to exposure and risk. The TTC will help us can change our lexicography from words like "zero" and "none," to "nothing found by analysis" and "no meaningful increase in risk." With chemists leading the way, accepting that detection does not always equal inevitable harm, will be easier for all.

Acknowledgement

The ILSI North America Technical Committee on Food and Chemical Safety has been instrumental in guiding the thinking for much of this paper. That said, the opinions herein are mine alone and do not necessarily reflect those of the Committee, Unilever, or any other group.

References

1. Food and Drug Administration. Food additives; Threshold of regulation of substances used in food-contact articles: Final rule. *Federal Register* **1995**, *60*, 36582-36596.
2. Oser, B. L. Insignificant levels of chemical additives in food. *Food Drug Cosmetic Law J.* **1958**, *13*, 477-479.
3. Frawley, J. P. Scientific evidence and common sense as a basis for food-packaging regulations. *Food Cosmet. Toxicol.* **1967**, *5*, 293-308.
4. Rulis, A. M.; Hattan, D. G. FDA's priority-based assessment of food additives II: General toxicity parameters. *Regul. Toxicol. Pharmacol.* **1985**, *5*, 152-174.
5. Rulis, A. M.; Hattan, D. G.; Morgenroth, III, V. M. FDA's priority based assessment of food additives. *Regul. Toxicol. Pharmacol.* **1984**, *4*, 37-56.
6. Rulis, A. M. De Minimis and the threshold of regulation. In *Food Protection Technology*; Felix, C. W. Ed.; Lewis Publishers Inc.: Chelsea, MI, 1986; pp 29-37.
7. Rulis, A. M. Establishing a threshold of concern. In *Risk Assessment in Setting National Priorities*, Bonin, J. J.; Stevenson, D.,E. Eds.; Plenum Press: New York, Vol. 7, 1989; pp 271-278.
8. Rulis, A. M. Threshold of regulation: Options for handling minimal risk situations. In *Food Safety Assessment.* Finley, J. W.; Robinson, S. F.; Armstrong D.J., Eds.; American Chemical Society Symposium Series 484; American Chemical Society: Washington, D.C., 1992. pp 132-139.
9. Cheeseman, M. A.; Machuga, E. J.; Bailey, A. B. A tiered approach to threshold of regulation. *Food Chem. Toxicol.* **1999**, *37*, 387-412.
10. Kroes, R.; Galli, C. L.; Munro, I.; Schilter, B.; Tran, L. A.; Walker, R.; Würtzen, G. Threshold of toxicological concern for chemical substances present in the diet: a practical tool for assessing the need for toxicity testing. *Food Chem. Toxicol.* **2000**, *38*, 255-312.
11. Kroes, R.; Muller, D.; Lambe, J.; Löwik, M. R .H.; van Klaveren, J.; Kleiner, J.; Massey, R.; Mayer, S.; Urieta, I.; Verger, P.; Visconti, A., Assessment of intake from the diet. *Food Chem. Toxicol.* **2002**, *40*, 327–385.
12. Kroes, R.; Renwick, A. G.; Cheeseman, M.; Kleiner, J.; Mengelsdorf, I.; Piersma, A.; Schilter, B.; Schlatter, J.; van Schothorst, F.; Vos, J. G.; Wurtzen, G. Structure-based thresholds of toxicological concern – guidance for application to substances present at low levels in the diet. *Food Chem. Toxicol.* **2004**, *42*, 65-83.
13. Kroes, R.; Kleiner, J.; Renwick A. The threshold of toxicological concern concept in risk assessment. *Toxicol. Sci.* **2005**, *86*, 226-230.
14. Barlow, S. *ILSI Europe Concise Monograph Series. Threshold of Toxicological Concern (TTC).* International Life Sciences Institute Europe: Brussels. 2005.

15. Kroes, R.; Renwick A. G.; Feron V.; Galli, L. L.; Gibney, M.; Greim, H.; Guy, R. H.; Lhuguenot, J. C.; van de Sandt, J. J. Application of the threshold of toxicological concern (TTC) to the safety of cosmetic ingredients. *Food Chem. Toxicol.* **2007**, *45*, 2533-2562.
16. Munro, I. C.; Ford, R. A.; Kennepohl, E.; Sprenger, J. G. Correlation of structural class with no-observed effect levels: a proposal for establishing a threshold of concern. *Food Chem. Toxicol.* **1996,** *34*, 829-867.
17. Cramer, G. M.; Ford, R. A.; Hall, R. L. Estimation of a toxic hazard – a decision tree approach. Review. *Food Cosmet. Toxicol.* **1978**, *16*, 255–276.
18. The Joint FAO/WHO Expert Committee on Food Additives. *A procedure for the safety evaluation of flavouring agents.* WHO Food Additive Series 35. Annex 5. World Health Organization, Geneva. 1996.
19. Müller, L.; Mauthe, R. J.; Riley, C. M.; Andino, M. M.; Antonis, D. D.; Beels, C.; DeGeorge, J.; De Knaep, A. G.; Ellison, D.; Fagerland, J. A.; Frank, R.; Fritschel, B.; Galloway, S.; Harpur, E.; Humfrey, C. D.; Jacks, A. S.; Jagota, N.; Mackinnon, J.; Mohan, G.; Ness, D. K.; O'Donovan, M. R.; Smith, M.D.; Vudathala, G.; Yotti, L. A rationale for determining, testing, and controlling specific impurities in pharmaceuticals that possess potential for genotoxicity. *Regul Toxicol Pharmacol.* **2006**, *44*, 198-211.
20. Blackburn, K.; Stickney, J.A.; Carlson-Lynch, H. L.; McGinnis, P. M.; Chappell, L.; Felter, S. P. Application of the threshold of toxicological concern approach to ingredients in personal and household care products. *Reg. Toxicol. Pharmacol.* **2005**, *43*, 249-259.
21. Dolan, D. G.; Naumann, B. D.; Sargent, E. V.; Maier, A.; Dourson, M. Application of the threshold of toxicological concern concept to pharmaceutical manufacturing operations. *Reg. Toxicol. Pharmacol.* **2005**, *43*, 1-9.
22. Humphrey, C. D. N. Recent developments in the risk assessment of potentially genotoxic impurities in pharmaceutical drug substances. *Toxicol. Sci.* **2007**, *100*, 24-28.
23. Environmental Protection Agency. Pesticide Registration (PR) Notice 96-8: Toxicologically Significant Levels of Pesticide Active Ingredients. 1996. http://www.epa.gov/PR_Notices/pr96-8.html.
24. Scholliers, P. Defining food risks and food anxieties throughout history. *Appetite.* **2008**, 51, 3-6.

25. Kaplan, G. A.; Pamuk, E. R.; Lynch, J. W.; Cohen, R. D.; Balfour, J. L. Inequality in income and mortality in the United States: analysis of mortality and potential pathways. *BMJ.* **1996**; *312*, 999-1003.
26. Kennedy, B. P.; Kawachi, I.; Glass, R.; Prothrow-Stith, D. Income distribution, socioeconomic status, and self rated health in the United States: multilevel analysis. *BMJ.* **1998**, *317*, 917-921.

Chapter 10

Effects of Pasteurization on Detection and Toxicity of the Beans from *Abrus precatorius*

Eric A.E. Garber

Division of Bioanalytical Chemistry, Office of Regulatory Science, Center for Food Safety and Applied Nutrition, FDA, College Park, MD 20740

During the past decade, major advances have been made in the development of novel technologies to detect toxins. However, not much is known about the effects of food processing on either toxicity or the ability of commonly used assays to detect the toxin. Further, most of the studies have employed highly purified toxin and not easily prepared crude extracts. Using an electrochemiluminescence (ECL) immunosorbent assay, the effects of pasteurization on the ability to detect abrin in a crude extract of the beans from *Abrus precatorius* (rosary peas) in milk were examined and compared to the effects of pasteurization on toxicity, oral and intraperitoneal (i.p.), in female BALB/c mice. Neither form of pasteurization examined, Low Temperature Long Time Treatment (LTLT, 30 minutes at 145 °F) or High Temperature Short Time (HTST, 15 seconds at 161 °F), had an effect on the ability to detect abrin or the toxicity of the extract in milk.

Research during the past few years on the detection and stability of toxins in food has focused primarily on purified toxins. Numerous methods have been developed with limits of detection less than 1 ng/g. Despite a large body of data, very little work has been done using easily prepared crude extracts. An extensive body of biochemical literature has described the stabilizing effects of sugars and other solutes on proteins (*1, 2*). Many complex enzymes are more stable in their natural milieu than following purification and removal of associated auxiliary cofactors viewed as extraneous to the known catalytic functions being studied. It is important to examine the effects of food processing and preparation on the ability to detect and on the toxicity of crude extracts of proteinaceous toxins. This is particularly important for toxins derived from ubiquitous plants. Abrin is derived from the seeds of *Abrus precatorius*, rosary peas, which are commonly found throughout Florida and routinely used in ornamental jewelry.

Abrin is a class II ribosome inactivating protein (RIP-2) derived from the seeds of *Abrus precatorius*, rosary peas, which are commonly found throughout Florida and routinely used in ornamental jewelry (*3*). As an RIP-2, abrin consists of two subunits held together by a disulfide bond (*4, 5*). The A chain is a 30 kDa N-glycosidase that cleaves adenine-4324 of 28s rRNA and thereby inhibits ribosomal activity (*6*). The B chain is a 33 kDa lectin that specifically recognizes galactosyl residues and serves to target the toxin to cells and facilitate uptake (*6, 7*). Depending on the cultivar, numerous immunologically indistinguishable variants (isozymes) of abrin have been detected and classified into three groups (I, II, and III). The three groups have mouse LD50 intraperitoneal (i.p.) values of 22, 2.4, and 10 μg/kg body weight, respectively, with abrin II being the most toxic (*8*). Though abrin resembles ricin, the RIP-2 isolated from castor beans (*9, 10, 11*), antibodies raised against abrin do not cross react with ricin and visa versa (*12, 13*).

Microcalorimetric measurements of thermal transitions in the heat capacitance of abrin II by Krupakar et al. (*14*) demonstrated the existence of two irreversible transitions (115 °F and 125 °F) in the temperature range routinely employed in the pasteurization of milk. The second transition (125 °F) was implicated as being derived from the A chain based on an analogous transition observed with purified A chain. Both transitions were shifted to higher temperatures by either the addition of salt or lactose. The addition of 50 mM lactose caused a 16°F increase in both transitions. The effects of pasteurization at 145 °F or 161 °F on the ability to detect abrin using a novel immunosorbent assay and on the oral and intrapertoneal toxicity of crude rosary pea extracts in milk, which contains 140 mM lactose, were examined.

Materials and Methods

Seeds of *Abrus precatorius* were purchased from B & T World Seeds (Aigues-Vives, France). Raw, un-pasteurized whole milk was provided by Henry Njapau, Ph.D. (FDA). All other reagents were of the best technical grade available.

Preparation of *Abrus precatorius* (Rosary Pea) Extract

The two most commonly used methods for the purification of abrin entail overnight extraction at 4 °C of ground peas with either 5% acetic acid or sodium chloride. Extraction using acetic acid typically gave overall yields of purified abrin of 0.075 to 0.25% (*15, 16*) while sodium chloride resulted in overall yields of 0.18 to 0.5% purified abrin (*8, 17*). The rosary pea extract used in this study was prepared by manually grinding rosary peas to mesh size >4, and mixing gently overnight with PBST (Sigma Chemical Co., PBS P3813 supplemented with 0.1 % v/v Tween-20). The abrin plus agglutinin content of the extract was determined, as described below, to be 0.05% of the starting material. Acetic acid extraction of a separate sample of the ground rosary peas according to a published method (*15*) indicted an abrin plus agglutinin content of 0.03%.

The PBST derived abrin extract was divided into aliquots and either concentrated against PEG (FW 14000) in 6 mm dialysis tubing (mol wt cut-off 12000) at 4 °C or immediately frozen at – 80 °C for later use.

Protein concentration was determined using the Bradford Assay (Sigma Chemical Co., Catalog #B 6916) and abrin plus agglutinin content determined as described below using an ECL-based immunosorbent assay.

Pasteurization

According to International Dairy Foods Association (IDFA), the most commonly employed method in the Unites States for pasteurizing milk is High Temperature Short Time (HTST) pasteurization followed by rapid cooling (*18*). HTST pasteurization consists of heating milk at 161 °F for 15 s. A second form of pasteurization, Low Temperature Long Time Treatment (LTLT), also known as Vat Pasteurization, entails heating the milk for 30 min at 145 °F. LTLT is routinely used in the dairy industry for preparing milk for starter cultures used to make cheese, yogurt, buttermilk and some ice cream mixes.

The pasteurized samples used in this study were prepared by heating 1 mL aliquots in 3 mL reaction vials, sealed with Teflon coated septa, and equipped with triangular stirring bars. The temperature of the samples and the silicon oil baths used to heat the vials were monitored using needle thermocouples. The reaction vials were typically heated to the desired temperature within 2 min. The temperatures of the samples were maintained to within 2 °F and upon completion of the heating process, the samples were immediately cooled in an ice bath prior to storing at – 80 °C. Detection and toxicity studies were initiated within 3 days of sample preparation.

Immunosorbent Assay for the Detection of Abrin

An electrochemiluminescence (ECL)-based immunosorbent assay was used to detect and quantify the presence of abrin plus agglutinin (*12, 19*). The assay entailed a single 20 min simultaneous incubation of the sample with mouse derived, ruthenium labeled monoclonal detector antibody (5F6) in streptavidin coated plates containing rabbit derived, biotinylated polyclonal (capture) antibodies. The ECL signal was generated by the oxidation/reduction of the ruthenium measured using a Sector™ PR100 ECL plate reader manufactured by Meso Scale Diagnostics, (Gaithersburg, MD). As with other immunoassays for the detection of RIPs, the assay cannot distinguish between abrin and the agglutinin present in rosary peas. Thus, the concentration of abrin measured represents an upper limit, with the literature reporting a ratio of approximately 2:1 for abrin:agglutinin in rosary peas (*8*).

Toxicity Studies

Oral and i.p. toxicity measurements were made using 6 week old BALB/c female mice. The handling, care, and administration of samples to the mice were conducted by Biocon, Inc. (Gaithersburg, MD) through a contract administered by Tetracore, Inc. (Rockville, MD). The toxicity study was divided into two phases. The first phase determined the lethal dose range of the extract and the second phase examined the effects of pasteurization. Oral toxicity measurements entailed administering a single sample by gavage to non-fasted mice and monitoring the effects for ten days. Oral toxicity measurements were followed by i.p. toxicity measurements. The mice from the oral toxicity studies were regrouped for the i.p. studies such that no group consisted of the same mice as in the oral toxicity measurements. Further, the mice that received the lowest amount of toxin in the oral part were used to assemble the highest dose group for the i.p. measurements. Each group of mice consisted of three animals except in those cases in which the survivorship from the oral toxicity part was insufficient.

Statistical Analysis of Data

Pairwise chi-square (X^2) and Cochran-Mantel-Haenszel (CMH) analyses of the toxicity data were performed with the Bonferroni inequality used to adjust chi-square-derived P values to account for multiple tests of the same data (*20*). CMH analyses were performed for stratified subgroups of the number of fatalities and the explanatory variables (method of pasteurization and route of administration). Analysis of variance (row mean square) was used when the explanatory variables were not on an ordinal scale and the ordinal mean number of fatalities was the column variable. The chi-square and CMH analyses gave similar results, with SAS system (SAS Institute, Cary, NC) used to obtain more accurate P values for the CMH analysis.

Results and Discussion

Effects of Pasteurization on the Ability to Detect Abrin in Milk

The two forms of pasteurization examined, HTST and LTLT, had no effect on the ability to detect abrin from rosary pea extract using an immunosorbent assay (Table I). This indicated that the heat treatments did not affect either the binding of the capture or detector antibodies to their respective epitopes. It is possible that major changes could have occurred in the conformation of the protein provided the epitopes were intact and accessible for antibody binding.

Table 1 Immunosorbent Detectable Abrin [a]

Pasteurization	*Protein Content* [b]
no heat treatment	1.0 ± 0.3 mg/mL
161 °F 15 sec (HTST) [c]	0.9 ± 0.2 mg/mL
145 °F 30 min (LTLT)	1.0 ± 0.2 mg/mL
milk w PBS (no extract)	0.0 ± 0.0 mg/mL

[a] abrin+agglutinin, the two proteins are indistinguishable using the immunosorbent assay.

[b] averages of $\geq$ 3 samples, representing more than a 100-fold concentration range with each sample analyzed in duplicate.

[c] temperatures were maintained within 2 °F.

Effects of Pasteurization on the Toxicity of Abrin in Milk

The oral and i.p. toxicities of rosary pea extract in raw milk and following HTST and LTLT pasteurization are presented in Figures 1 and 2 for BALB/c female mice. The limited number of mice per group (n = 3) along with the high apparent oral LD50, estimated at > 1 mg per kg body weight, precluded an extensive titration and detailed estimation of the oral LD50. In contrast, the higher i.p. toxicity of rosary pea extract enabled a more precise estimation of an LD50 and comparison between the effects of pasteurization on toxicity. As indi cated in Figure 2, the i.p. toxicity of the rosary pea extracts was not significantly affected by pasteurization ($P > 0.25$), and was between 3 and 9 μg/kg body weight and comparable to published LD50 values (*13*). The individual mortalities of mice administered by gavage 0.049 mg/kg HTST treated extract (Figure 1) and 0.33 μg/kg no heat treatment extract (Figure 2) probabably represent random mortalities.

Microcalorimetric Transitions

The data presented in this study did not contradict microcalorimetry studies (*14*) which described two irreversible transitions (changes in heat capacitance) for abrin II below the temperatures used for pasteurization. There are several possible explanations why no comparable changes were observed in either the ability to detect or the toxicity of abrin. One possible explanation is that the cultivar used in this study did not contain sufficient abrin II to result in measurable changes in either the toxicity or level of detectable abrin in the extracts and the abrin present (I and/or III) did not display analogous thermal transitions. Though this possibility could not be conclusively ruled out, abrin purification from another lot of rosary peas purchased from the same supplier conducted under contract to the FDA yielded abrin I, abrin II, and abrin III at a ratio of 1 : 4.3 : 8.6. Inasmuch as the i.p. LD50 values for abrin I, abrin II, and abrin III were 22, 2.4, and 10 μg/kg body weight (*8*), abrin II should be the primary source of toxicity in the rosary pea extract used and contribute 31% to the detectable abrin. A second possible explanation would entail the thermal transitions indicative of conformational changes that did not affect the biological activity of abrin. Alternatively, the thermal transitions could reflect conformational changes whose effects on catalytic activity were not kinetically significant in regards to whole organism toxicity. Consistent with this later explanation were the observed effects of lactose on the microcalorimetric transitions (*14*) and ricin deposition in organisms (*21*) while toxicity did not change until extremely high levels of galactose were present (*17*). Detailed analyses of the effects of temperature on the conformation of abrin and the functions of the A and B chains would be required before the transitions observed by Krupakar et al. (*14*) could be fully interpreted.

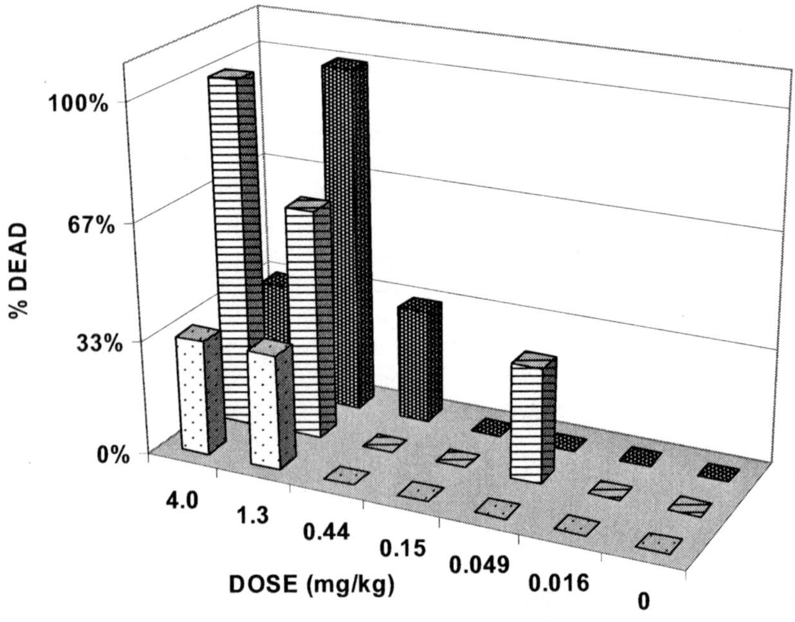

Figure 1. Oral Toxicity of milk samples containing rosary pea extract subjected to LTLT pasteurization (145 °F, 30 min); HTST pasteurization (161 °F, 30 min); or no heat treatment. Samples were administered as a single dose by gavage to BALB/c female mice at 4.0, 1.3, 0.44, 0.15, 0.049, 0.016, and 0 mg/kg body weight and monitored for 10 days. Each group consisted of 3 mice.

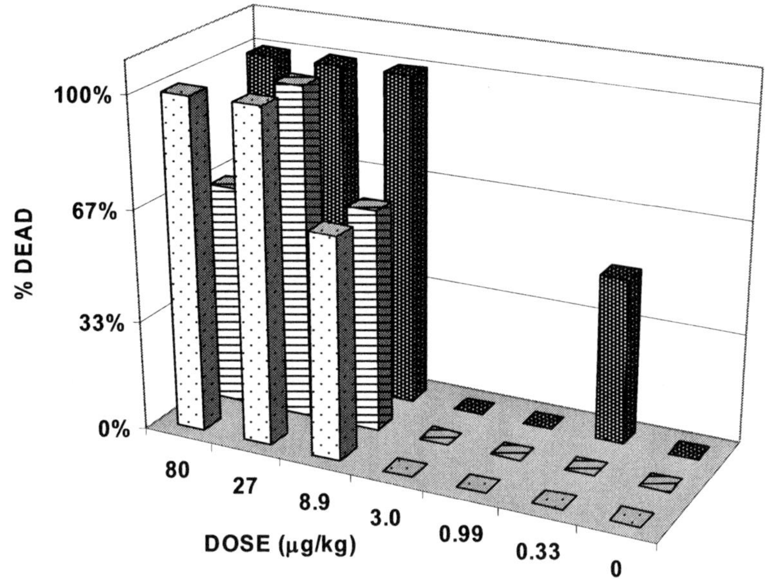

Figure 2. IP toxicity of milk samples containing rosary pea extract subjected to LTLT pasteurization (145 °F 30 min); HTST pasteurization (161 °F 30 min); or no heat treatment. Samples were administered by gavage to BALB/c female mice at 80, 27, 8.9, 3.0, 0.99, 0.33, and 0 µg/kg body weight and monitored for 6 days. Each test group consisted of 3 mice except for two groups which consisted of 2 mice each (mice administered 33 µg/kg body weight HTST treated extract and mice administered 33 µg/kg body weight of extract not heat treated).

Conclusions

The results indicate that HTST and LTLT pasteurization did not affect the biological activity of abrin from either a functional or antigenic perspective (for the antibodies employed). Whole animal toxicity requires the lectin (B chain) component of abrin to recognize, bind, and catalyze uptake of the protein into the cell. Once inside the cell, the toxin (A chain) catalyzes the de-adenylkation of 28s rRNA, resulting in ribosomal inhibition and death. From a functional perspective, the A and B chains were not significantly altered by either form of pasteurization. The possibility that the toxicity observed represented other components in rosary peas is inconsistent with the literature which describes abrin as the primary toxin in rosary peas and the agglutinin as non-toxic (*22*).

Acknowledgments

Gratitude is expressed to George Sigal, Ph.D. (Meso Scale Diagnostics), Jennifer Aldrich, M.S. and Thomas W. O'Brien, Ph.D. (Tetracore, Inc.) for providing the ECL Instrumentation and reagents used in the development of the immunosorbent assay. Gratitude is also expressed to Marc S. Boyer (FDA) for the CMH statistical analyses and Frank Klotz (Biocon, Inc.) for providing technical expertise in the handling, care, and administration of the samples to the BALB/c mice. Appreciation is expressed to George Ziobro, Ph.D., Benjamin Canas, Michael A. McLaughlin, Ph.D., Mitchell J. Smith, Ph.D., and Douglas L. Park, Ph.D. of the FDA along with Gregory Orlowski (JIFSAN) and Lynn L.B. Rust, Ph.D. (NIH) for helpful discussions.

References

1. Arakawa, T.; Timasheff, S. N. Stabilization of protein structure by sugars. *Biochem.* **1982,** 21, 6536-6544.
2. Xie, G.; Timasheff, S. N. The thermodynamic mechanism of protein stabilization by trehalose. *Biophys. Chem.* **1997,** 64, 25-43.
3. Dickers, K. J.; Bradberry, S. M.; Rice, P.; Griffiths, G. D.; Vale, J. A. Abrin poisoning. *Toxicol. Rev.* **2003,** 22, 137-142.
4. Reddy, G. B.; Narayanan, S.; Reddy, Y.; Surolia, I. Suppression of DTT-induced aggregation of abrin by αA- and αB-crystallins: A model aggregation assay for α-crystallin chaperone activity in-vitro. *FEBS Lett.* **2002,** 522, 59-64.
5. Tahirov, T. H.;Lu, T-H.; Liaw, Y-C.; Chen, Y-L.; Lin, J-Y. Crystal structure of abrin-a at 2.14 Å. *J. Mol. Biol.* **1995,** 50, 354-357.
6. Olsnes, S. The history of ricin, abrin and related toxins. *Toxicon* **2004,** 44, 361-370.
7. Hudson, T.H.; Grillo, F. G. Brefeldin-A enhancement of ricin A-chain immunotoxins and blockade of intact ricin, modeccin, and abrin. *J. Biol. Chem.* **1991,** 266, 18586-18592.
8. Hedge, R.; Maiti, T. K.; Podder, S. K. Purification and characterization of three toxins and two agglutinins from *Abrus precatorius* seed by using lactamyl-sepharose affinity chromatography. *Anal. Biochem.* **1991,** 194, 101-109.
9. Endo, Y.; Mitsui, K.; Motizuki, M.; Tsurugi, K. The mechanism of action of ricin and related toxic lectins on eukaryotic ribosomes: The site and the characteristics of the modification in 28 S ribosomal RNA caused by the toxins. *J. Biol. Chem.* **1987,** 262, 5908-5912.

10. Collins, E. J.; Robertus, J. D.; LoPresti, M.; Stone, K. L.; Williams, K. R.; Wu, P.; Hwang, K.; Piatak, M. Primary amino acid sequence of α-trichosanthin and molecular models for abrin A-chain and α-trichosanthin. *J. Biol. Chem.* **1990,** 265, 8665-8669.
11. Olsnes, S.; Fernandez-Puentas, C.; Carrasco, L.; Vazquez, D. Ribosome inactivation by toxic lectins abrin and ricin: Kinetics of the enzymic activity of the toxin A-chains. *Eur. J. Biochem.* **1975,** 60, 281-288.
12. Garber, E. A.E.; Walker, J. L.; O'Brien, T. W. Detection of abrin in food using elisa and electrochemiluminescence (ECL) technologies. *J. Food Prot.* **2008,** *in press.*
13. Hedge, R; Podder, S. K. Studies on the variants of the protein toxins ricin and abrin. *Eur. J. Biochem.* **1992,** 204, 155-164.
14. Krupakar, J.; Swaminathan, C. P.; Das, P. K.; Surolia, A.; Podder, S. K. Calorimetric studies on the stability of the ribosome-inactivating protein abrin II: Effects of pH and ligand binding. *Biochem. J.* **1999,** 338: 273-279.
15. Olsnes, S. Toxic and nontoxic lectins from *Abrus precatorius*. *Methods Enz.* **1978,** 50, 323-330.
16. Wei, C. H.; Hartman, F. C.; Pfuderer, P.; Yang, W-K. Purification and characterization of two major toxic proteins from seeds of *Abrus precatorius*. *J. Biol. Chem.* **1974,** 249, 3061-3067.
17. Griffiths, G. D.; Lindsay, C. D.; Upshall, D. G. Examination of the toxicity of several protein toxins of plant origin using bovine pulmonary endothelial cells. *Toxicology* **1994,** 90, 11-27.
18. International Dairy Foods Association (IDFA). Pasteurization: definition and methods. Available at: http://www.idfa.org/facts/milk/pasteur.cfm Accessed: 10, July **2008**.
19. Garber, E. A.E.; O'Brien, T. W. Detection of ricin in food using electro-chemiluminescence (ECL)-based technology. *J. AOAC Intl.* **2008,** 91, 376-382.
20. Glantz, S. A. Primer of Biostatistics; McGraw-Hill; New York, NY, 1997.
21. Fodstad, G.; Olsnes, S.; Pihl, A. Toxicity, distribution and elimination of the cancerostatic lectins abrin and ricin after parenteral injection into mice. *Br. J. Cancer* **1976,** 34, 418-425
22. Saltvedt, E. *Ricinus* and *abrus* agglutinin. *J. Oslo City Hospital* **1977,** 27, 53-68.

Chapter 11

Detection and Confirmation of Food Allergen using Mass Spectrometric Techniques: Characterization of Allergens in Hazelnut using ESI and MALDI Mass Spectrometry

Dorcas Weber, Gustavo Polenta, Benjamin P.-Y. Lau and Samuel Benrejeb Godefroy

Bureau of Chemical Safety, Food Directorate, Sir Frederick Banting Research Centre, Health Canada, Ottawa, Ontario K1A 0K9

Mass spectrometric techniques were used to identify and characterize allergens in a hazelnut protein extract either directly in solution or after separation on 1D and 2D gels. Five major proteins were identified including *Cor a 8* (Lipid transfer protein precursor), *Cor a 1*, *Cor a 9* (11S globulin-like protein), *Cor a 11* (48 kDA glycoprotein precursor) and Oleosin. The protein extract was either digested with trypsin followed by analysis using electrospray and MALDI for peptides, or the intact proteins were analyzed directly with MALDI. Complementary results were obtained from both methods. A new data acquisition method using alternating low and high energy acquisition was also investigated.

Introduction

Food allergens, a class of unintentional food contaminants, are proteins that cause immunological responses that are expressed as food allergies. These reactions only affect selected individuals, mostly children under the age of three, and can vary from a mild discomfort to life threatening anaphylaxis. There is no known treatment and the only effective prevention is avoidance. Therefore, it is important for health officials to constantly monitor food commodities for undeclared allergens. There is a large body of research on developing analytical methods for allergen detection, most of which are immunologically based (*1*). There are two main approaches to the allergen assays: the direct method and the indirect method. The direct methods are clinical methods that detect the allergen and involve the use of extracted blood and test for human IgE. These methods depend on the availability and the standardization of human IgE. The indirect methods are based on the detection of markers which indicate the presence of potentially allergenic food products. These methods are divided into two main categories – detection of proteins and detection of DNA. DNA detection uses polymerase chain reaction (PCR) techniques (*2*). Protein detection uses immunological approaches such as ELISA and Western Blot or physicochemical approaches using chromatography and mass spectrometry. Currently the most common method for food allergen detection is the ELISA method which is target specific and involves at least one antibody with specificity for a particular antigen. Simple methodology, specificity and good sensitivity give this approach great potential for standardization. Many commercial kits are available for various allergens, but they are far from comprehensive. Also, cross reactivity may be a problem and lead to false positives. Therefore, immunological methods need other methods for confirmation. Mass spectrometry is a well established technique accepted by regulatory agencies as a detection and confirmation method for contaminants. The technique is specific, sensitive and can be multi-targeted. The proteomic approach using mass spectrometry for the analysis of allergens has been reported for peanuts (*3,4*), milk (*5*), and gluten (6,7,8,9).

This paper describes the various mass spectrometric techniques used to characterize food allergens using hazelnut as an example. Hazelnut belongs to the tree nut class of allergens and is one of the nine major allergen groups that are monitored by health agencies to warn consumers who may have adverse immunoreactions to these proteins (*10*). Tree nuts cause potentially life-threatening food allergies, most of which are lifelong problems. Some of the species such as the pecans and walnuts show great homology between them, but others are quite different. They often are found undeclared in food like chocolate bars which is probably due to cross contamination during manufacturing. The major protein identified in hazelnut that causes allergic reactions was *Cor a 9*, an 11S globulin which is a seed storage protein. This family of proteins includes other known food allergens such as *Ara h 3*(in peanuts) and *glycine max* (soybean). The homology between these proteins ranges from 45% to 50%. These proteins can be identified and characterized using either a top down or bottom up mass spectrometry based proteomic approach (*11*). Peptide markers for these proteins can also be used for quantification.

Experimental

Reagents

LC-MS Chromasolv grade water and acetonitrile, formic acid (FA), trifluoroacetic acid (TFA), dithiothreitol (DTT), iodoacetamide (IA), leucine enkephalin, ammonium bicarbonate, polyethylene glycol (PEG) 1000, 2000 and 3000, sodium iodide (NaI), adrenocorticotropic hormone clip 18-39 (ACTH), protein calibration standards [insulin, ubiquitin, cytochrome C, myoglobin, trypsinogen and bovine serum albumin (BSA)] were all obtained from Sigma Chemical Co. (St. Louis, MO). OmniSolv HPLC grade acetone, methanol and 2-propanol were obtained from EMD Chemicals Inc. (Gibbstown, NJ). Anhydrous ethyl alcohol was obtained from Commercial Alcohols Inc. (Brampton, Ontario). Sequencing grade trypsin was obtained from Promega, (Madison, WI). Matrices for MALDI sample preparation α-cyano-4-hydroxy cinnamic acid (CHCA) and sinapinic acid (SA) were obtained from Waters (Milford, MA).

Protein Extraction

For protein extraction, hazelnuts were first defatted by homogenizing 50 g of nuts in an Ultra Turrax homogenizer (IKA Works, Germany) with 250 ml of hexane. The insoluble fraction was separated by filtration and re-extracted four times by repeating the procedure, and then the defatted extract was left overnight at room temperature to dry. For protein extraction, 10 g of defatted hazelnut extract were mixed with 100 ml of PBS and stirred for one hour at 45°C. After centrifuging the mixture at 10,000 g, the supernatant was left overnight at 4°C which was then centrifuged the next day at 12,000 g for 30 min at 4°C. Protein concentration of the resultant supernatant was determined by the Lowry method.

Electrophoretic Analysis

For the 1D electrophoresis, proteins were separated using 10% NuPAGE Bis-Tris polyacrylamide gels from Invitrogen (Invitrogen Corp., Carlsbad, Ca.). Equal amounts (30 ug) of total protein were loaded onto each lane of the gel. The electrophoresis was run with NuPAGE MES (2-(N-morpholino) ethane sulfonic acid) SDS (sodium dodecyl sulphate) running buffer containing 50 mM

MES, 50 mM Tris base, 3.5 mM SDS and 1 mM EDTA. The samples were reduced with NuPAGE sample buffer (0.293 M sucrose, 141 mM Tris base, 106 mM Tris HCl, 69.5 mM SDS, 0.51 mM EDTA, 0.22 mM Serva blue G 250, 0.175 mM Phenol red) and NuPAGE reducing agent according to the manufacturer's instructions. Prior to sample loading, samples were heated for 10 min at 90°C. Additionally, NuPAGE antioxidant was added to the NuPAGE MES SDS running buffer in the upper chamber to prevent reduced proteins from reoxidizing during electrophoresis. The gels were run at a constant voltage (130 V) followed by staining with 0.1 % (w/v) Coomassie Blue solution.for 90 min. Marker proteins were used to estimate the MW of the different protein bands.

2D electrophoresis used the ZOOM IPGRunner System from Invitrogen (Invitrogen Corp., Carlsbad, CA.) to perform isoelectric focusing (IEF) as the first dimension. One hundred µg of protein dissolved in a buffer solution were used to rehydrate the strips (pH 3-10 non linear). IEF strips were run according to the following conditions: 175 V for 15 min, 175-2000 V ramp for 45 min, and 2000 V for 30 min. Strips were loaded onto 4-12% gradient ZOOM gels (Invitrogen). Second dimension was the 1D SDS electrophoresis method described previously.

In-gel Digestion

For 1D gel, the bands were excised with a Band Picker (5.0mm, from The Gel Company, San Francisco, CA). The band was then cut into 10 1 mm x1 mm pieces and loaded into a well on a 96 well microtiter plate. For 2D gel, the spots were excised with a Spot Picker (1.5mm, from The Gel Company, San Francisco, CA) and each spot was loaded in individual wells of the microtiter plate. The gels were destained using a 50:50 (v/v) mixture of 100mM ammonium bicarbonate and acetonitrile. After the gels were washed and dried, DTT (10 mM in 100 mM ammonium bicarbonate) was added for reduction. The mixture was incubated at 37 °C for 30 min. IA (55mM in 100mM ammonium bicarbonate) was added for alkylation and was incubated at room temperature in the dark for 20 min. The gels were then washed with 100mM ammonium bicarbonate buffer and dehydrated with acetonitrile. Trypsin (78 ng) in 100 mM ammonium bicarbonate was added to each well and incubated at 37 °C for 4.5 h. The peptides were extracted with an aqueous solution containing 2% acetonitrile and 1% formic acid and then transferred to 250 µl autosampler vials for mass spectrometric analysis.

In-solution Digestion of the Hazelnut Protein Extract

A 30 µL aliquot of the protein extract was digested with trypsin. The detailed protocol was reported in our previous work on gluten (*6*). Briefly, reduction and alkylation of the sample was carried out using DTT and IA respectively. Digestion with trypsin followed. After 1% FA was added, the mixture was transferred to autosampler vials

ESI LC-MS/MS

LC-MS/MS was performed on a Waters MALDI QTOF Premier mass spectrometer (Waters, Milford, MA) with analyzer in V configuration and interfaced to Waters nanoAcquity UPLC system (Waters, Milford, MA) configured with a binary solvent manager for delivering the gradient and an auxiliary solvent manager for delivering the lock mass reference compound through the LockSpray probe for accurate mass correction. The LC system consisted of a trap column [Symmetry C18 5 μm 180 μm x 20 mm (Waters)] for sample concentration and precolumn cleanup and an analytical column [BEH 130 C18 1.7 μm 100 μm x 100 mm (Waters)] for peptide separation. Solvent A was H_2O with 0.1% FA and solvent B was ACN with 0.1%FA. The sample was injected with the initial mobile phase at 1% B onto the trap column where the trap valve was opened for 3 min with a flow of 5 μL/min. The trap valve was closed and the sample was eluted into the analytical column at a flow of 0.4 μL/min with the following gradient: 1% B to 50% B in 30 min, held at 50% B for 1 min, from 31 min to 40 min, ramped to 85% B, stayed at 85% B until 46 min, then reduced back to 1% B. The column was allowed to recondition at 1% B for 14 min. Total run time for this gradient was 60 min.

The mass spectrometer was operated in nanoESI positive ion mode with LockSpray enabled. The capillary voltage was set to 2.8 kV. A cone voltage of 45 V and a resolution of >10000 (FWHH) were used. Leucine enkephalin at a concentration of 200 ng/μL in 50% acetonitrile and 0.1% FA with a flow of 0.4 μL/min was used in the LockSpray as reference for accurate mass correction. The mass spectrometric data was acquired using MassLynx v4.1 data system (Waters).

For data survey analysis, the acquisition mode was data directed analysis (DDA). MS survey data was acquired from *m/z* 400 to 1600 with switching to MS/MS acquisition when the intensity of individual ion rising above a threshold of 10 counts/sec. The survey scans were carried out at a scan time of 1 second and an inter-scan delay of 0.02 second. The MS/MS mass spectra were acquired over the range *m/z* 50 to 1700 with a scan time of 1 second and an inter-scan delay of 0.02 second. The maximum number of ions selected for MS/MS from a single MS survey scan was set to 3 ions. The system returned to MS survey when the TIC rose above 3000 counts/second. The reference mass (lock mass) scanning conditions were 1 second scan time; a frequency of 10 second; sampling cone 40 volts and collision energy of 5 volts.

For total peptide analysis using MS^E, two alternating functions were set for acquisition. Both are TOFMS functions with the first using collision energy set to a value of 5 volts. The second function used collision energy ramp with initial energy of 30 volts to a final energy of 45 volts. Both functions scanned from m/z 100 to 1800 with a 1 second scan time and 0.02 second inter-scan delay. The reference mass setting was the same as in DDA.

MALDI TOF MS of Tryptic Digest

MALDI TOF MS was carried out using a Waters MALDI micro MX mass spectrometer (Waters, Milford, MA) which incorporates a MALDI source and axial time of flight mass analyzer with reflectron detector for recording MS data. An equal aliquot of the tryptic digest (typically 2-3 μL) was mixed with a 10 mg/ml solution of purified CHCA in a 50:50 (v/v) ethanol/acetonitrile 0.1% TFA and 1 μL of the mixture was spotted on a MALDI target plate. For calibration of the instrument, a PEG/NaI/CHCA solution was prepared by mixing 3 μL of PEG 1000, PEG 2000, PEG 3000 (10mg/mL each, prepared in 50:50 (v/v) acetonitrile / water), 9 μL of CHCA and 4 μL of 2 mg/mL of sodium iodide (prepared in 50:50 (v/v) 2-propanol / water). One μL of the mix was spotted onto the target plate. The plate was allowed to air dry and then loaded into the instrument. Data was acquired in positive ion reflectron mode with automated software control of the scanning pattern and laser energy. In MS mode, the PEG/NaI mix was used to generate a multi-point external calibration. ACTH was used as either external or internal lock mass correction.

Off-line LC MALDI TOF MS of Tryptic Digest

The tryptic digest mixture was injected and separated by reverse phase chromatography on a nanoAcquity UPLC system as described in the ESI LC-MS/MS section. The eluent from the LC column was combined with a 3 mg /mL CHCA solution (50:50 (v/v) ethanol/acetonitrile 0.1% TFA) delivered by a syringe pump (Harvard Apparatus, Holliston, MA) at 1μL/min and spotted directly onto a MALDI target plate by the CTC PAL spot collecting system. The spots were allowed to air dry and the sample plate containing the fractions collected were analyzed as described in the MALDI TOF MS section

MALDI TOF MSMS of Tryptic Digest

MALDI TOF MSMS work was carried out using a Waters MALDI QTOF Premier mass spectrometer (Waters, Milford, MA) with the MALDI source option. Data were acquired in positive ion detection mode with analyzer in V mode using DDA survey with MS function scanning from *m/z* 800 to 3000 with a scan time of 1 second and an inter-scan delay of 0.1 second. The target is scanned in a spiral pattern with a step rate of 12 Hz, a firing rate of 200 Hz and laser energy at 250. The criteria for switching from MS to MSMS were based on peak selection from MS survey data. The maximum number of MS/MS peaks was set at 30 from the MS survey data and only peaks that were in the range of *m/z* 800 to 3000 with intensity above 15 counts were considered for MS/MS. In MS/MS mode, the scan time was 1 second with a 0.1 second inter-scan delay. Collision energy profile was used and the acquisition mass range was set to automatic. Peak detection was used to stop the acquisition with the criteria that when TIC was greater than 2000 counts and each MS/MS was stopped when intensity rose above 500 counts. The data from the top 5% of the mass range was also discarded and the scan was completed after 10 seconds regardless.

MALDI TOF MS of Intact Protein

The intact protein profile of the hazelnut extract was obtained using the Waters MALDI micro MX mass spectrometer (Waters, Milford, MA) in positive ion linear mode. Sinapinic acid prepared at a concentration of 10 mg/mL in 60:40 (v:v) aqueous 0.1% TFA:acetonitrile was used as matrix (refer to as Matrix 2 in later discussion). Two techniques for the preparation of the sample target were experimented with. The dry drop method involved the mixing of the sample and matrix 2 in a 3:1 (v:v) matrix:sample ratio. The thin film method is a two step method. The plate is first coated with 1μL of the 10 mg / mL sinapinic acid in acetone (Matrix 1). Then the sample and Matrix 2 was mixed 1:1 (v:v) and spotted on top of the Matrix 1 thin film. The plate was allowed to air dry. A multi-point calibration of the instrument was generated from a spot containing a mixture of proteins [insulin, ubiquitin, cytochrome C, myoglobin, trypsinogen and BSA (concentration of 10 pmol/μL each)].

Bioinformatic Software for Protein Database Search

Mass spectral data collected were analyzed using three bioinformatics suites. ProteinLynx global server v 2.3 (PLGS) (Waters, Milford, MA) was the software developed by the instrument manufacturer and was used to process all the data acquired by Waters instruments. In addition, the bioinformatics software package from Matrix Science (Matrix Science Inc., Boston, MA) which included Mascot Server (an in-house licensed database search engine), Mascot Distiller (a peak detecting software which processed the raw data into de-isotoped peak lists) and Mascot Deamon (a client application which automated the submission of data files to Mascot server allowing batch mode, real time monitoring) was used for processing DDA data to provide additional information. Peaks Studio v 4.5 (BSI Bioinformatics Solutions Inc., Waterloo, Ontario) was also used for DDA data processing. This suite provides peptide sequencing using a process known as *de novo* sequencing and identifies the protein with the aid of a protein sequence database. Survey (or DDA) data was processed with all three software and results were compared.

General Protocol

A general approach for the protein and peptide marker discovery for proteins in the extract is illustrated in Figure 1. The sample was homogenized followed by extraction with a buffer solution and de-fatting with hexane. The extract was further cleaned up using acetone precipitation and re-dissolving the protein in an appropriate buffer. 1D and 2D gels were prepared for the characterisation of the selected proteins including those that were found immunogenic by the Western Blot technique. However, a faster protocol was also used to identify the major proteins found in the extract without further cleanup. This was done by treating the extract to an in-solution proteolytic digestion protocol using trypsin. Protein identification was achieved by analysing the digest using either LC-MS/MS, LC-MS^E of the peptides followed by database search or using MALDI TOF for peptide mass finger printing (PMF). This is similar to bottom up proteomics, a common method used to sequence and identify proteins using mass spectrometry. For complementary information on the protein, the intact proteins in the mixture were analyzed with MALDI TOF MS operating in the linear mode to obtain the molecular profile of the proteins found.

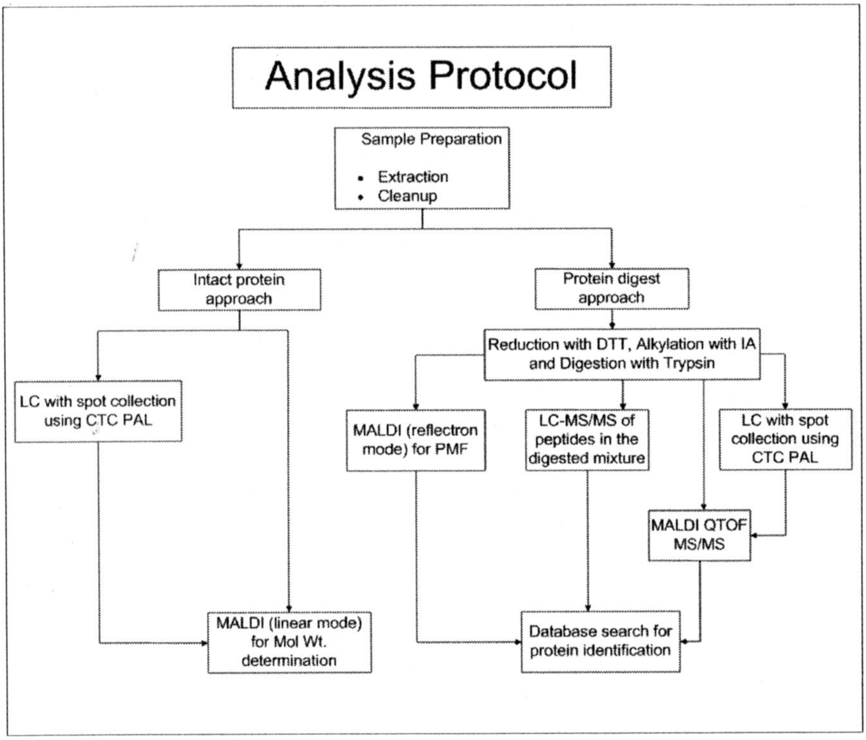

Figure 1. Typical work flow chart for protein and peptide marker discovery in food allergen analysis using mass spectrometry

Results and Discussion

From the in-solution digestion of the hazelnut protein extract, three major proteins were identified using nanoESI LC-MS/MS with DDA acquisition. A typical chromatogram obtained by this approach is shown in Figure 2. Database search results from the MS/MS data using the three bioinformatics software packages namely PLGS, Mascot and PEAKS Studio are shown in Figure 3. Three major proteins from *Corylus avellana* (hazelnut) were identified. They were 11S globulin-like protein (also known as *Cor a 9*), 48-

kDa glycoprotein precursor (*Cor a 11*) and the Lipid transfer protein precursor (*Cor a 8*). Since all three independent database searches returned with the same proteins, one can draw the conclusion that these proteins are truly present in the sample. Major peptides identified from the 11S globulin-like protein (see Figure 4a) and the 48-kDa glycoprotein precursor (see Figure 4b) can be selected as peptide markers to set up for LC-MS/MS with MRM (multiple reaction monitoring) using a triple quadrupole instrument to achieve lower detection limit and also produce reliable quantification data. Peptides that were identified in the in-solution digest from these two proteins are shown in Table I. An example of the MS/MS fragmentation of one of the potential peptide markers from the 11S globulin-like protein is shown in Figure 5. Most of these peptides have very rich fragmentation in the MS/MS spectra and three or more transitions can be easily selected for MRM experiments.

DDA is a criteria dependent method where a decision has to be made to switch from MS mode to MSMS mode based on some preset threshold or conditions. This presents a problem when the sample is a complex mixture with many co-eluting peptides. While the system is switched to the MSMS mode, newly eluted peptides could be missed. This can be remedied by a newly developed scanning method, MS^E, for modern Q-TOF instruments. With the fast scanning technology of the newer instruments, it is possible to acquire data using two rapidly alternating MS functions; with the first as low collision energy (5 volts) and the second at elevated collision energy (either fixed at 30 volts or with a ramp setting of 30-45 volts) resulting in a data set containing a total time resolved record of all detectible precursor and product ions. The precursor and product ions are associated by both retention time alignment and peak shape, allowing overlapping ion clusters to be analyzed. Data processing using $Identity^E$ software (part of the PLGS suite) involves an algorithm which includes retention time alignment, peak deconvolution and peak depletion resulting in the digital captured of all MS and MS/MS data of the sample in one single run. This usually results in more peptides detected, hence higher protein coverage. Evaluation of this method for a high-coverage peptide mapping study using a tryptic digest of yeast enolase was discussed (*12*). A typical MS^E run is shown in Figure 6 and the peptides identified for 11S globulin-like protein are shown in Table II. The protein coverage using this technique was 31% versus 26% using the DDA method thus illustrating the point discussed above. One drawback of using this approach is that the data acquired this way is not compatible with the other two bioinformatics software packages and so the analysis is dependent solely on the $Identity^E$ software.

An additional feature of the MS^E analysis is the ability to do label-free quantification of the proteins (*13,14,15*). This is accomplished by adding a known amount of a calibration protein digest to the final digest of the unknown sample before the LC-MS analysis. The IdentityE software module from PLGS obtained the response factor for several (generally 3) peptides from the calibration digest and used them to calculate the quantity of the targeted protein in the sample (using 3 or more of the response of identified peptides from the protein). It has been shown that an average MS signal response for the three most intense tryptic peptides per mole of protein is constant within a coefficient of variation of less than ±10% (*15*). Hence, this can be used to establish a universal response factor with a calibration protein. Initial work on quantification of the proteins found in the hazelnut extract using ADH as calibration protein was very promising. Quadruplet injections of the same extract also yield reproducible results as shown in Table III. Future work will include comparing the results obtained using a triple quadrupole instrument and also checking the reproducibility with digestions of the same sample.

In order to characterize more proteins in the solution extract, a 1D and a 2D gel of the extracts were prepared. Bands (1D) and spots (2D) from the gels were excised and an in-gel digest were carried out. The digests were analyzed using LC-MS/MS in both survey mode and MS^E mode for total peptide discovery. Two more proteins from hazelnut were identified. Also, the 11S globulin-like protein was identified in 3 bands (1D) and 3 regions (2D) of the gels. The bands and regions of the proteins in the 1D and the 2D gel excised and analyzed are shown in Figure 7a and 7b. In the 2D gels, individual spots were analyzed and the regions were grouped with similar proteins found. For the 1D gel, additional analysis using the Western Blot technique indicated that bands 2, 3, 5 and 6 shows immunogenic activities. The identity of these bands and regions are shown in Table IV. The results from both gels are correlated as shown in the table. A total of 5 unique proteins were reported from the protein database sequence search using the three different bioinformatics software packages listed earlier. All the proteins identified are either allergenic or potentially allergenic. Band 1 or region 1 was found to be lipid transfer protein precursor (*Corylus avellana*) and was also known as *Cor a 8* and has been suggested to be a relevant allergen for patients with hazelnut allergy (*16,17*). Band 2 or region 2 was identified as hazelnut oleosin which is an oil body associated protein and was recently reported as a hazelnut allergen (*18,19*). Another major allergen *Cor a 1.0402* was found in band 3 or region 3. There were several studies to compare the immuno-reactivity with another major hazel pollen allergen *Cor a 1.01* using *Cor a 1.04* and its variants as recombinant allergens in hope of developing specific immunotherapy of hazelnut allergy (*20, 21*). Band 6 or region 6 identified the 48-kDa glycoprotein precursor [*Corylus avellana*] as the major protein present. This protein is also known as *Cor a 11* and has been reported as a minor allergen (*22*). Bands 4, 5 with regions 4, 5 and 7 all reported the presence of the 11S globulin by the database search results. On close examination of the peptides identified in band 4 and region 4, all the peptides found were in the last 1/3 of the protein sequence proposed (see Figure 8a), and coverage for band 5 was from the top part of the sequence (Figure 8b) and finally, sequence coverage for region 7 is spread throughout the protein (see Figure 8c). The 11S globulin-like protein is also known as the *Cor a 9* and has been identified as a major hazelnut food allergen (*23,24*). It belongs to the 11S seed storage globulin protein family. Although it has a calculated mass of 59 kDa, there is evidence that the experimental mass was 35-40 k Da which is from the acidic subunit. It is also known to have an oligomeric mass of 360 kDa. The acid subunit has also been identified as the main IgE binding region. The protein was proposed as a hexameric seed storage protein with subunits synthesized as a precursor which is then cleaved into two disulphide linked subunits with masses around 20 and 40 kDa before the hexamer is formed. This explained the heterogeneity of the proteins observed especially in the 2D gel.

Band 5 and region 5 are protein fragment or variants from the acid subunit of the 11S globulin-like protein and is a major contributor to the allergenicity cause by this protein since it also exhibit immunogenic activity shown in the Western blot. This was in accordance with findings by other workers (*25*). Band 4 and region 4 contains a protein with pI value about 9.5 (based on the 2D gel data) and from the peptide distribution results from the database search, the theoretical pI of the sequence from the region 337-515 is 9.68 with an expected molecular weight of about 20 kDa which agrees well with the experimental data. This protein fragment is probably the complementary unit of the acid subunit found in Band 5 or region 5. The complete protein is found in region 7, and also in band 6 where it was not separated from the 48 kDa glycoprotein. All the proteins characterized in this work were reported as allergens for hazelnut (see Table V), and their sequences are included in the Protein AllergenOnline Database which is hosted by the Food Allergy Research and Resource Program (FARRP) from the University of Nebraska-Lincoln (*26*).

Although electrospray LC-MS is regarded as the most common approach to protein and peptide marker discovery in proteomics, MALDI also plays a role in providing additional information and offer other advantages. In this work, the use of MALDI for both the digests and intact proteins for the hazelnut protein extract was explored. Using the digest approach, results from the PMF (Peptide Mass Fingering) search of the resultant mass spectra are shown in Figures 9 and 10. Depositing the samples onto the target plate allows the sample to be archived and used for further analysis. The sample turnaround time is very fast, therefore suitable for preliminary screening for a large batch of samples. However, in complex matrices, the peptides can be heavily populated making the database search difficult to identify minor components. This situation was remedied by the use of an offline-LC method. This effectively provides a cleaned-up fraction resulting in a much better search score. This is demonstrated in Figure 11 comparing the PLGS search for the total digest versus that from a CTC PAL spot collected during the elution of the targeted peptides on the MALDI plate. Using the MALDI QTOF Premier for analysis adds another dimension of selectivity. With this hybrid instrument, peptides from the laser ionization can be further fragmented in the collision cell resulting in MS/MS data. This can be searched against the database for protein identification. Results are shown in Table VI. More peptides were identified giving higher protein coverage from the MALDI DDA survey data of the hazelnut protein solution digest compared to MALDI MS results.

The last mass spectrometric method explored in this work used MALDI to obtain a protein profile of the intact protein present in the extract. Figure 12 shows the mass profile for a hazelnut protein extract directly spotted onto a target plate and analyzed with the MALDI micro MX in linear mode. LTP (*Cor a 8*) was observed as a major peak at ~12 kDa. When the laser energy was increased, the larger proteins indentified as the acid subunit of *Cor a 9*, *Cor a 11* and *Cor a 9* were observed. The mass profile was fairly broad indicating these proteins are quite heterogeneous, which corroborate with the results observed from the 2D gel.

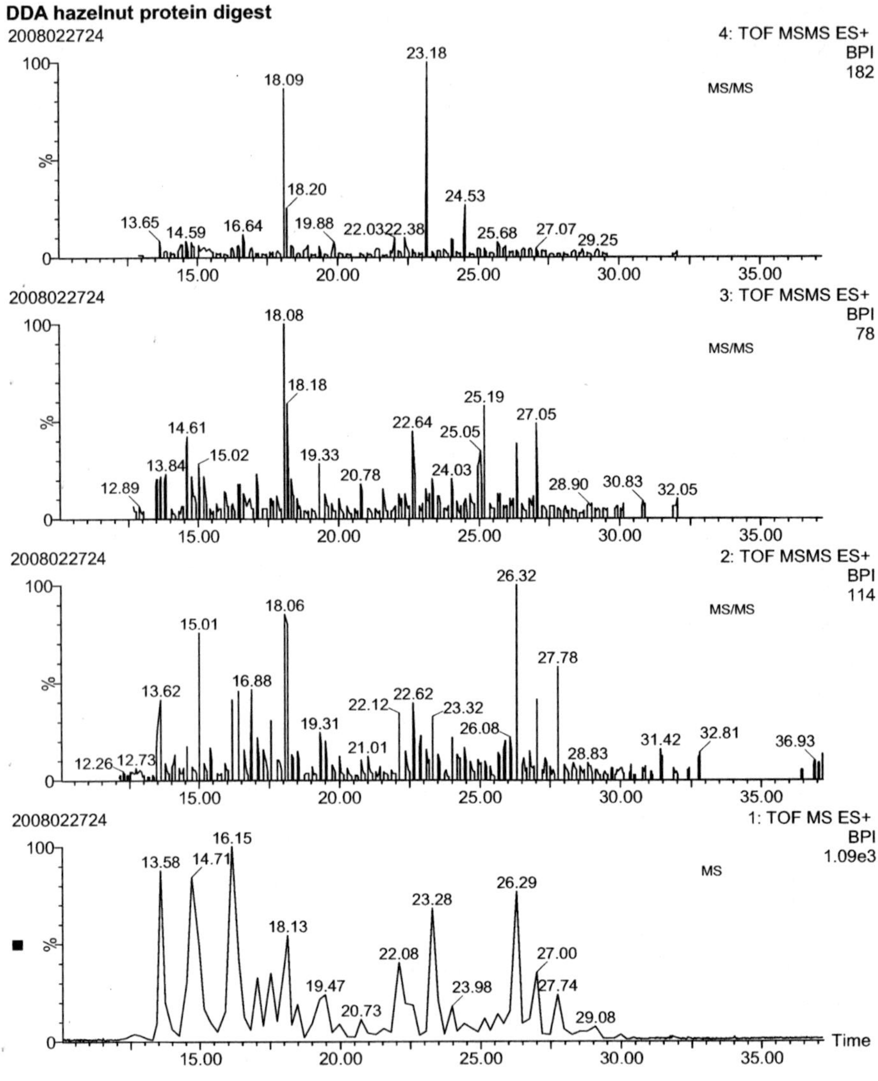

Figure 2. Raw data display of the hazelnut protein extract in-solution trypsin digest showing the MS scan (bottom trace) and three channels of MS/MS data (top three traces) from a survey DDA acquisition.

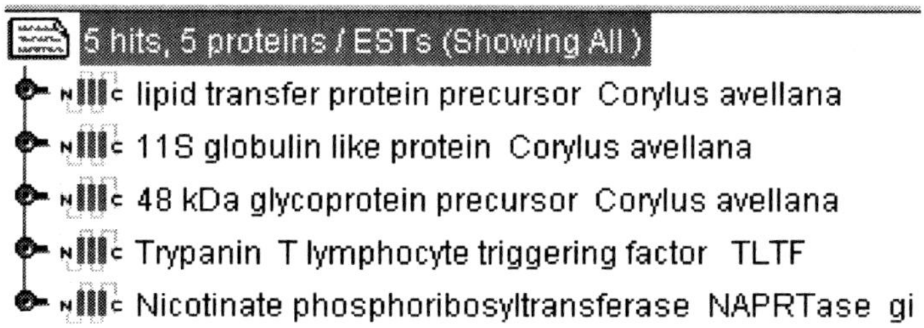

(a)

```
MS data file    : D:\2008 Data\200802 DDA\2008022724.raw
Database        : NCBInr 20070904 (5424415 sequences; 1877877880 residues)
Timestamp       : 28 Feb 2008 at 13:34:02 GMT
Protein hits    : gi|18479082  11S globulin-like protein [Corylus avellana]
                  gi|19338630  48-kDa glycoprotein precursor [Corylus avellana]
                  gi|120674    Glyceraldehyde-3-phosphate dehydrogenase, cytosolic
                  gi|600108    legumin A precursor [Vicia narbonensis]
                  gi|13507262  lipid transfer protein precursor [Corylus avellana]
```

(b)

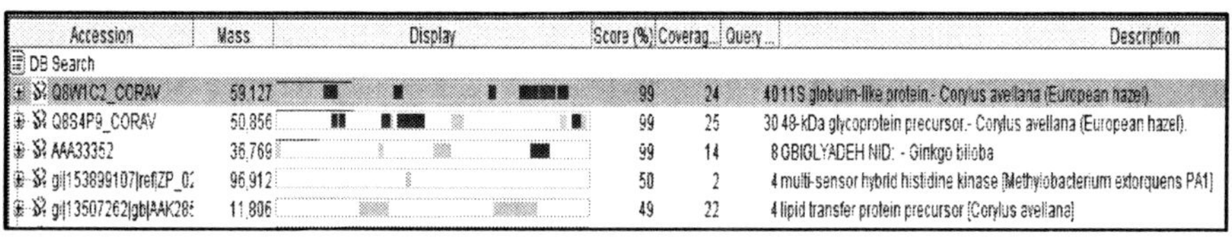

(c)

Figure 3. Database search results for the hazelnut protein digests' LC-MS/MS DDA data using (a) ProteinLynx Global Server v2.3 (b) MASCOT and (c) PEAKS Studio v4.5

```
Sequence coverage:  26%

  1 MAKLILVSFS LCLLVLFNGC LGINVGLRRQ QQRYFGECNL DRLNALEPTN
 51 RIEAEACQIE SWDHNDQQFQ CAGVAVIRRT IEPNGLLLPQ YSNAPELIYI
101 ERGRGITGVL FPGCPETFED PQQQSQQGQR QGQGQSQRSE QDRHQKIRHF
151 REGDIIALPA GVAHWCYNDG DSPVVTVSLL HTNNYANQLD ENPRHFYLAG
201 NPDDEHQRQG QQQFGQRRRQ QQHSHGEQGE QEQQGEGNNV FSGFDAEFLA
251 DAFNVDVDTA RRLQSNQDKR RNIVKVEGRL QVVRPERSRQ EWERQERQER
301 ESEQERERQR RQGGRGRDVN GFEETICSLR LRENICTRSR ADIYTEQVGR
351 INTVNSNTLP VLRWLQLSAE RGDLQREGLY VPHWNLNAHS VVYAIRGRAR
401 VQVVDDNGNT VFDDELRQGQ VLTIPQNFAV AKRAESEGFE WVAFKTNDNA
451 QISPLAGRTS AIRALPDDVL ANAFQISREE ARRLKYNRQE TTLVRSSRSS
501 SERKRRSESE GRAEA
```

(a)

```
Sequence coverage:  21%

  1 MLPKEDPELK KCKHKCRDER QFDEQQRRDG KQICEEKARE RQQEEGNSSE
 51 ESYGKEQEEN PYVFQDEHFE SRVKTEEGRV QVLENFTKRS RLLSGIENFR
101 LAILEANPHT FISPAHFDAE LVLFVAKGRA TITMVREEKR ESFNVEHGDI
151 IRIPAGTPVY MINRDENEKL FIVKILQPVS APGHFEAFYG AGGEDPESFY
201 RAFSWEVLEA ALKVRREQLE KVFGEQSKGS IVKASREKIR ALSQHEEGPP
251 RIWPFGGESS GPINLLHKHP SQSNQFGRLY EAHPDDHKQL QDLDLMVSFA
301 NITKGSMAGP YYNSRATKIS VVVEGEGFFE MACPHLSSSS GSYQKISARL
351 RRGVVFVAPA GHPVAVIASQ NNNLQVLCFE VNAHGNSRFP LAGKGNIVNE
401 FERDAKELAF NLPSREVERI FKNQDQAFFF PGPNKQQEEG GRGGRAFE
```

(b)

Figure 4. Protein sequence of (a) 11S globulin-like protein(corylus avellana) and (b) 48-kDa glycoprotein precursor(corylus avellana) found in the tryptic digest with the peptides identified highlighted

Table I. Peptides from the 11S globulin-like protein and the 48-kDa glycoprotein precursor of Corylus avellana identified in the protein digest of the hazelnut with its corresponding precursor mass and charge

Peptide mass and charge	*Peptide sequence*
11S globulin-like protein	
538.7658 (2+)	QGQQQFGQR
678.8475 (2+)	TNDNAQISPLAGR
700.3306 (2+)	AESEGFEWVAFK
720.9126 (2+)	INTVNSNTLPVLR
807.4547 (2+)	QGQVLTIPQNFAVAK
815.4339 (2+)	ALPDDVLANAFQISR
967.9587 (2+)	VQVVDDNGNTVFDDELR
780.0751 (3+)	EGLYVPHWNLNAHSVVYAIR
881.8063 (3+)	TIEPNGLLLPQYSNAPELIYIER
1322.2056 (2+)	TIEPNGLLLPQYSNAPELIYIER
933.8401 (3+)	RTIEPNGLLLPQYSNAPELIYIER
48-kDa glycoprotein precursor	
666.3612 (2+)	IPAGTPVYMINR
981.1318 (3+)	ILQPVSAPGHFEAFYGAGGEDPESFYR
682.3670 (2+)	AFSWEVLEAALK
617.9971 (3+)	IWPFGGESSGPINLLHK
523.7856 (2+)	ELAFNLPSR
755.3602 (2+)	NQDQAFFFPGPNK
765.3582 (3+)	NQDQAFFFPGPNKQQEEGGR

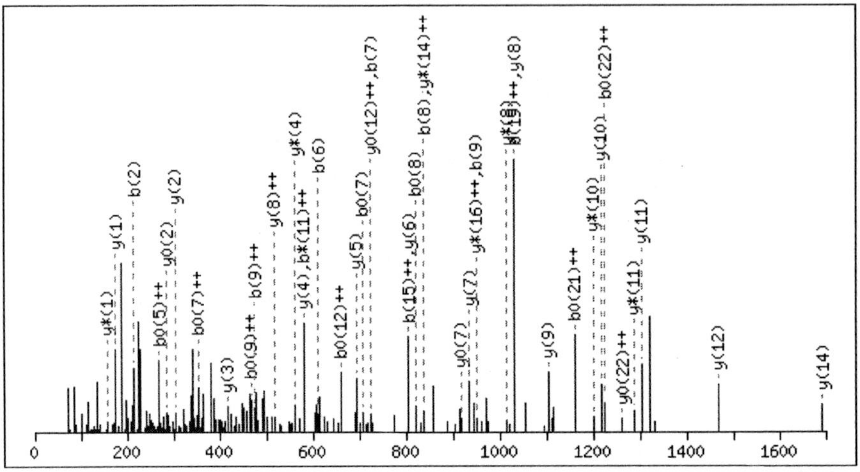

#	b	b^{++}	b^{*}	b^{*++}	b^{0}	b^{0++}	Seq.	y	y^{++}	y^{*}	y^{*++}	y^{0}	y^{0++}	#
1	102.0550	51.5311			84.0444	42.5258	T							23
2	**215.1390**	108.0731			197.1285	99.0679	I	2542.3552	1271.6812	2525.3286	1263.1679	2524.3446	**1262.6759**	22
3	344.1816	172.5944			326.1710	163.5892	E	2429.2711	1215.1392	2412.2446	1206.6259	2411.2605	1206.1339	21
4	441.2344	221.1208			423.2238	212.1155	P	2300.2285	1150.6179	2283.2020	1142.1046	2282.2179	1141.6126	20
5	555.2773	278.1423	538.2508	269.6290	537.2667	**269.1370**	N	2203.1757	1102.0915	2186.1492	1093.5782	2185.1652	1093.0862	19
6	**612.2988**	306.6530	595.2722	298.1397	594.2882	297.6477	G	2089.1328	1045.0700	2072.1063	1036.5568	2071.1223	1036.0648	18
7	**725.3828**	363.1951	708.3563	354.6818	707.3723	**354.1898**	L	2032.1114	1016.5593	2015.0848	1008.0460	2014.1008	1007.5540	17
8	**838.4669**	419.7371	821.4403	411.2238	**820.4563**	410.7318	L	1919.0273	960.0173	1902.0007	**951.5040**	1901.0167	951.0120	16
9	**951.5510**	**476.2791**	934.5244	467.7658	933.5404	**467.2738**	L	1805.9432	903.4753	1788.9167	894.9620	1787.9327	894.4700	15
10	1048.6037	524.8055	1031.5772	516.2922	1030.5932	515.8002	P	**1692.8592**	846.9332	1675.8326	**838.4199**	1674.8486	837.9279	14
11	1176.6623	588.8348	1159.6358	**580.3215**	1158.6517	579.8295	Q	1595.8064	798.4068	1578.7799	789.8936	1577.7958	789.4016	13
12	1339.7256	670.3665	1322.6991	661.8532	1321.7151	**661.3612**	Y	**1467.7478**	734.3775	1450.7213	725.8643	1449.7373	**725.3723**	12
13	1426.7577	713.8825	1409.7311	705.3692	1408.7471	704.8772	S	**1304.6845**	652.8459	**1287.6579**	644.3326	1286.6739	643.8406	11
14	1540.8006	770.9039	1523.7740	762.3907	1522.7900	761.8986	N	**1217.6525**	609.3299	**1200.6259**	600.8166	1199.6419	600.3246	10
15	1611.8377	**806.4225**	1594.8112	797.9092	1593.8271	797.4172	A	**1103.6095**	552.3084	1086.5830	543.7951	1085.5990	543.3031	9
16	1708.8905	854.9489	1691.8639	846.4356	1690.8799	845.9436	P	**1032.5724**	**516.7898**	**1015.5459**	508.2766	1014.5619	507.7846	8
17	1837.9331	919.4702	1820.9065	910.9569	1819.9225	910.4649	E	**935.5197**	468.2635	918.4931	459.7502	**917.5091**	459.2582	7
18	1951.0171	976.0122	1933.9906	967.4989	1933.0066	967.0069	L	**806.4771**	403.7422	789.4505	395.2289	788.4665	394.7369	6
19	2064.1012	**1032.5542**	2047.0746	1024.0410	2046.0906	1023.5489	I	**693.3930**	347.2001	676.3665	338.6869	675.3824	338.1949	5
20	2227.1645	1114.0859	2210.1380	1105.5726	2209.1539	1105.0806	Y	**580.3089**	290.6581	**563.2824**	282.1448	562.2984	281.6528	4
21	2340.2486	1170.6279	2323.2220	1162.1147	2322.2380	**1161.6226**	I	**417.2456**	209.1264	400.2191	200.6132	399.2350	200.1212	3
22	2469.2912	1235.1492	2452.2646	1226.6359	2451.2806	**1226.1439**	E	**304.1615**	152.5844	287.1350	144.0711	**286.1510**	143.5791	2
23							R	**175.1190**	88.0631	**158.0924**	79.5498			1

Figure 5. MS/MS fragmentation of a triply charged peptide (TIEPNGLLLPQYSNAPELIYIER) from 11S globulin-like protein

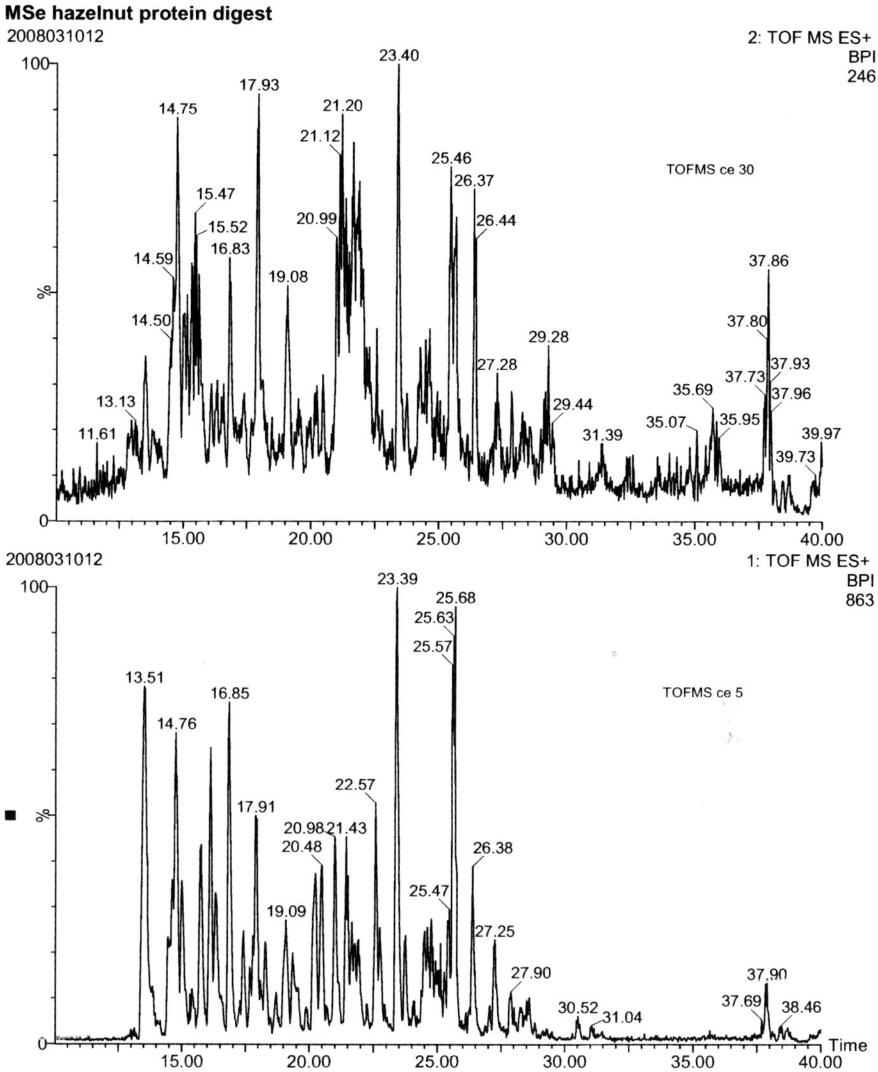

Figure 6. Typical raw data from a MS^E acquisition of a hazelnut protein digest with the bottom trace from function 1 TOFMS at low collision energy (5 v) and top trace function 2 TOFMS at elevated collision energy (30 v)

Table II. Peptides identified in the hazelnut protein digest from data acquired using MS^E and processed with Identity^E software.

Peptide MH+ (Da)	*Sequence*
1934.909	(R)VQVVDDNGNTVFDDELR(Q)
1356.6866	(K)TNDNAQISPLAGR(T)
1440.8169	(R)INTVNSNTLPVLR(W)
1629.8595	(R)ALPDDVLANAFQISR(E)
2644.387	(R)TIEPNGLLLPQYSNAPELIYIER(G)
1613.901	(R)QGQVLTIPQNFAVAK(R)
1151.5691	(R)ADIYTEQVGR(I)
1027.5531	(R)LNALEPTNR(I)
2904.3582	(R)GITGVLFPGCPETFEDPQQQSQQGQR(Q)
1441.8009	(R)INTVNSNTLPVLR(W)
1002.5367	(R)WLQLSAER(G)
1357.6707	(K)TNDNAQISPLAGR(T)
1445.7383	(L)PDDVLANAFQISR(E)
1399.6528	(R)AESEGFEWVAFK(T)
1630.8435	(R)ALPDDVLANAFQISR(E)
1394.7023	(R)SRADIYTEQVGR(I)
800.4261	(N)ALEPTNR(I)
1133.5538	(R)ADIYTEQVGR(I)
965.5051	(D)IYTEQVGR(I)
852.421	(I)YTEQVGR(I)
985.506	(R)WLQLSAER(G)
914.469	(L)NALEPTNR(I)

Table III. Quantification result for the three major proteins identified from MS^E data for 4 injections of a hazelnut protein digest using ADH as calibration protein

	fmol / µL			
	inj 1	inj 2	inj 3	inj 4
lipid transfer protein precursor [Corylus avellana]	5.39	5.97	5.41	5.02
11S globulin-like protein [Corylus avellana]	37.81	37.59	40.53	37.4
48-kDa glycoprotein precursor [Corylus avellana]	10.26	10.76	10.89	10.41

Table IV. Proteins identified in the 1D and 2D gel of the hazelnut protein

Band from 1D gel	*Region from 2D gel*	*Protein identified from NCBI protein sequence database*	*Mol. Wt (Da)*	*Protein Coverage (%)*
1	1	lipid transfer protein precursor [Corylus avellana] Cor a 8	12368	42
2	2	oleosin [Corylus avellana]	14723	27
3	3	major allergen variant Cor a 1.0402 [Corylus avellana]	17614	72
4	4	11S globulin-like protein [Corylus avellana] Cor a 9	59605	32
5	5	11S globulin-like protein [Corylus avellana] Cor a 9	59605	32
6*	6	48-kDa glycoprotein precursor [Corylus avellana] Cor a 11	51110	47
	7	11S globulin-like protein [Corylus avellana] Cor a 9	59605	32

**in addition 11S globulin-like protein [Corylus avellana] was also found in this band*

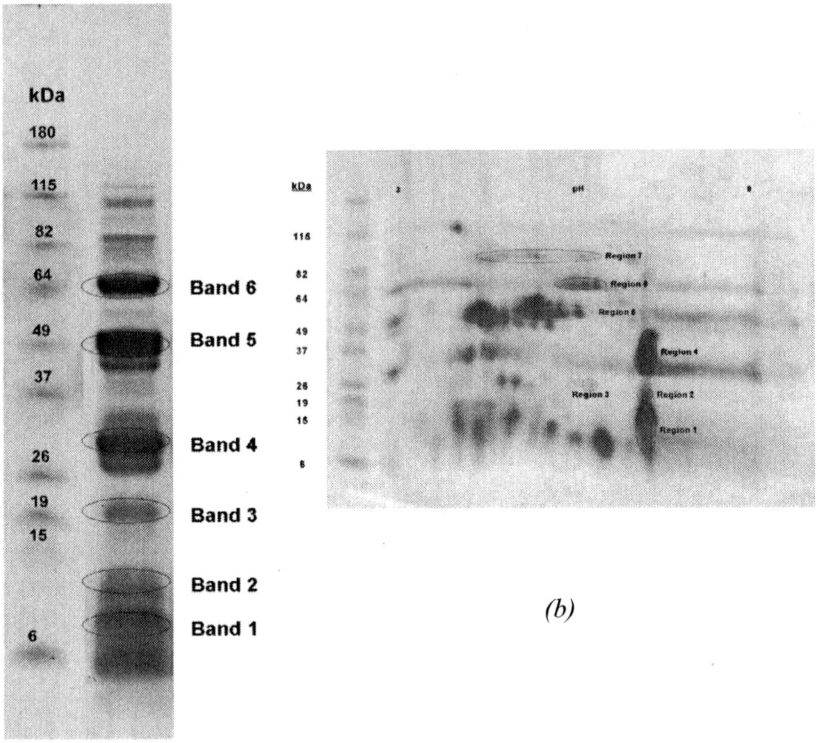

(a)

Figure 7. (a) 1D gel and (b) 2D gel of the hazelnut protein extract showing the bands and regions characterized by LC-MS/MS after in-gel digestion

(a)

Sequence coverage 21%

```
  1 MAKLILVSFS LCLLVLFNGC LGINVGLRRQ QQRYFGECNL DRLNALEPTN
 51 RIEAEACQIE SWDHNDQQFQ CAGVAVIRRT IEPNGLLLPQ YSNAPELIYI
101 ERGRGITGVL FPGCPETFED PQQQSQQGQR QGQGQSQRSE QDRHQKIRHF
151 REGDIIALPA GVAHWCYNDG DSPVVTVSLL HTNNYANQLD ENPRHFYLAG
201 NPDDEHQRQG QQQFGQRRRQ QQHSHGEQGE QEQQGEGNNV FSGFDAEFLA
251 DAFNVDVDTA RRLQSNQDKR RNIVKVEGRL QVVRPERSRQ EWERQERQER
301 ESEQERERQR RQGGRGRDVN GFEETICSLR LRENICTRSR ADIYTEQVGR
351 INTVNSNTLP VLRWLQLSAE RGDLQREGLY VPHWNLNAHS VVYAIRGRAR
401 VQVVDDNGNT VFDDELRQGQ VLTIPQNFAV AKRAESEGFE WVAFKTNDNA
451 QISPLAGRTS AIRALPDDVL ANAFQISREE ARRLKYNRQE TTLVRSSRSS
501 SERKRRSESE GRAEA
```

(b)

Sequence coverage 20%

```
  1 MAKLILVSFS LCLLVLFNGC LGINVGLRRQ QQRYFGECNL DRLNALEPTN
 51 RIEAEACQIE SWDHNDQQFQ CAGVAVIRRT IEPNGLLLPQ YSNAPELIYI
101 ERGRGITGVL FPGCPETFED PQQQSQQGQR QGQGQSQRSE QDRHQKIRHF
151 REGDIIALPA GVAHWCYNDG DSPVVTVSLL HTNNYANQLD ENPRHFYLAG
201 NPDDEHQRQG QQQFGQRRRQ QQHSHGEQGE QEQQGEGNNV FSGFDAEFLA
251 DAFNVDVDTA RRLQSNQDKR RNIVKVEGRL QVVRPERSRQ EWERQERQER
301 ESEQERERQR RQGGRGRDVN GFEETICSLR LRENICTRSR ADIYTEQVGR
351 INTVNSNTLP VLRWLQLSAE RGDLQREGLY VPHWNLNAHS VVYAIRGRAR
401 VQVVDDNGNT VFDDELRQGQ VLTIPQNFAV AKRAESEGFE WVAFKTNDNA
451 QISPLAGRTS AIRALPDDVL ANAFQISREE ARRLKYNRQE TTLVRSSRSS
501 SERKRRSESE GRAEA
```

(c)

Sequence coverage 32%

```
  1 MAKLILVSFS LCLLVLFNGC LGINVGLRRQ QQRYFGECNL DRLNALEPTN
 51 RIEAEACQIE SWDHNDQQFQ CAGVAVIRRT IEPNGLLLPQ YSNAPELIYI
101 ERGRGITGVL FPGCPETFED PQQQSQQGQR QGQGQSQRSE QDRHQKIRHF
151 REGDIIALPA GVAHWCYNDG DSPVVTVSLL HTNNYANQLD ENPRHFYLAG
201 NPDDEHQRQG QQQFGQRRRQ QQHSHGEQGE QEQQGEGNNV FSGFDAEFLA
251 DAFNVDVDTA RRLQSNQDKR RNIVKVEGRL QVVRPERSRQ EWERQERQER
301 ESEQERERQR RQGGRGRDVN GFEETICSLR LRENICTRSR ADIYTEQVGR
351 INTVNSNTLP VLRWLQLSAE RGDLQREGLY VPHWNLNAHS VVYAIRGRAR
401 VQVVDDNGNT VFDDELRQGQ VLTIPQNFAV AKRAESEGFE WVAFKTNDNA
451 QISPLAGRTS AIRALPDDVL ANAFQISREE ARRLKYNRQE TTLVRSSRSS
501 SERKRRSESE GRAEA
```

Figure 8. Protein sequence coverage of the 11S globulin-like protein from (a) band 4 and region 4, (b) band 5 and region 5 and (c) region 7

Table V. Currently sequenced allergen protein database entries for species Corylus avellana and common name European hazelnut obtained from AllergenOnline v. 8.0 (farrp allergen protein database, University of Nebraska-Lincoln)

Allergen	Type	comment	GI #	Length
Cor a 1.0201	Aero Plant	Cor a 1 (Bet v 1 homologue)	1321731	160
Cor a 1.0403	Food Plant	Cor a 1 (Bet v 1 homologue)	11762104	161
Cor a 1.0103	Aero Plant	Cor a 1 (Bet v 1 homologue)	22684	160
Cor a 1.0401	Food Plant	Cor a 1 (Bet v 1 homologue)	5726304	161
Cor a 1.0404	Food Plant	Cor a 1 (Bet v 1 homologue)	11762106	161
Cor a 1.0104	Aero Plant	Cor a 1 (Bet v 1 homologue)	22686	160
Cor a I	Aero Plant	Cor a 1 (Bet v 1 homologue)	584968	160
Cor a 1.0301	Aero Plant	Cor a 1 (Bet v 1 homologue)	1321733	160
Cor a 1.0402	Food Plant	Cor a 1 (Bet v 1 homologue)	11762102	161
Unassigned	Food Plant	Corylus Oleosin	29170509	140
Cor a 10	Aero Plant	Putative luminal binding protein	10944737	668
Cor a 11	Food Plant	48 kDa glycoprotein (vicilin like)	19338630	448
Cor a 2	Aero Plant	Profilin	12659206	131
Cor a 2	Aero Plant	Profilin	12659208	131
Cor a 8	Food Plant	Lipid transfer protein	13507262	115
Cor a 9	Food Plant	11S globulin-like protein	18479082	515
Cor a 1.0102	Aero Plant	Cor a 1 (Bet v 1 homologue)	22690	160

m/z	Charge	Sequence
2904.2678	1	(R)GITGVLFPGCPETFEDPQQQSQQGQR(Q)
1698.7787	1	(R)HFYLAGNPDDEHQR(Q)
1076.5072	1	(R)QGQQQFGQR(R)
990.4998	1	(R)SRQEWER(Q)
1440.8247	1	(R)INTVNSNTLPVLR(W)
1002.5568	1	(R)WLQLSAER(G)
1571.8529	1	(R)WLQLSAERGDLQR(E)
2338.1702	1	(R)EGLYVPHWNLNAHSVVYAIR(G)
1934.9268	1	(R)VQVVDDNGNTVFDDELR(Q)
1613.9385	1	(R)QGQVLTIPQNFAVAK(R)
1555.7596	1	(K)RAESEGFEWVAFK(T)
1399.6559	1	(R)AESEGFEWVAFK(T)
1356.6730	1	(K)TNDNAQISPLAGR(T)
1884.8417	1	(K)TNDNAQISPLAGRTSAIR(A)
1629.8643	1	(R)ALPDDVLANAFQISR(E)

```
  1 MAKLILVSFS LCLLVLFNGC LGINVGLRRQ QQRYFGECNL DRLNALEPTN
 51 RIEAEACQIE SWDHNDQQFQ CAGVAVIRRT IEPNGLLLPQ YSNAPELIYI
101 ERGRGITGVL FPGCPETFED PQQQSQQGQR QGQGQSQRSE QDRHQKIRHF
151 REGDIIALPA GVAHWCYNDG DSPVVTVSLL HTNNYANQLD ENPRHFYLAG
201 NPDDEHQRQG QQQFGQRRRQ QQHSHGEQGE QEQQGEGNNV FSGFDAEFLA
251 DAFNVDVDTA RRLQSNQDKR RNIVKVEGRL QVVRPERSRQ EWERQERQER
301 ESEQERERQR RQGGRGRDVN GFEETICSLR LRENICTRSR ADIYTEQVGR
351 INTVNSNTLP VLRWLQLSAE RGDLQREGLY VPHWNLNAHS VVYAIRGRAR
401 VQVVDDNGNT VFDDELRQGQ VLTIPQNFAV AKRAESEGFE WVAFKTNDNA
451 QISPLAGRTS AIRALPDDVL ANAFQISREE ARRLKYNRQE TTLVRSSRSS
501 SERKRRSESE GRAEA
```

Figure 9. Typical PMF (peptide map fingerprinting) search result for the identified 11S globulin-like protein from a hazelnut protein digest analyzed on a MALDI target plate showing the peptide markers and the protein sequence coverage

m/z	Charge	Sequence
950.4208	1	(R)QFDEQQR(R)
1106.5370	1	(R)QFDEQQRR(D)
1077.5552	1	(R)VQVLENFTK(R)
1233.6976	1	(R)VQVLENFTKR(S)
1415.6962	1	(R)ESFNVEHGDIIR(I)
1331.7300	1	(R)IPAGTPVYMINR(D)
1220.5967	1	(R)ALSQHEEGPPR(I)
1157.5557	1	(K)HPSQSNQFGR(L)
1202.5292	1	(K)GSMAGPYYNSR(A)
1046.5297	1	(K)ELAFNLPSR(E)
1509.6915	1	(K)NQDQAFFFPGPNK(Q)

```
  1 MLPKEDPELK KCKHKCRDER QFDEQQRRDG KQICEEKARE RQQEEGNSSE
 51 ESYGKEQEEN PYVFQDEHFE SRVKIEEGRV QVLENFTKRS RLLSGIENFR
101 LAILEANPHT FISPAHFDAE LVLFVAKGRA TITMVREEKR ESFNVEHGDI
151 IRIPAGTPVY MINRDENEKL FIVKILQPVS APGHFEAFYG AGGEDPESFY
201 RAFSWEVLEA ALKVRREQLE KVFGEQSKGS IVKASREKIR ALSQHEEGPP
251 RIWPFGGESS GPINLLHKHP SQSNQFGRLY EAHPDDHKQL QDLDLMVSFA
301 NITKCSMACP YYNSRATKIS VVVECECFFE MACPHLSSSS CSYQKISARL
351 RRGVVFVAPA GHPVAVIASQ NNNLQVLCFE VNAHGNSRFP LAGKGNIVNE
401 FERDAKELAF NLPSREVERI FKNQDQAFFF PGPNKQQEEG GRGGRAFE
```

Figure 10. Typical PMF (peptide map fingerprinting) search result for the identified 48 kDa glycoprotein from a hazelnut protein digest analyzed on a MALDI target plate showing the peptide markers and the protein sequence coverage

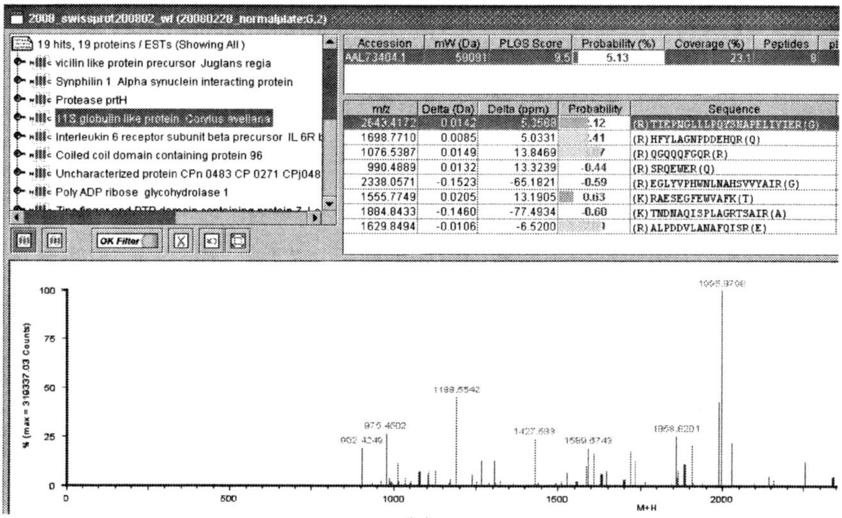

(a)

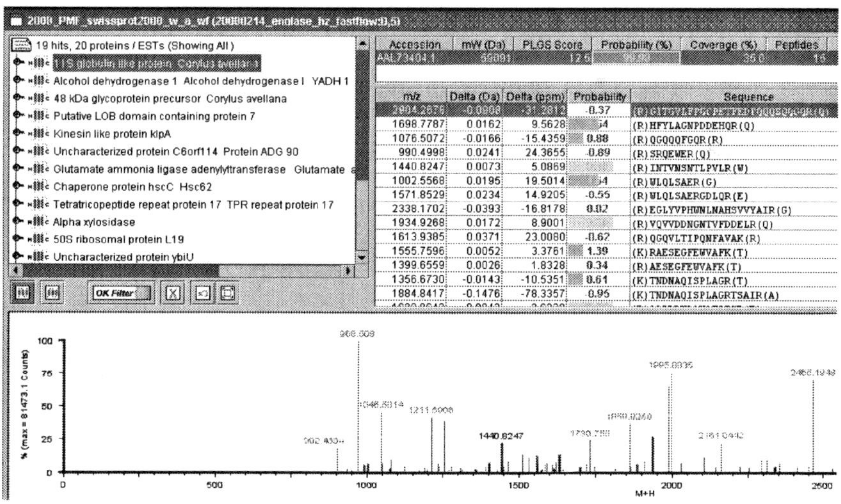

(b)

Figure 11 . Comparison of MALDI-TOF PMF results from (a) total digest and (b) selected CTC-PAL spot collected on target plate

Table VI. PLGS search result from MALDI DDA survey data on a hazelnut protein solution digest using the QTOF Premier in MALDI mode

Protein identified	No. of peptides	% coverage
11S globulin like protein Corylus avellana	19	40

m/z	Charge	Sequence
2799.5390	1	(R)RTIEPNGLLLPQYSNAPELIYIER(G)
2643.4160	1	(R)TIEPNGLLLPQYSNAPELIYIER(G)
2904.3977	1	(R)GITGVLFPGCPETFEDPQQQSQQGQR(Q)
1698.7451	1	(R)HFYLAGNPDDEHQR(Q)
1076.5073	1	(R)QGQQQFGQR(R)
832.3732	1	(R)LQSNQDK(R)
990.4656	1	(R)SRQEWER(Q)
1394.6890	1	(R)SRADIYTEQVGR(I)
1151.5070	1	(R)ADIYTEQVGR(I)
1440.8093	1	(R)INTVNSNTLPVLR(W)
1002.5261	1	(R)WLQLSAER(G)
2338.1882	1	(R)EGLYVPHWNLNAHSVVYAIR(G)
1934.8930	1	(R)VQVVDDNGNTVFDDELR(Q)
1555.7373	1	(K)RAESEGFEWVAFK(T)
1399.6260	1	(R)AESEGFEWVAFK(T)
1356.6738	1	(K)TNDNAQISPLAGR(T)
1884.8310	1	(K)TNDNAQISPLAGRTSAIR(A)
1629.8391	1	(R)ALPDDVLANAFQISR(E)
2115.0867	1	(R)ALPDDVLANAFQISREEAR(R)

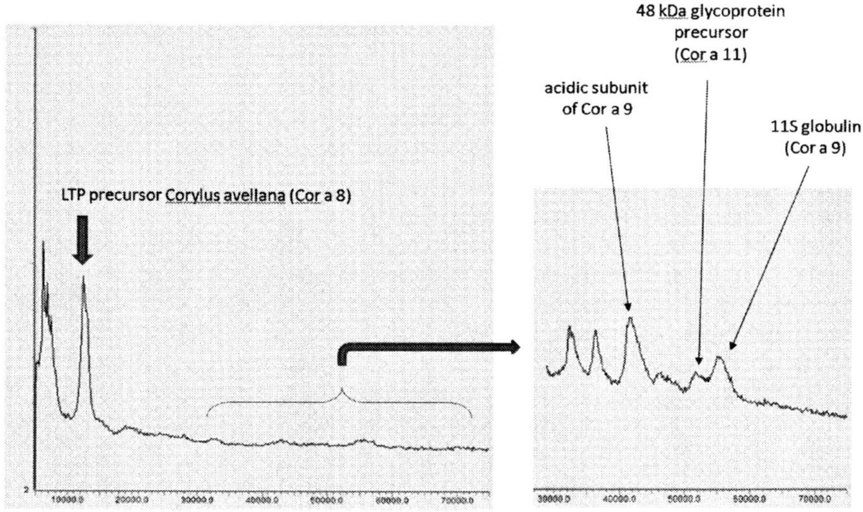

(a) *(b)*

Figure 12. Typical MALDI TOF (linear mode) intact protein profile from a hazelnut protein extract with laser energy at (a) 196 and (b) 220

Conclusion

In this paper, we have presented a number of approaches to the characterization of allergens in a food commodity using hazelnut as a model. We have shown that for a crude analysis of the protein present; in solution digestion followed by either ESI or MALDI give reasonably reliable protein identification in the mixture. However, the coverage of the protein found can sometimes be incomplete due to the complexity of the mixture. 1D and 2D gel separation can provide a much better protein coverage since the sought for proteins are isolated or separated from each other. The intact protein approach enables more accurate measurement of a relative molecular weight value for the intact proteins and complements the data from the analysis of the protein digest experiments. We have shown that mass spectrometry is a powerful and versatile technique for the identification and characterization of food allergens. It can also serve as a confirmatory method and complement results obtained from other approaches such as ELISA.

References

1. *Detecting allergens in food.*; Koppelman, S.; Hefle, S., Eds.; Woodhead Publishing Limited: Cambridge, UK, **2006**.

2. Poms, R.E.; Anklam, E.; Kuhn, M. Polymerase chain reaction techniques for food allergen detection : Methodological aspects of food allergens. *J. AOAC Int.***2004,**87, 1391-1397.

3. Careri,M.; Costa,A.; Elviri, L.; Lagos,J.-B.; Mangia,A.; Terenghi,M.; Cereti, A.; Perono Garoffo, L. Use of specific peptide biomarkers for quantitative confirmation of hidden allergenic peanut proteins Ara h 2 and Ara h 3/4 for food control by liquid chromatography–tandem mass spectrometry. *Analytical and Bioanalytical Chemistry*, **2007**,389, 1901-1907.

4. Shefcheck,K.J.; Musser, S.M. Confirmation of the Allergenic Peanut Protein, Ara h 1, in a Model Food Matrix Using Liquid Chromatography/ Tandem Mass Spectrometry (LC/MS/MS). *J. Agric. Food Chem.* **2004,** 52 (10), 2785 -2790.

5. Weber,D.; Raymond,P.; Ben-Rejeb, S.; Lau, B. Development of a Liquid Chromatography-Tandem Mass Spectrometry Method Using Capillary Liquid Chromatography and Nanoelectrospray Ionization-Quadrupole Time-of-Flight Hybrid Mass Spectrometer for the Detection of Milk Allergens. *J. Agric. Food Chem.* **2006**, 54 (5), 1604 -1610.

6. Weber,D.; Lau, B. P.-Y.; Benrejeb Godefroy,S. Characterization of Gluten Proteins in Grain Flours by Liquid Chromatography – Tandem Mass Spectrometry using a Quadrupole Time-of-Flight Mass Spectrometer. *ACS Symposium Series No. 1001/Food Contaminants,Mycotoxins and Food Allergens, 482-500* **2008**.

7. Weber, D.; Cyr, T.; Lau, B. P.-Y. Benrejeb-Godefroy, S. Gluten determination in beer using mass spectrometric techniques. 55th ASMS Annual Conference, Indianapolis, Indiana, **2007**.

8. Méndez,E.; Valdes,I.; Camafeita, E. Analysis of Gluten in Foods by MALDI-TOFMS. *Mass Spectrometry of Proteins and Peptides: Methods in Molecular Biology,* **2000**, 146, 355-367.

9. Akagawa, M.; Handoyo, T.; Ishii, T.; Kumazawa, S.; Morita, N.; Suyama K. Proteomic Analysis of Wheat Flour Allergens. *J. Agric. Food Chem.* **2007**, 55, 6863-6870.

10. Crespol, J. F.; James, J. M.; Fernandez-Rodriguez, C.; Rodriguez, J. Food allergy: nuts and tree nuts. *British Journal of Nutrition*, **2006**, 96 Suppl. 2, S95–S102.

11. Reid, G.E.; McLuckey, S.A., Top down protein characterization via tandem mass spectrometry. *J. Mass Spectrom.* **2002**,37, 663–675.

12. Chakraborty, A.B.; Berger, S.J.; Gebler, J.C. Use of an integrated MS - multiplexed MS/MS data acquisition strategy for high-coverage peptide mapping studies. *Rapid Commun Mass Spectrom.* **2007**,21(5),730-44.

13. Schwarz, E.; Levin, Y.; Wang, L.; Leweke, F.M.; Bahn, S. Peptide correlation: A means to identify high quality quantitative information in large-scale proteomic studies. *J Sep Sci.* **2007**, 30(14), 2190-2197.

14. Levin, Y.; Schwarz, E.; Levin, Y.; Wang, L.; Leweke, F.M.; Bahn, S. Label-free LC-MS/MS quantitative proteomics for large-scale biomarker discovery in complex samples. *J Sep Sci.* **2007**, 30(14), 2198-2203.

15. Silva, J.C.; Gorenstein, M.V.; Li, G.Z.; Vissers, J.P.C.; Geromanos, S.J. Absolute Quantification of Proteins by $LCMS^{E}$. *Mol Cell Proteomics.* **2006**, 5(1),144-156.

16. Schocker, F.; Lüttkopf, D.; Scheurer, S.; Petersen, A.; Cisteró-Bahima, A.; Enrique, E.; San Miguel-Moncín, M.; Akkerdaas, J.; van Ree, R.; Vieths, S.; Becker, W.M. Recombinant lipid transfer protein Cor a 8 from hazelnut: a new tool for in vitro diagnosis of potentially severe hazelnut allergy. *J Allergy Clin Immunol.* **2004**,113(1):141-7.

17. Flinterman, A.E.; Akkerdaas, J.H.; den Hartog Jager, C.F.; Rigby, N.M.; Fernandez-Rivas, M.; Hoekstra, M.O.; Bruijnzeel-Koomen, C.A.; Knulst, A.C.; van Ree, R.; Pasmans, S.G. Lipid transfer protein-linked hazelnut allergy in children from a non-Mediterranean birch-endemic area. *J Allergy Clin Immunol.* **2008**, 121(2):423-428.

18. Zuidmeer, L.; Winter, M.; Akkerdaas, J.; Versteeg, S.; Summers, C.; Knulst, A.; Brettlova, V.; Wensing, M.; Schilte, P.; van Ree, R. Hazelnut oleosin: identification and characterization of a novel hazelnut allergen. *2nd International Symposium on Molecular Allergology, Allergens: from Identification to Therapy,* **2007,** Rome, Italy.

19. Akkerdaas, J.H.; Schocker, F.; Vieths, S.; Versteeg, S.; Zuidmeer, L.; Hefle, S.L.; Aalberse, .RC.; Richter, K.; Ferreira, F.; van Ree, R. Cloning of oleosin, a putative new hazelnut allergen, using a hazelnut cDNA library. *Mol Nutr Food Res.* **2006,** 50(1):18-23.

20. Lüttkopf, D.; Müller, U.; Skov, P.S.; Ballmer-Weber, B.K.; Wüthrich, B.; Skamstrup Hansen, K.; Poulsen, L.K.; Kästner, M.; Haustein, D.; Vieths, S.

Comparison of four variants of a major allergen in hazelnut (Corylus avellana) Cor a 1.04 with the major hazel pollen allergen Cor a 1.01. *Mol Immunol.* **2002,** 38(7):515-25.

21. Breiteneder, H.; Ferreira, F.; Hoffmann-Sommergruber, K.; Ebner, C.; Breitenbach, M.; Rumpold, H.; Kraft, D.; Scheiner, O. Four recombinant isoforms of Cor a I, the major allergen of hazel pollen, show different IgE-binding properties. *Eur J Biochem.* **1993,** 212(2):355-62.

22. Lauer, I.; Foetisch, K.; Kolarich, D.; Ballmer-Weber, B.K.; Conti, A.; Altmann, F.; Vieths, S.; Scheurer, S. Hazelnut (Corylus avellana) vicilin Cor a 11: molecular characterization of a glycoprotein and its allergenic activity. *Biochem J.* **2004**, 15; 383(Pt 2): 327–334.

23. Beyer, K.; Grishina, G.; Bardina, L.; Grishin, A.; Sampson, H.A. Identification of an 11S globulin as a major hazelnut food allergen in hazelnut-induced systemic reactions. *J Allergy Clin Immunol.* **2002,** 110(3):517-23.

24. Pastorello, E.A.; Vieths, S.; Pravettoni, V.; Farioli, L.; Trambaioli, C.; Fortunato, D.; Lüttkopf, D.; Calamari, M.; Ansaloni, R.; Scibilia, J.; Ballmer-Weber, B.K.; Poulsen, L.K.; Wütrich, B.; Hansen, K.S.; Robino, A.M.; Ortolani, C.; Conti, A. Identification of hazelnut major allergens in sensitive patients with positive double-blind, placebo-controlled food challenge results. *J Allergy Clin Immunol.* **2002**, 109(3):563-70.

25. http://foodallergens.ifr.ac.uk/

26. http://www.allergenonline.com/

Chapter 12

Inactivation of Microbial Contaminants in Fresh Produce

Brendan A. Niemira, Bassam A. Annous, Xuetong Fan, Ching-Hsing Liao, Joseph Sites

Food Safety Intervention Technologies Research Unit, Eastern Regional Research Center, U.S. Department of Agriculture, Agricultural Research Service, 600 E. Mermaid Lane, Wyndmoor, PA 19038

With the microbial safety of fresh produce of increasing concern, conventional sanitizing treatments need to be supplemented with effective new interventions to inactivate human pathogens. The Produce Safety research project at the United States Department of Agriculture's Eastern Regional Research Center develops and validates new interventions to improve the safety of fresh and fresh-cut fruits and vegetables. Inoculation with suppressive microbial communities inhibits the growth of *Salmonella* on vegetable surfaces by up to 99% during the course of storage. Rapid thermal treatments and gaseous chlorine dioxide can achieve reductions of Salmonella on cantaloupe of more than 99.99%. Irradiation can reduce *E. coli* O157:H7, *Salmonella* and *Listeria monocytogenes* on leafy vegetables and other produce by more than 99.99% while preserving product quality. Chemical and sensory analysis has demonstrated the safety and wholesomeness of irradiated foods. A novel processing technology, cold plasma, has shown promising results, with 99.9% reductions of *Salmonella* and *E. coli* O157:H7 on apple surfaces. This chapter will summarize the advances made in these areas, as well as research results on the process of scaling up effective interventions from laboratory scale to pilot plant scale, including the critical process of evaluating the effects of the various interventions on sensory and nutritional quality attributes, yield, physiology, and shelf-life.

Consumers in the United States have experienced increases in the incidence of foodborne illness (FBI) outbreaks associated with contaminated fruits, vegetables, salads, and juices in recent decades (*1*). While it is recognized that pre-harvest (good agricultural practices, GAP), post-harvest (good manufacturing practices, GMP) and supply-chain (good handling practices, GHP) controls can help to reduce risk, these practices have not been able to prevent product recalls of tomatoes, leafy greens, melons, sprouts and other fresh produce, and the associated repeated FBI outbreaks. The lack of a broadly applicable antimicrobial process (a "kill step") is hampering the food safety efforts of the fresh produce industry (*2, 3*).

In 2005, the U.S. Department of Agriculture initiated a new five year research project within the Agricultural Research Service to address this critical need (*4*). The project, entitled "Intervention Technologies for Enhancing the Safety and Security of Fresh and Minimally Processed Produce and Solid Plant-Derived Foods", builds on a notable history of produce safety research projects at the Eastern Regional Research Center in Wyndmoor, PA. The objective of the project is to develop more effective means for decontaminating organic and conventionally grown fresh and minimally processed fruits and vegetables, including sprout seeds. Food safety and security will be more completely assured by assessing the efficacy of new and/or improved intervention technologies. The researchers determine the effectiveness of treatment combinations (multiple hurdle approach), assess factors that might limit treatment efficacy, and transfer effective decontamination technology to the produce industry in order to reduce the risk of foodborne illness. The produce safety treatments being developed are broadly grouped into physical, chemical and biological interventions. This research group is a representative sample of the type of research activities being conducted across the entire field. In order to give an overview of the kind of research inquiries being pursued in this area at many different locations, the latest results from this particular research project will be highlighted in this chapter within the context of a broader review.

Cold Plasma

Cold plasma is a relatively new sanitizing technology in the field of food processing. Various types of plasma-generating technologies have been used for non-food applications, such as lighting, electronics, and materials processing. As a food processing technology, however, a number of technological hurdles must be overcome before widespread implementation. For all practical purposes, cold plasma may be regarded as an energetic form of gas, although it is technically a distinct state of matter. As energy is added to materials, they change state, going from solid to liquid to gas, with large-scale inter-molecular structure breaking down. In general, as additional energy is added, the intra-atomic structures of the components of the gas break down, yielding plasmas - concentrated collections of ions, radical species and free electrons (*5-8*). According to Niemira and Sites *(8)*, cold plasma technologies used to treat foods fall into three general categories: *electrode contact* (in which the target is in contact with or between electrodes), *direct treatment* (in which active plasma is deposited directly on the target) and *remote treatment* (in which active plasma is generated at some distance, and plasma is moved to the target).

When applied to foods and inert surfaces, electrode contact systems have been shown to achieve reductions of *Escherichia coli, Staphylococcus aureus, Bacillus subtillis,* and *Saccharomyces cerevisiae* as high as 5 logs (*9, 10*). Direct treatment of *E. coli* placed within the 1 mm gap spacing of the plasma reactors reduced the pathogens populations by 4.6 and 5.1 log cfu/ml after treatment of 10 s and 60 s, respectively (*11*). As the space between the plasma emitter and the treated culture was increased, antimicrobial efficacy was reduced, until at 10 mm spacing, no reductions were observed at any power level tested. Remote treatment reactors reduced *E. coli* and *St. aureus* inoculated on polypropylene by 4 or 2 log cfu/ml, respectively, after a 10s treatment (*12*). The one atmosphere uniform glow discharge plasma system (OAUGDP) produced D-values of 22 s (*Shigella flexneri* and *Vibrio parahaemolyticus*) to 51s (*E. coli* O157:H7) for pathogens on agar (*13*). On food surfaces, a 2 min treatment with the OAUGDP reduced *E. coli* O157:H7 on red delicious apples by ca. 3 log cfu, reduced *Salmonella* Enteritidis on cantaloupe by ca. 3 log cfu, and reduced *Listeria monocytogenes* on Iceberg lettuce by ca. 2 log cfu (*14*).

Cold plasma generated by a gliding arc emitter system (Figure 1) inactivated *E. coli* O157:H7 and *Salmonella* on agar plates and on the surface of golden delicious apples (*8*). Higher flow rates of plasma (30 or 40 L/min) were more effective than lower flow rates (10 or 20 L/min). Longer exposures also yielded greater reductions in pathogen population. With plasma flowing from the gliding arc emitter at 40 L/min, treatments of 3 minutes reduced *Salmonella* by 2.9 to 3.7 log cfu, and reduced *E. coli* O157:H7 by 3.4 - 3.6 log cfu. In that study, the plasma tended to increase the temperature of the treated apples. The maximum temperature of any plasma treated apple (50.8°C, 28°C above ambient) was obtained after a lower plasma flow rate (20 L/min) for 3 min. As this temperature is not high enough to kill either *Salmonella* or *E. coli* O157:H7, the antimicrobial effects obtained in this study were not the result of heat.

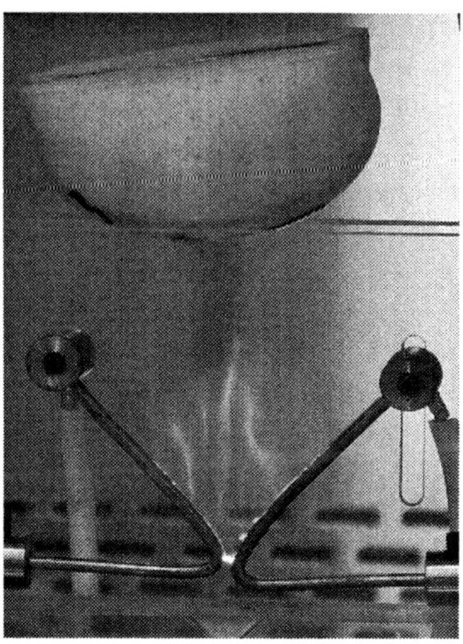

Figure 1. Gliding arc cold plasma applied as an antimicrobial treatment to *Escherichia coli* O157:H7 on the surface of a golden delicious apple.

Cold plasma is a promising new nonthermal process that can effectively reduce human pathogens on fresh produce. Research is ongoing to advance the state of the art in plasma emitter design, and to improve the operational application of the technology to fruits and vegetables.

Irradiation of Fresh Produce

Irradiation is a nonthermal process in which high-energy electrons or photons are applied to foods, resulting in the inactivation of associated pathogens (*15*). An extensive body of research has demonstrated that this technology is safe and effective. In 2008, the FDA approved the use of irradiation up to 4.0 kGy on fresh Iceberg lettuce and fresh spinach, to improve food safety and to extend shelf life (*16*). More recent research has focused on the ability of irradiation to inactivate pathogens within the interior spaces of a leaf, fruit or vegetable. These populations of internalized pathogens are isolated from conventional antimicrobial treatments. The inefficient uptake of bacteria via roots and vasculature make microbiological analysis problematic, complicating research aimed at developing effective interventions. Penetrating processes such as irradiation may be uniquely suited for dealing with this type of contamination, but the literature is as yet scant. In a study by Nthenge et al. (*17*), irradiation was shown to eliminate pathogenic bacteria internalized within leaf tissues as a result of root uptake. Lettuce plants grown in hydroponic solutions inoculated with *E. coli* O157:H7 contained the pathogen in the leaf tissue. Irradiation effectively killed the pathogen while a treatment with 200 ppm aqueous chlorine was ineffective.

In other studies which used a direct inoculation method that introduces inoculum into the leaf intracellular spaces, irradiation was shown to be similarly effective in eliminating internalized *E. coli* O157:H7 from baby spinach and various types of lettuce (Romaine, Iceberg, Boston, green leaf, red leaf), while 300 or 600 ppm sodium hypochlorite was generally ineffective (*18, 19*). D_{10} values for internalized cells (0.30-0.45 kGy) were shown to be 2- to 3-fold higher than for surface associated cells (0.12-0.14 kGy) (*18*). This suggests that the context of the intercellular space may provide a chemical or structural protection for these internalized bacteria. Additional information is necessary to develop a complete understanding of this phenomenon. Pathogen populations within the leaf are expected to be very low in a commercial setting; near-complete elimination of internalized pathogens may therefore be practically achieved using irradiation doses that do not cause undue sensory damage.

Related research has recently begun to assess the ability of irradiation to inactivate biofilm-associated pathogens. Living within the tightly-knit exopolysaccharide matrix of a biofilm serves to protect pathogens from chemical antimicrobial treatments (*20, 21*). Irradiation is a penetrating process, but the data on the efficiency of irradiation in killing biofilm-associated pathogens is very limited. The particular isolate and the biofilm culture conditions (growth temperature, medium, time of cultivation, etc.) can influence irradiation efficacy. Biofilm-associated cells of *Salmonella* were as sensitive or significantly more sensitive to ionizing radiation than respective planktonic cells (*22*). Biofilms of *L. monocytogenes* and *L. innocua* grown at various temperatures were equally or more sensitive to irradiation as planktonic cells

(*23*). For *E. coli* O157:H7, in vitro biofilms cultivated for varying times up to 72 h were either more sensitive or less sensitive to irradiation, depending on the isolate examined (*24*). That study concluded that the response of *E. coli* O157:H7 in the form of a biofilm was influenced by growth conditions, but did not extend the conclusion to address the probable response of in vivo biofilms. Thus, information is not yet available on the specific effect of leaf-surface biofilms on the efficacy of irradiation. The information available in this emerging field of inquiry suggests a complex difference between the two physiological states of these cells, planktonic and biofilm-associated, in their response to irradiation (*24*). Further research is expected to improve our understanding of how biofilms, in particular leaf-, fruit- or vegetable-surface biofilms, may alter the efficacy of irradiation.

Quality of Irradiated Fresh Produce

Recent studies have demonstrated that most fresh-cut fruits and vegetables irradiated at doses of 1 kGy or less did not cause any significant change in appearance, texture, flavor or nutrient quality. Shelf-life of some fresh-cut fruits and vegetables can be extended by low dose irradiation due to the reduction of spoilage microorganisms. For example, Koorapati et al. (*25*) showed irradiation at doses above 0.5 kGy prevented microbial-induced browning and blotching of sliced mushroom. Studies have also shown that irradiated fresh produce may have higher antioxidant content than non-irradiated controls as irradiation increased synthesis of phenolic compounds (*26*).

In some fresh-cut fruits and vegetables, irradiation may cause tissue softening and loss of ascorbic acid (*27*). The losses in quality due to irradiation can be minimized by combination with other sanitizers or techniques such as modified atmosphere packaging (MAP), heat treatment, calcium infiltration and antibrowning agents (*15, 28*). For example, Boynton et al. (*29*) showed that fresh-cut cantaloupes irradiated at 1 kGy in MAP of 4% O_2, 10% CO_2 had the highest rating in sweetness and cantaloupe flavor intensity and lowest in off-flavor after 17 days storage compared to the control and the 0.5 kGy samples.

Formation of Furan due to Irradiation

Furan (C_4H_4O) is regarded as a possible carcinogen according to the Department of Health and Human Services and the International Agency for Research on Cancer, because it causes cancer in animals in studies where the animals are exposed to furan (*30, 31*). This compound is commonly found in foods that have been treated with traditional heating techniques, such as cooking, jarring, and canning (*32, 33*). As a result, both the U.S. Food and Drug Administration (FDA) and the European Food Safety Authority are requesting data and information about furan formation and its toxicity (*34, 35*). Studies showed that ionizing radiation can induce furan formation in solutions of simple sugars and ascorbic acid as well as in fruit juice (*36, 37*), suggesting that the sources of furan upon irradiation are simple sugars such as fructose, sucrose, glucose, and ascorbic acid. Many fruits and vegetables are rich in sugars and ascorbic acid. Fan and Sokorai (*38*) irradiated nineteen fruits and vegetables and measured furan formation in those irradiated produce. Their results showed that irradiation produced low levels of furan in grape and pineapple. In all other fruits and vegetables, furan levels were not detectable or below 1 ng/g. Soluble solids content, titratable acidity and pH varied among the fruits. It appears that the presence of high amount of sugars and low pH are prerequisites for furan formation in fresh-cut produce. Considering the low ng/g of furan detected in grape and pineapple as well as the volatility of furan, irradiation-induced furan is unlikely to be a concern for fresh-cut produce.

Use of Irradiation and Hot Water Treatment on Cantaloupes

Consumption of fresh produce has been linked to outbreaks of foodborne illness and recalls in the U.S. due to contamination with human pathogens. Melons (mostly cantaloupes) are one of the groups of produce that are most frequently associated with outbreaks and contamination with foodborne pathogens (*39, 40*). Between 1990 and 2000, more than 700 cases of salmonellosis were reported in the U.S. and Canada (*41*). The high rates of pathogen contamination associated with melon highlight the need for effective interventions for both whole and cut melons.

Over the last decade, many chemical antimicrobials have been investigated for their effectiveness against human pathogens. However, most chemical interventions have limited effectiveness for reducing the microbiological population on the surface of cantaloupes (*42, 43*), partially due to the rough surface (netting), which provides a protective environment to microbes.

Certain hot water treatments have been shown to effectively reduce human pathogens and native microflora on whole cantaloupes (*44, 45*). Ukuku et al. (*46*) demonstrated that immersion of inoculated cantaloupe in hot water or 5% hydrogen peroxide solution at 70 °C for 1 min resulted in up to a 3.8 log cfu/cm^2 reduction in *Salmonella*. Annous et al. (*44*) reported that surface pasteurization with hot water at 76 °C for 3 min resulted in more than 5 log CFU/cm^2 reduction in *S. enterica serovar* Poona and *E. coli* populations with inoculated cantaloupes.

Fan et al. (*47*) submerged whole cantaloupes into water in the following three conditions: 10 °C water for 20 min (control), 20 ppm chlorine at 10 °C for 20 min, and 76 °C water for 3 min. The hot water significantly reduced both total plate count (TPC) and yeast and mold counts on rind of whole fruits while chlorine or cold water wash did not result in a significant reduction of microbial population. Fresh-cut pieces prepared from hot water-treated cantaloupes had lower TPC than the other two treatments in the later storage periods (day 13-20) (Table I). Lower yeast and mold count on rind of whole fruits due to hot water treatment did not always result in lower count of the microorganism on fresh-cut fruit prepared from the hot-water-treated cantaloupes. Soluble solids content, ascorbic acid content, fluid loss, and aroma and appearance scores were not consistently affected by either hot water or chlorine treatment. Therefore, hot water treatment of whole fruit was superior to chlorine in reducing microbial population of both whole and fresh-cut cantaloupe.

Table I. The Effect of Hot Water Pasteurization of Whole Cantaloupe on Total Plate Count (Log CFU/g) of Fresh-Cut Cantaloupes Stored at 4 °C

Storage time (day)	*Treatments* Cold water control	Chlorine	Hot water
		Trial 1	
1	3.1±0.6 a	2.9±0.7 a	2.9±0.6 a
6	3.8±0.8 a	3.5±0.3 ab	2.3±0.7 b
8	4.2±0.6 a	4.3±0.2 a	3.9±0.5 a
10	4.9±0.5 a	4.8±0.7 a	2.9±0.8 b
13	6.3±0.7 a	5.6±0.8 a	3.6±0.3 b
16	6.9±0.7 a	6.9±0.7 a	2.8±0.3 b
20	7.9±0.7 a	7.8±0.7 a	5.0±1.3 b
LSD	1.1	1.1	1.2
		Trial 2	
1	2.8±1.0 a	2.6±0.9 a	3.2±1.7 a
6	4.3±0.8 a	3.9±1.1 a	2.4±2.1 a
8	4.8±0.9 a	4.9±2.2 a	3.0±0.6 a
10	5.8±0.9 a	6.0±2.4 a	2.7±2.0 a
13	7.3±0.5 a	6.5±1.6 a	3.2±0.8 b
16	7.4±0.5 a	7.4±0.8 a	3.8±0.4 b
20	8.0±0.6 a	7.3±0.5 ab	4.6±2.8 b
LSD	1.4	2.5	2.8

[a]Means with same letters within the same rows are not significantly different ($P>0.05$).
[b]The least significant difference at $P<0.05$ levels for the storage effect.

Fan et al. (*48*) also investigated the feasibility of using hot water treatment in combination with low dose irradiation to reduce native microbial populations while maintaining the quality of fresh-cut cantaloupe. Whole cantaloupes were washed in tap water at 20 or 76°C for 3 min. Fresh-cut cantaloupe cubes, prepared from the washed fruit, were then packaged in clamshell containers, and half the samples were exposed to 0.5 kGy of gamma radiation. Results showed that hot water surface pasteurization reduced the microflora population by 3 logs on the surface of whole fruits, resulting in a lower microbial load on fresh-cut cubes, compared to those from cold 20°C water treated fruit. Irradiation of cubes prepared from untreated fruit to an absorbed dose of 0.5 kGy achieved similar low microbial load of the cubes as those prepared from hot water treated fruit. The combination of the two treatments was able to further reduce the microflora population. Color, titratable acidity, pH, ascorbic acid, firmness, and drip loss were not consistently affected by treatment with irradiation, hot water or the combination of the two. The results demonstrated that the combination of hot water pasteurization of whole cantaloupe and low dose irradiation of packaged fresh-cut melon can reduce the population of native microflora while maintaining quality of this product.

Surface Pasteurization of Fresh Produce using Chemical and Thermal Treatments

Numerous outbreaks of foodborne illness associated with consumption of cantaloupes contaminated with *Salmonella* (*49, 50*) and the detection of *Salmonella* in surveys of imported and domestic cantaloupes (*39, 40*) have focused attention of regulatory agencies and researchers on the problem of melon contamination and disinfection. A large number of sanitizing treatments for cantaloupe have been investigated (*42, 51, 52*), all with limited success achieving no more than 2-3 log (99-99.9%) reductions in pathogen levels and fall short of meeting the FDA's target of 5 log (99.999%) reductions. Therefore, more effective decontamination technologies are needed.

Chlorine is the most widely used sanitizer by the fresh fruits and vegetables industry. Chlorine is only partially effective in reducing populations of *Salmonella* on melons (*51, 53*) and it has the potential of forming harmful byproducts (*54*). But, the benefit of continuing to use chlorine by the produce industry for the prevention of potential cross contamination far outweighs the concern for the potential formation of harmful byproducts. A concern by the produce industry for the potential regulatory constraints on using chlorine in its present form has increased efforts to identify and evaluate alternative sanitation agents.

Electrolyzed water (EW) has been studied as an alternative to chlorine for reducing pathogens and/or spoilage microorganisms on surfaces of fruits and vegetables (*55-57*). Major advantages of using EW over sodium hypochlorite are 1) EW is produced on site by the electrolysis of 1% sodium chloride solution with the help of an electrolysis flow generator, and 2) there is no need for the handling or storage of potentially dangerous sodium hypochlorite in liquid or solid form (*58*).

While chemical sanitation wash treatments are only capable of inactivating bacterial cells attached to the produce surfaces, hot water wash treatments can inactivate bacterial cells below the produce surface (*59*), and thus is a potential alternative to chemical washes (*44, 60*). Unlike steam treatment, hot water immersion provides superior heat transfer between produce and heating medium (*61*) and can quickly establish a uniform temperature profile in produce (*44, 61*). Surface pasteurization of fresh produce using hot water immersion has been used to control insects and is the most effective method for destroying microorganisms, including postharvest plant pathogens that cause spoilage. While hot water or steam surface pasteurization has been shown to be an effective treatment in reducing levels of human pathogens on the surface of meat and poultry (*62, 63*), intact eggs (*64*), and cantaloupes (*24*), it is not usually used in the fresh and fresh-cut produce industry. Fresh fruits and vegetables investigated for surface pasteurization include apples, melons, mango, lemon, orange, cucumber, pear, tomato, and alfalfa seeds.

Annous (*65*) reported that hot water surface pasteurization of cantaloupes at 76°C for 3 min resulted in significantly lower *S.* Poona cell densities as compared to controls (Table II). Although the data indicated that this process did not result in complete inactivation of *S.* Poona cells on cantaloupe surface, five of the six samples tested showed no growth on selective and recovery media. Washing treatments with chlorine, acidic electrolyzed water (AEW), or tap water at RT for 20 min resulted in non-significant reductions of up to 1.2 logs as compared to controls (Table II). These results indicated that these aqueous sanitizers were not able to dislodge/remove *S.* Poona cells attached to the surface of cantaloupes. The same effect was previously seen with cantaloupes inoculated with *S.* Poona (*44, 51, 53*).

Resistance to washing treatments with aqueous sanitizers was reported to be due to the attachment of *S.* Poona cells to inaccessible sites (net-like structure) on the rind of the cantaloupe, thus avoiding contact with the washing solution (*44, 66*). Annous et al. (*44, 66*) reported that *S.* Poona cells started initiating biofilm formation through attachment to the rind of the cantaloupe, using fimbriae, following inoculation and drying for 2 h. Also, these researchers reported that once attached to the rind of the cantaloupe, *S.* Poona cells developed biofilm through growth and excretion of exopolysaccharides (Figure 2). The data presented in Table I indicated that *S.* Poona cells on the surface of cantaloupes were thermally inactivated using the hot water surface pasteurization treatment. Therefore, hot water surface pasteurization of cantaloupes can decrease the risk of foodborne illnesses associated with this commodity as previously reported (*44, 45, 65*).

Table II. Residual Populations of *Salmonella* Poona RM 2350[a] on Artificially Inoculated Cantaloupes Stored for 24 h at 4°C or Room Temperature Prior to Surface Pasteurization and Sanitizing Wash Treatments[b]

	(log CFU/cm²)			
	4°C		*Room temperature*	
Treatment[c]	*XLT-4*	*TSA with XLT-4 overlay*	*XLT-4*	*TSA with XLT-4 overlay*
2 h Control	4.81 ± 0.43a	5.40 ± 0.42a	5.11 ± 0.44ab	5.76 ± 0.50ab
24 h Control	4.18 ± 0.56ac	4.95 ± 0.33ac	6.19 ± 0.31d	6.59 ± 0.50ad
24 h 200 ppm chlorine (RT for 20 min)	3.36 ± 0.15c	3.79 ± 0.03c	5.30 ± 0.39bd	5.74 ± 0.42bd
24 h Acidic electrolyzed water (RT for 20 min)	3.20 ± 0.22c	3.85 ± 0.10c	5.70 ± 0.04bd	5.88 ± 0.08bd
24 h tap water (76°C for 3 min)[d]	0.04 ± 0.02e	0.04 ± 0.02e	1.15 ± 0.63e	1.48 ± 0.53e
24 h tap water (RT for 20 min)	3.22 ± 0.37c	4.59 ± 0.38ac	5.87 ± 0.51bd	6.15 ± 0.56bd

[a] *Salmonella* Poona populations were enumerated on XLT-4 agar medium (non-injured), and on TSA with XLT-4 overlay medium (injured). Cell concentrations were reported as log CFU/cm^2 rind.

[b] Data is reported as the mean ± standard deviation for two separate runs with each run having consisted of three separate cantaloupes. Across all rows and columns, means

with the same letter (a, b, c, d or e) in common are not different ($p < 0.05$) by the Bonferroni T-tests.

[c] Cantaloupes were dip inoculated with *Salmonella* Poona for 5 min, allowed to air dry under a biosafety cabinet for 2 h, and were stored at either room temperature or 4°C for up to 24 h prior to washing treatment at the indicated temperature and time.

[d] Although five of six cantaloupes tested for this treatment showed no growth, 0.04 log CFU/cm^2 (minimum detection level) was taken as the minimum population for determining the mean and standard deviation.

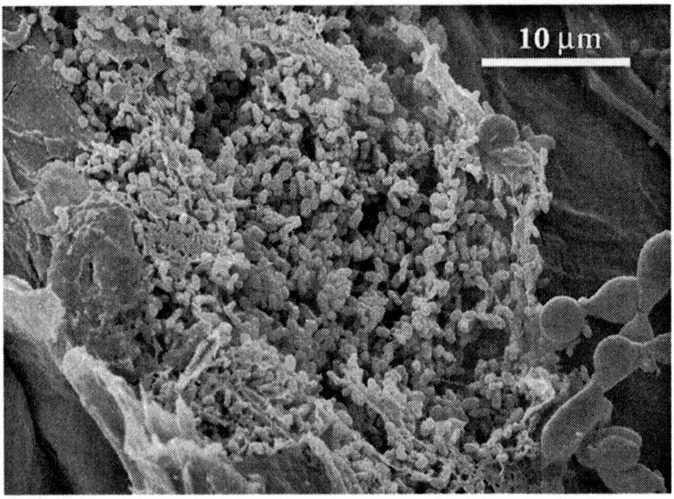

Figure 2. Scanning electron microscopy image showing attachment and biofilm formation by *Salmonella* Poona cells inside the netting of inoculated cantaloupe. Cantaloupes were inoculated and allowed to dry at 20 °C for 72 hours prior to imaging. Note the extracellular matrix encapsulating cells.

Annous (*65*) reported that hot water surface pasteurization treatment resulted in a significant reduction, up to 3.3 logs, in yeast and mold populations on the rind of cantaloupes (Table III). Also, chlorine wash treatment resulted in a significant reduction, up to 1.8 logs, in yeast and mold populations on the rind of cantaloupes (Table III). There was no significant difference between the controls and the remaining washing treatments tested in this study (Table III). Also, fresh-cut cantaloupe prepared from hot water treated melons had an extended shelf-life in excess of 4 weeks as compared to chlorine based treatments (*65*). Fan et al. (*47, 48*) reported that hot water surface pasteurization was able to significantly reduce total aerobic microorganisms on the surface of the rind and on the fresh-cut fruit prepared from the treated cantaloupes. Also, these researchers reported that the total aerobic microorganisms on the packaged fresh-cut fruit were significantly lower than the controls following seven days of storage at 4°C. These results demonstrate the utility of hot water for the inactivation of *Salmonella* on cantaloupes, extending the shelf life of the fresh and fresh-cut cantaloupes, and providing a framework to producers of fresh-cut melon for the potential use of hot water as an intervention treatment for enhancing the microbiological safety of this commodity.

Table III. Residual Populations of Yeast and Molds[a] on Cantaloupes Artificially Inoculated with *Salmonella* Poona RM 2350 and Stored for 24 h at 4°C or Room Temperature Prior to Surface Pasteurization and Sanitizing Wash Treatments[b]

	(log CFU/cm²)	
Treatment[c]	*4°C*	*Room temperature*
2 h Control	4.32 ± 0.47ac	5.00 ± 0.46ac
24 h Control	4.66 ± 0.67ac	5.06 ± 0.37a
24 h 200 ppm Chlorine (RT for 20 min)	2.90 ± 0.57bd	3.67 ± 0.47bc
24 h Acidic electrolyzed water (RT for 20 min)	3.79 ± 0.15abc	3.87 ± 0.64abc
24 h Tap water (76°C for 3 min)	1.89 ± 1.16d	1.79 ± 0.45d
24 h Tap water (RT for 20 min)	3.63 ± 0.29ab	4.84 ± 0.24ac

[a] Native yeast and mold populations were enumerated using 3M yeast and mold Petrifilm, and reported as log CFU/cm^2 rind.

[b] Data is reported as the mean ± standard deviation for two separate runs with each run having consisted of three separate cantaloupes. Across all rows and columns, means with the same letter (a, b, c, d or e) in common are not different ($p < 0.05$) by the Bonferroni T-tests.

[c] Cantaloupes were dip inoculated with *Salmonella* Poona for 5 min, allowed to air dry under biosafety cabinet for 2 h, and were stored at either room temperature or 4°C for 24 h prior to washing treatment at the indicated temperature and time.

Annous and Burke (*65*) reported that washing apples in different sanitizing solutions at 60°C resulted in up to 2.6 log CFU/g more reductions in *E. coli* O157:H7 populations on the apple surface as compared to wash treatments at 25°C (Table IV). However, they reported that these same treatments were not effective in inactivating cells attached to inaccessible sites (stem and calyx) of an apple (Tables V and VI). Similar results related to efficacy of sanitation washes on bacterial cells attached to inaccessible sites (calyx and/or stem) of an apple were previously reported by Annous et al. (*67*). Fleischman et al. (*68*) reported similar results for surface pasteurization of apples using water at 95°C for up to 60 s. Hot water immersion of apples can result in heat damage to the apple resulting in browning of the skin at temperatures above 60°C and softening of the sub-surface flesh above 70-80°C (*69, 70*).

Table IV. Effect of Washing Treatment on Log Reduction[a] in *Escherichia coli* O157:H7 Cell Concentration Applied to the Skin Region of the Apple (Adopted from 4)

Washing solution	*Inoculated control[b]*	*log_{10} CFU/g* — *Washing temperature* 25 °C	60 °C
Tap water	6.37	3.71 ± 0.25ab	4.23 ± 1.24ab
5% Hydrogen peroxide	5.24	3.97 ± 1.20ab	3.74 ± 0.68ab
1200 ppm sanova[c]	5.49	4.38 ± 0.45ab	4.83 ± 0.75a
400 ppm chlorine (pH[d] = 6.5)	5.39	3.00 ± 1.23abc	4.84 ± 0.15a
Acidified electrolyzed water	4.65	1.64 ± 0.19c	4.07 ± 0.37ab

[a] Mean cell population following (duplicate samples) washing treatment minus mean cell population (duplicate samples) of untreated inoculated control. Means with the same letter (a, b or c) in common are not significantly different at $p<0.05$.

[b] Mean populations of untreated inoculated control samples.

[c] Acidified sodium chlorite solution, prepared as per manufacturer's specifications.

[d] pH of the chlorine solution adjusted to 6.5 using concentrated hydrochloric acid.

Table V. Effect of Washing Treatments on Log Reduction[a] in *Escherichia coli* O157:H7 Cell Concentration Applied to the Calyx Region of the Apple (Adopted from 4)

Washing solution	*Inoculated control[b]*	*log_{10} CFU/g* — *Washing temperature* 25 °C	60 °C
Tap water	6.71	0.19 ± 0.18ab	0.43 ± 0.15ab
5% Hydrogen peroxide	5.64	0.39 ± 0.08ab	0.80 ± 0.44ab
1200 ppm sanova[c]	5.80	0.48 ± 0.09ab	1.06 ± 0.14a
400 ppm chlorine (pH[d] = 6.5)	6.11	0.66 ± 0.37ab	0.95 ± 0.28a
Acidified electrolyzed water	5.18	-0.04[d] ± 0.20b	-0.09[e] ± 0.28b

[a] Mean of cell population (duplicate samples) following washing treatment minus means of cell population (duplicate samples) of untreated inoculated control. Means with the same letter (a or b) in common are not significantly different at $p<0.05$.

[b] Mean populations of untreated inoculated control samples.

[c] Acidified sodium chlorite prepared as per manufacturer's specifications.

[d] pH of the chlorine solution was adjusted to 6.5 using concentrated hydrochloric acid.

[e] Negative numbers indicate no reduction in cell populations was detected following washing treatment.

Table VI. Effect of Washing Treatment on Log Reduction[a] in *Escherichia coli* O157:H7 Cell Concentration Applied to the Stem Region of the Apple (Adopted from 4)

		(log₁₀ CFU/g	
		Washing temperature	
Washing solution	*Inoculated control[b]*	*25 °C*	*60 °C*
Tap water	6.37	-0.10[c] ± 0.12d	0.11 ± 0.0.12d
5% Hydrogen peroxide	5.50	1.83 ± 0.17ab	0.96 ± 0.72bc
1200 ppm sanova[d]	5.66	2.24 ± 0.68a	2.04 ± 0.62ab
400 ppm chlorine (pH[e] = 6.5)	6.53	0.49 ± 0.51cd	1.56 ± 0.26abc
Acidified electrolyzed water	5.19	-0.20[c] ± 0.27d	-0.30[c] ± 0.30d

[a] Mean cell population (duplicate samples) following washing treatment minus mean cell population (duplicate samples) of untreated inoculated control. Means with the same letter (a, b, c or d) in common are not significantly different at $p<0.05$.

[b] Mean populations of untreated inoculated control samples.

[c] Negative numbers indicate no reduction was detected following washing treatment.

[d] Acidified sodium chlorite prepared as per manufacturer's specifications.

[e] pH of the chlorine solution was adjusted to 6.5 using concentrated hydrochloric acid.

Hot water treatments of a wide variety of fruits and vegetables can greatly improve microbiological qualities and increase shelf life, while maintaining sensory qualities of the produce. However, over processing the produce can result in thermal injury to apples and juice extracted from treated oranges, and significantly reduces seed germination. These adverse effects can be controlled by the process temperature and time. Since individual commodities have different thermal tolerance, the hot water immersion treatment should be tailored to each commodity. While the rind of a cantaloupe (*44*) and the peel of an orange (*71*) effectively insulated the flesh from thermal damage at temperatures above 70°C, the peel of an apple did not protect the flesh from thermal damage at temperatures above 60°C (*69*). Thus, the tolerance to hot water immersion over a range of temperatures must be determined for individual commodities at different maturity stages (*70*).

Biological Controls

Produce surfaces harbor a wide variety of microorganisms including bacteria, yeasts and fungi. On fresh produce populations total aerobic bacteria ranging from 10^2 to 10^9 cfu per gram of tissue have been reported, with the highest level found on sprouting seeds (*72*). The majority of bacteria recovered are Gram-negative rods and most belong to the genera *Pseudomonas*, *Enterobacter*, *Erwinia* and *Pantoea* (*73*). Other Gram-negative rods identified include *Alcaligenes*, *Chromobacterium*, *Chryseomonas*, *Citrobacter*, *Flavobacterium*, *Klebsiella*, *Serratia*, and *Xanthomonas*. Presence of Gram-positive bacteria on produce is less common and the genera often recovered include *Bacillus*, *Micrococcus*, *Paenibacillus*, *Sarcina* and *Leuconostoc* spp (*74*). Lactic acid bacteria as a group constitute a minor component of microbial community on leafy vegetables (*75*).

Diverse groups of bacteria as outlined above are believed to play an important role in maintaining the quality and safety of fresh produce (*76*). They can enhance, restrict, or show no effect on the survival and growth of enteric pathogens on produce before or after harvest (*75*). Some of them can degrade plant cell walls and alter plant environment to make it more suitable for colonization of enteric pathogens (*61, 77*). The others can suppress the growth of pathogens by producing an array of antimicrobial compounds or by depleting the nutrients or spaces essential for the multiplication of contaminated pathogens (*74*). Enhancing the growth of native or introduced antagonists on fresh produce thus represents a potential strategy to control the proliferation of undesirable microbes on produce.

Native microbial complexes originating from different types of produce including seed sprouts (*78*), cantaloupe (*79*) and baby carrot (*80*) have been shown to inhibit the growth of enteric pathogens both in vitro and in situ. In our laboratory, microflora naturally associated with baby carrot has been the focus of our investigation for at least two reasons. First, raw carrot was one of a few produce not found naturally contaminated with *L. monocytogenes* and other pathogens in a number of field and market surveys (*81*). Secondly, an earlier study by Beuchat and Bracket (*82*) showed that the population of *L. monocytogenes* could be reduced by 2 to 3 log units following a brief exposure of this pathogen to shredded carrot. Results from a series of experiments recently conducted in our laboratory conclude that anti-Listeria activity of fresh carrot was in part due to the antagonistic action of native microflora.

Data as summarized in Figure 3 show that resident microbial complex recovered from baby carrot can inhibit the growth of four major foodborne pathogens grown in sterile carrot extracts. On bell pepper disks co-inoculated with approximately 10^4 cfu/disk of a pathogen and 10^5 to 10^6 cfu/disk of carrot microflora, the growth of each pathogen (*Salmonella*, *L. monocytogenes*, *E. coli* O157:H7, or *Yersinia enterocolitica*) can be reduced by up to 3.9 log units as compared to the growth of pathogen on disks not inoculated with carrot microflora (*80*). It should be noted however that direct use of native microbial complex for control of human pathogens on produce may not be suitable. A substantial proportion of fluorescent pseudomonads associated with raw carrot were pectolytic and capable of producing an array of cell wall-degrading enzymes for induction of tissue maceration or spoilage (*73*). The deleterious effect of these pseudomonads on the quality or shelf life of baby carrot can not be ignored. To avoid the undesirable effect of a sub-proportion of native microbial complex, efforts have been made to isolate individual strains of non-pectolytic antagonists that can be used as candidates for development into biological control agents.

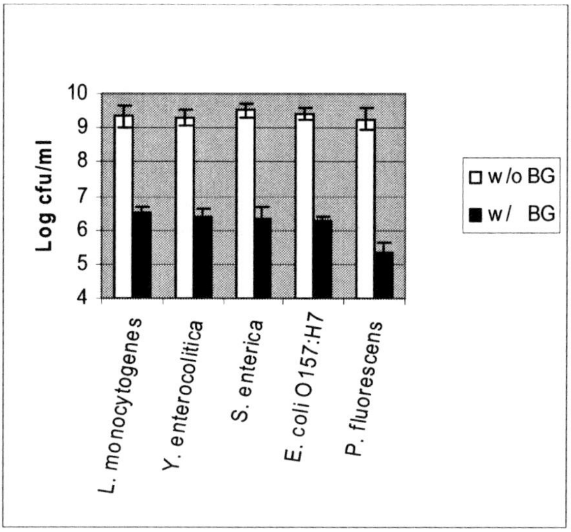

Figure 3. Comparison of the populations of *Listeria monocytogenes, Yersinia enterocolitica, Salmonella enterica, Escherichia coli* O157:H7, and *Pseudomonas fluorescens* grown at 20 °C for 2 days in sterile carrot extract supplemented with or without native microflora (BG) designated respectively, as w/BG or w/o BG as shown above.

Forty-two representative strains were isolated and characterized. Two of them exhibiting the highest degree of anti-Listeria activity by agar spot bioassays were identified as *P. fluorescens* AG3A and *Bacillus* YD1. These two strains were evaluated in conjunction with another known antagonist, *P. fluorescens* 2-79 (or *Pf* 2-79), for their activity to inhibit the growth of pathogens (*L. monocytogenes*, *Y. enterocolitica*, *Salmonella* and *E. coli* O157:H7) and soft-rot bacteria (*Erwinia carotovora* subsp. *carotovora*, *P. marginalis*, and *P. viridiflava*) on three different agar media and also on bell pepper slices. Results show that *Pf* 2-79 was the most effective among the three antagonists tested in producing inhibition zones up to 20 mm in diameter against the growth of human pathogens and soft rot bacteria on three different agar media including Pseudomonas agar F (Difco), Pseudomonas agar F supplemented with 5 mM of FeCl3, and tryptic soy agar.

Using *Pf* 2-79 and bell pepper disks as a model, the effectiveness of this antagonist as a biocontrol agent was largely dependent on the ratio of the number of antagonist to the number of pathogen tested (Table VII). The greatest inhibition was observed when 100-fold higher number of the antagonist than the initial number of pathogen on pepper disks tested. The initial number of pathogen (approximately 10^4 to 10^5 cfu/disk) on pepper disks tested in this study is much higher than that to be expected on naturally contaminated produce (less than 1 cfu in 10 g of seed) (*83*). Dipping of pepper disks in a suspension containing 10^7 cfu/ml of *Pf* 2-79 or Bacillus YD1 for minutes should provide an adequate control of the growth of pathogens to a level that is clinically insignificant.

All three bacterial antagonists examined in recent studies exhibit a broad spectrum of antimicrobial activity but vary with their ability to grow and to suppress pathogens on produce at different temperatures. *Pf* AG3A and Bacillus YD1 inhibited the growth of foodborne pathogens on pepper disks at 20° C but not at 10°C. However, *Pf* 2-79 inhibited the growth of *L. monocytogenes* and *Y. enterocolitica* at either 20 or 10°C by up to 4 log units. Treatment of pepper disks with *Pf* 2-79 could also reduce the incidence of soft rot by 40 to 70%. *Pf* 2-79 is also effective in controlling the growth of psychrotrophic pathogens (such as *L. monocytogenes*, *Y. enterocolitica*, and *Aeromonas* spp.) and spoilage bacteria (such as *P. marginalis* and *P. viridiflava*) on produce that are often stored at refrigeration temperature.

Pf 2-79 was also evaluated for its potential as a biocontrol agent for control of *Salmonella* on sprouting seeds (*84*). Alfalfa seeds containing different concentrations of *Salmonella* ($\leq 10^1$ to 10^3 cfu/g) were co-inoculated with or without *Pf* 2-79 (10^5 cfu/g) and were then subject to sprout at room temperature for 6 days. The changes in the population of *Salmonella* and total bacterial count were monitored daily to determine the growth kinetics of *Salmonella* under the influence of native bacteria. Results show that the population of *Salmonella* on germinating seeds pre-treated with or without *Pf* 2-79 reached the maximum 2 to 3 days after sprouting when total bacterial count on sprouting seeds also reached the maximum of 10^9 cfu/g. The final population of Salmonella on germinating seeds not treated with *Pf* 2-79 showed a net increase of 3 to 4 log units. However, the final population of Salmonella on germinating seeds treated with *Pf* 2-79 showed a net increase of only 1 to 2 log units. Treatment of seeds with *Pf* 2-79 before sprouting can reduce the growth of *Salmonella* on sprouting seed by 2 to 3 log units and the degree of inhibition does not seem to be affected by the initial level of pathogens ($\leq 10^1$ to 10^3 cfu/g) on seeds (Figure 4).

As sanitizer treatment becomes a standard procedure in produce processing, the potential of increased safety risk due to a drastic reduction in the number of competitive native microflora and post contamination of processed produce with human pathogens is an issue of serious concern. Application of biocontrol agents to fresh produce can provide an additional hurdle to inhibit the re-growth of survivor pathogens following the chemical or physical treatment. Interaction with native microflora is important (*85, 86*). Application of antagonists which are psychrotrophic, such as strains *Pf* 2-79 and *Pf* AG3A, are especially useful for control of cold-tolerant pathogens such as *L. monocytogenes*, *Aeromonas* spp. and *Y. enterocolitica* and spoilage bacteria such as *P. marginalis* on fresh produce.

Table VII. Inhibition of *Yersinia enterocolitica, Listeria monocytogenes, Salmonella enterica,* and *Escherichia coli* O157:H7 as Affected by the Ratio of the Number of Antagonists (*Pseudomonas fluorescens* 2-79) to the Number of Pathogens Tested

	Final population of the pathogen (log cfu/disk) [log reduction compared to control C]			
P. fluorescens 2-79: pathogen ratio	*Yersinia enterocolitica*	*Listeria mono.*	*Salmonella enterica*	*E. coli* O157:H7
0 : 1[a]	[b]8.7 ± 0.3a	8.3 ± 0.4a	8.5 ± 0.2a	8.8 ± 0.4a
1 : 1	6.6 ± 0.4b [2.1]	5.4 ± 0.5b [2.9]	6.2 ± 0.4b [2.3]	6.3 ± 0.3b [2.5]
10 : 1	5.0 ± 0.4c [3.7]	5.1 ± 0.1b [3.1]	5.6 ± 0.3c [2.9]	6.0 ± 0.2b [2.8]
100 : 1	5.2 ± 0.3c [3.5]	4.5 ± 0.1c [3.8]	4.4 ± 0.2d [4.1]	5.2 ± 0.5c [3.6]
1,000 : 1	4.9 ± 0.2c [3.8]	4.3 ± 0.4c [4.0]	4.7 ± 0.2d [3.8]	4.9 ± 0.3c [3.9]

[a] Disks inoculated with the pathogen (0:1 ratio) alone were used as control values for calculation. Log reduction from control is presented in brackets for each pathogen/ratio combination.
[b]Values represents the mean of six determinations from two experiments and three duplications (n = 6) ± standard deviation. Values in the same column followed by the same letter (a, b or c) are not significantly different ($p \leq 0.05$).

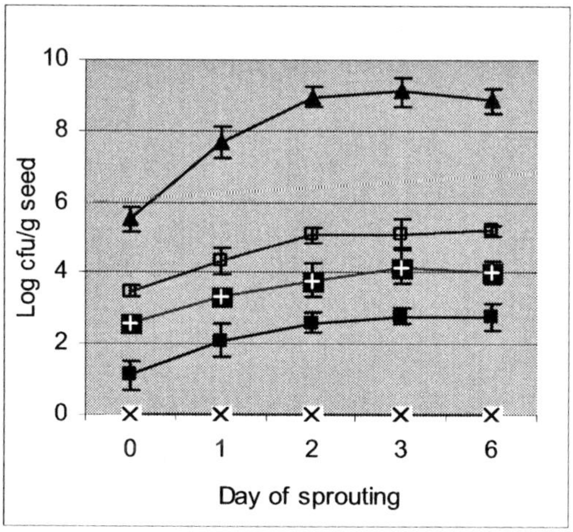

Figure 4. *Salmonella* growth kinetics on sprouting alfalfa seeds inoculated with different concentrations of *Salmonella* ranging from ≤ 10^0 (--×--), 10^1 (--■--), 10^2 (--+ --), to 10^3 (--□--) cfu g^{-1} and 10^5 cfu g^{-1} of *Pseudomonas fluorescens* 2-79 (or Pf 2-79). Change in Pf 2-79 population is shown as (--▲--).

Conclusions

The research described herein is driven by the critical need to develop procedures and processes that will result in safer produce. The fresh and fresh-cut produce industry requires new approaches and new tools to be able to continue to meet the needs of the consumer. These advances must integrate with the economics that exist within the marketplace, but must also offer new opportunities to change the dynamics of conventional production. In the coming years, the industry can take advantage of these technological advances to contribute to the production of fresh, nutritious produce that meets critical food safety goals for the consumer.

Acknowledgements

The authors would like to thank Drs. Nereus Gunther and Lihan Huang for their critical reviews of this chapter. Mention of trade names of commercial products in this publication is solely for the purpose of providing specific information and does not imply recommendation or endorsement by the U.S. Department of Agriculture.

References

1. Sivapalasingam, S.; Friedman, C. R.; Cohen, L.; Tauxe, R. V. Fresh produce: a growing cause of outbreaks of foodborne illness in the United States, 1973 through 1997. *J. Food Prot.* **2004**, *67*, 2342-2353.
2. UFPA (United Fresh Produce Association). Leafy Greens Food Safety Research Conference. **2007**. http://www.unitedfresh.org/newsviews/leafy_greens _food_safety_research (accessed Feb 6, 2008).
3. JIFSAN (Joint Institute for Food Safety and Applied Nutrition). Tomato Safety Research Needs Workshop. **2007**. http://www.jifsan.umd.edu/tomato_wkp2007.htm (accessed Feb 6, 2008).
4. ARS. **2008**. Agricultural Research Service CRIS Project 1935-41420-011-00: "Intervention Technologies for Enhancing the Safety and Security of Fresh and Minimally Processed Produce and Solid Plant-Derived Foods" (accessed September 2, 2008).
5. Birmingham, J. G.; Hammerstrom, D. J. Bacterial decontamination using ambient pressure nonthermal discharges. *IEEE Trans. Plasma Sci.* **2000,** *28*, 51-55.
6. Fridman, A.; Hirokov, A.; Gutsol, A. Non-thermal atmospheric pressure discharges. *J. Phys. D: Appl. Phys*. **2005**, *38*, R1 R24.
7. Gadri, R. B.; Roth, J. R.; Montie, T. C.; Kelly-Wintenberg, K.; Tsai, P.; Helfritch, D. J.; Feldman, P.; Sherman, D. M.; Karakaya, F.; Chen, Z. Sterilization and plasma processing of room temperature surfaces with a

one atmosphere uniform glow discharge plasma (OAUGDP). *Surface Coatings Technol.* **2000**, *131*, 528-542.

8. Niemira, B. A.; Sites, J. Cold plasma inactivates Salmonella Stanley and *Escherichia coli* O157:H7 inoculated on golden delicious apples. *J. Food Prot.* **2008**, *71*, 1357-1365
9. Deng, S. R.; Ruan, C. Y.; Mok, G.; Huang, X. L.; Chen, P. Inactivation of *Escherichia coli* on almonds using nonthermal plasma. *J. Food Sci.* **2007**, *72*, M62-M66.
10. Kelly-Wintenberg, K.; Hodge, A.; Montie, T. C.; Eleanu, L. D.; Sherman, D.; Roth, J. R.; Tsai, P.; Wadsworth, L. Use of a one atmosphere uniform glow discharge plasma to kill a broad spectrum of microorganisms. *J. Vac. Sci. Technol. A* **1999**, *17*, 1539-1544.
11. Sladek, R. E. J.; Stoeffels, E. Deactivation of *Escherichia coli* by the plasma needle. *J. Phys. D: Appl. Phys.* **2005**, *38*, 1716-1721.
12. Montie, T. C.; Kelly-Wintenberg, K.; Roth, J. R. An overview of research using the one atmosphere uniform flow discharge plasma (OAUGDP) for sterilization of surfaces and materials. *IEEE Trans Plasma Sci.* **2000**, *28*, 41-50.
13. Kayes, M. M.; Critzer, F. J.; Kelly-Wintenberg, K.; Roth, J. R.; Montie, T. C.; Golden, D. A. Inactivation of foodborne pathogens using a one atmosphere uniform glow discharge plasma. *Foodborne Path. Dis.* **2007**, *4*, 50-59.
14. Critzer, F. A.; Kelly-Wintenberg, K.; South, S. L.; Golden, D. A. Atmospheric plasma inactivation of foodborne pathogens on fresh produce surfaces. *J. Food Prot.* **2007**, *70*, 2290-2296.
15. Niemira, B. A.; Fan, X. Low dose irradiation of fresh produce: safety, sensory and shelf life. In *Food Irradiation Research and Technology;* Sommers, C. H.; Fan, X., Eds.; Blackwell Publishing: Ames, IA, 2006; pp. 169-184.
16. FDA. 2008. U.S Food and Drug Administration - Final Rule (73 FR 49593), Irradiation in the Production, Processing and Handling of Food 21 CFR Part 179 http://www.cfsan.fda.gov/~lrd/fr080822.html (accessed September 2, 2008).
17. Nthenge, A. K.; Weese, J. S.; Carter, M.; Wei, C. I.; Huang, T. S. Efficacy of gamma radiation and aqueous chlorine on *Escherichia coli* O157:H7 in hydroponically grown lettuce plants. *J. Food Prot.* **2007**, *70*, 748-752.
18. Niemira, B. A. Relative efficacy of sodium hypochlorite wash vs. irradiation to inactivate *Escherichia coli* O157:H7 internalized in leaves of romaine lettuce and baby spinach. *J. Food Prot.* **2007**, *70*, 2526-2532.
19. Niemira, B. A. Irradiation vs. chlorination for elimination of *Escherichia coli* O157:H7 internalized in lettuce leaves: influence of lettuce variety. *J. Food Sci.* **2008**, *73*, M208-M213.
20. Stewart, P. S.; Mukherjee, P. K.; Ghannoum, M. A. Biofilm antimicrobial resistance. In *Microbial Biofilms*; Ghannoum, M. A.; O'Toole, G. A. Eds.; ASM Press: Washington, DC, 2004; pp. 250-268.
21. Robbins, J. B.; Fisher, C. W.; Moltz, A. G.; Martin, S. E. Elimination of *Listeria monocytogenes* biofilms by ozone, chlorine and hydrogen peroxide. *J. Food Prot.* **2005,** *68*, 494-498.

22. Niemira, B. A.; Solomon, E. B. Sensitivity of planktonic and biofilm-associated Salmonella to ionizing radiation. *Appl. Environ. Microbiol.* **2005,** *71*, 2732-2736.
23. Niemira, B. A. Influence of growth temperature on irradiation sensitivity of planktonic and biofilm-associated *Listeria monocytogenes* and *L. innocua*. *Food and Bioprocess Technol.* [Online early access]. DOI 10.1007/s11947-008-0079-5. Published Online: April 2008

24. Niemira, B.A. Irradiation sensitivity of planktonic and biofilm-associated *Escherichia coli* O157:H7 isolates is influenced by culture conditions. *Appl. Environ. Microbiol.* **2007,** *73*, 3239-3244.
25. Koorapati, A.; Foley, D.; Pilling, R; Prakash, A. Electron-beam irradiation preserves the quality of white button mushroom (Agaricus bisporus) slices. *J. Food Sci.* **2004,** 69, S25-S29.
26. Fan, X. Antioxidant capacity of fresh-cut vegetables exposed to ionizing radiation. *J Sci Food Agric.* **2005,** *85,* 995-1000.
27. Fan, X; Niemira, B. A.; Prakash, A. Irradiation of fresh and fresh-cut fruits and vegetables. *Food Technol.* **2008**, *3*, 36-43.
28. Prakash, A.; Foley, D. Improving safety and extending shelf-life of fresh-cut fruits and vegetable using irradiation. In *Irradiation of Food and Packaging: Recent Developments;* Komolprasert, V.; Morehouse, K. M., Eds.; American Chemical Society: Washington, DC, 2004; pp. 90-106.
29. Boynton, B. B.; Welt, B. A.; Sims, C. A.; Balaban, M. O.; Brecht, J. K.; Marshall, M. R. Effects of low-dose electron beam irradiation on respiration, microbiology, texture, color, and sensory characteristics of fresh-cut cantaloupe stored in modified-atmosphere packages. *J. Food Sci.* **2006**, *71*, S149–S155.
30. International Agency for Research on Cancer (IARC). Monographs on the evaluation of carcinogenic risks to humans: dry cleaning, some chlorinated solvents and other industrial chemicals; *IARC*: Lyon, France, **1995**, *63,* 393-407.
31. National Toxicology Program, Furan CAS No.110-00-9. Report on carcinogens, 11th ed., U. S. Department of Health and Human Services, Public Health Service, **2004**. Also available: http://ntp.niehs.nih.gov/ntp/roc/eleventh/reason.pdf (accessed February 21, 2008).
32. FDA. Exploratory data on furan in food data. U.S. Food and Drug Administration. **2004.** http://vm.cfsan.fda.gov/~dms/furandat.html (accessed February 21, 2008).
33. Locas, C. P.; Yaylayan, V. A. Origin and mechanistic pathways of formation of the parent furan-a food toxicant. *J. Agric. Food Chem.* **2004,** *52,* 6830-6836.
34. FDA. Furan in food, thermal treatment; request for data and information. *Fed. Regist.* **2004,** *69*, 25911-25913.
35. European Food Safety Authority. Invitation to submit data on furan in food and beverages. **2006**. http://www.efsa.europa.eu/en/science/data_collection/furan.html (accessed February 1, 2007).

36. Fan, X. Formation of furan from carbohydrates and ascorbic acid following exposure to ionizing radiation and thermal processing. *J. Agric. Food Chem.* **2005,** *53,* 7826-7831.
37. Fan, X. Impact of ionizing radiation and thermal treatments on furan levels in fruit juice. *J. Food Sci.* **2005**, *71,* E409-E414.
38. Fan, X.; Sokorai, K. S. B. Effects of ionizing radiation on furan formation in fresh-cut fruits and vegetables. *J. Food Sci.* **2008**, *73*, C79-C83.
39. FDA. FDA survey of imported fresh produce. FY 1999 field assignment. U.S. Food and Drug Administration. Center for Food Safety and Applied Nutrition. Office of Plant and Dairy Foods and Beverages. **2001**. Available at: http://www.cfsan.fda.gov/~dms/prodsur6.html (accessed June 27, 2008).
40. FDA. Survey of domestic fresh produce: interim results. FY 2000/2001 filed assignment. U.S. Food and Drug Administration. Center for Food Safety and Applied Nutrition; Office of Plant and Dairy Foods and Beverages. **2003**. Available at: http://www.cfsan.fda.gov/~dms/prodsu10.html (accessed June 27, 2008).
41. FDA. Program information manual. **2001**. Available at: http://www.cfsan. fda.gov/~ear/ret-mln.html (accessed November 4, 2008).
42. Sapers, G. M.; Miller, R. L.; Pilizota, V.; Mattrazzo, A. M. Antimicrobial treatments for minimally processed cantaloupe melon. *J. Food Sci.* **2001,** *66,* 345-349.
43. Ukuku, D. O.; Fett, W. F.; Sapers, G. M. Inhibition of Listeria monocytogenes by native microflora of whole cantaloupe. *J. Food Safety* **2004**, *24*, 129-146.
44. Annous, B. A.; Burke, A..; Sites, J. E. Surface pasteurization of whole fresh cantaloupes inoculated with Salmonella Poona or *Escherichia coli*. *J. Food Prot.* **2004,** *67*, 1876-1885.
45. Solomon, E. B.; Huang, L.; Sites, J. E.; Annous, B. A.. Thermal inactivation of Salmonella on cantaloupes using hot water. *J. Food Sci.* **2006,** *71*, M25-M30.
46. Ukuku, D. O.; Pilizota, V.; Sapers, G. M. Effect of hot water and hydrogen peroxide treatment on survival of Salmonella and microbial quality of whole cantaloupe and fresh-cut cantaloupe. *J. Food Prot.* **2004,** *67,* 432-437.
47. Fan, X.; Annous, B. A.; Beaulieu, J.; Sites, J. Effect of hot water surface pasteurization of whole fruit on shelf life and quality of fresh-cut cantaloupes. *J. Food Sci.* **2008,** *73*, M91-98.
48. Fan, X.; Annous, B. A.; Sokorai, K. J.; Burke, A. M.; Mattheis, J. P. Combination of hot water surface pasteurization of whole fruit and low dose irradiation of fresh-cut cantaloupe. *J. Food Prot.* **2006,** *69,* 912-919.
49. Centers for Disease Control and Prevention. Multistate outbreaks of Salmonella serotype Poona infections associated with eating cantaloupe from Mexico---United States and Canada, 2000--2002. *MMWR* **2002,** *51*, 1044-1047.
50. Tamplin, M. Salmonella and cantaloupes. *Dairy Food Environ. Sanit.* **1997,** *17,* 287-286.

51. Annous, B. A.; Sapers, G. M.; Jones, D. M.; Burke, A. Improved recovery procedure for evaluation of sanitizer efficacy in disinfecting contaminated cantaloupes. *J. Food Sci.* **2005**, *70*, M242-M247.
52. Ukuku, D. O.; Fett, W. F. Effect of nisin in combination with EDTA, sodium lactate, and potassium sorbate for reducing Salmonella on whole and fresh-cut cantaloupe. *J. Food Prot.* **2004,** *67*, 2143-2150.
53. Ukuku, D. O.; Sapers, G. M. Effect of sanitizer treatment on Salmonella Stanley attached to the surface of cantaloupe and cell transfer to fresh-cut pieces during cutting practices. *J. Food Prot.* **2001,** *64,* 1286-1291.
54. Richardson, S. D.; Thurston Jr., A. D.; Caughran, T. V.; Collette, T. W; Patterson, K. S.; Lykins Jr., B. W. Chemical by-products of chorine and alternative disinfectants. *Food Technol.* **1998,** *52,* 58-61.
55. Izumi, H. Electrolyzed water as disinfectant for fresh-cut vegetables. *J. Food Sci.* **1999,** *64,* 536-539.
56. Koseki, S.; Isobe, S.; Itoh, K. Efficacy of acidic electrolyzed water ice for pathogen control on lettuce. *J. Food Prot.* **2004,** *67,* 2544-2549.
57. Wang, H.; Feng, H.; Luo, Y. Dual-phasic inactivation of *Escherichia coli* O157:H7 with peroxyacetic acid, acidic electrolyzed water and chlorine on cantaloupes and fresh-cut apples. *J. Food Safety* **2006,** *26,* 335-347.
58. Kim, C.; Hung, Y. C.; Brackett, R. E.; Lin, C. S. Efficacy of electrolyzed oxidizing water in inactivating Salmonella on alfalfa seeds and sprouts. *J. Food Prot.* **2003,** *66,* 208-214.
59. Breidt, E.; Hayes, J. S.; Fleming, H. P. Reduction of microflora of whole pickling cucumbers by blanching. *J. Food Sci.* **2000**, *65*, 1354.
60. Lichter, A.; Dvir, O; Rot, I.; Akerman, M.; Regev, R.; Weisblum, A. Hot water brushing: an alternative method to SO2 fumigation for color retention of litchi fruits. *Postharvest Biol. Technol.* **2000**, *18*, 235.
61. Couey, H.M. Heat treatment for control of postharvest diseases and insect pests of fruits. *HortScience.* **1989** *24, 198-202.*
62. Purnell, G.; Mattick, K.; Humphrey, T. The use of 'hot wash' treatments to reduce the number of pathogenic and spoilage bacteria on raw retail poultry, *J. Food Eng.* **2004**, *62*, 29.
63. Whyte, P.; McGill, K.; Collins, J. D. An assessment of steam pasteurization and hot water immersion treatments for the microbiological decontamination of broiler carcasses *Food Microbiol.* **2003,** *20*, 111.
64. Schuman, J. D.; Sheldon, B. W.; Vandepopuliere, J. M.; Ball Jr., H. R. Immersion heat treatments for inactivation of *Salmonella enteritidis* with intact eggs, *J. Appl. Microbiol.* **1997,** *83*, 438.
65. Annous, B. A. Efficacy of hot water surface pasteurization vs. chlorine wash treatments for reducing populations of Salmonella Poona on inoculated cantaloupe melons. *J. Food Safety* (submitted). **2008**
66. Annous, B. A.; Solomon, E. B.; Cooke, P. H.; Burke, A. Biofilm formation by Salmonella spp. on cantaloupe melons. *J. Food Safety* **2005,** *25,* 276-287.
67. Annous, B. A.; Sapers, G. M.; Mattrazzo, A. M.; Riordan, D. C. R. Efficacy of washing with a commercial flatbed brush washer, using conventional and experimental washing agents, in reducing populations of *Escherichia coli* on artificially inoculated apples. *J. Food Prot.* **2001**, *64*, 159.

68. Fleischman, G. J.; Bator, C.; Merker, R.; Keller, S. E. Hot water immersion to eliminate *Escherichia coli* O157:H7 on the surface of whole apples: thermal effects and efficacy. *J. Food Prot.* **2001**, *64*, 451.
69. Sapers, G. M.; Miller, R. L.; Annous, B. A.; Burke, A. M. Improved antimicrobial wash treatments for decontamination of apples. *J. Food Sci.* **2002**, *67*, 1886-1891.
70. Lurie, S. Postharvest heat treatments. *Postharv. Biol. Technol.* **1998**, *14*, 257–269.
71. Pao, S.; Davis, C. L. Enhancing microbiological safety of fresh orange juice by fruit immersion in hot water and chemical sanitizers. *J. Food Prot.* **1999**, *62*, 756.
72. Fett, W. F. Inhibition of Salmonella enterica by plant-associated pseudomonads in vitro and on sprouting alfalfa seed. *J. Food Prot.* **2006**, *69,* 719-728.
73. Liao, C. H.; Fett, W. F. Analysis of native bacteria and selection of strains antagonistic to human pathogens on fresh produce. *J. Food Prot.* **2001**, *64,* 1110-1115.
74. Nguyen-the, C.; Carlin, F. The microbiology of minimally processed fresh fruits and vegetables. *Crit. Rev. Food Sci. Nutr.* **1994**, *34,* 371-401.
75. Liao, C. H.; McEvoy, J. L.; Smith, J. L. Control of bacterial soft rot and foodborne human pathogens on fresh fruits and vegetables. In *Advances in Plant Disease Management;* Huang, H. C.; Aharya, S. N., Eds.; Research Signpost: Kerala, India; 2003; pp. 165-193.
76. Lund, B. M. Ecosystems in vegetable foods. *J. Appl. Baceriol.* (Symp. Suppl.) **1992,** *73,* 115S-126S.
77. Liao, C. H.; Sapers, G. M. Influence of soft rot bacteria on growth of Listeria monocytogenes on potato tuber slices. *J. Food Prot.* **1999,** *62,* 343-348.
78. Matos, A.; Garland, J. L. Effects of community versus single strain of inoculants on the biocontrol of Salmonella and microbial community dynamics in alfalfa sprouts. *J. Food Prot.* **2005**, *68,* 40-48.
79. Ukuku, D. O.; Fett, W. F.; Sapers, G. M. Inhibition of Listeria monocytogenes by native microflora of whole cantaloupe. *J. Food Safety* **2004**, *24,* 129-146.
80. Liao, C. H. Inhibition of foodborne pathogens by native microflora recovered from fresh peeled baby carrot and propagated in cultures. *J. Food Sci.* **2007**, *72,* M134-M139.
81. Beuchat, L. R. Pathogenic microorganisms associated with fresh produce. *J. Food Prot.* **1996**, *59*, 204-216.
82. Beuchat, L. R.; Brackett, R. E. Inhibitory effect of raw carrots on *Listeria monocytogenes*. *Appl. Environ. Microbiol.* **1990,** *56,* 1734-1742.
83. FDA (U.S. Food and Drug Administration). Microbiological safety evaluations and recommendations on sprouted seeds. **1999,** www.cfsan.fda.gov/mow/sprouts2.html (accessed 7/28/2006).
84. Liao, C. H. Growth of Salmonella on sprouting seeds as affected by the inoculum size, native microbial load and Pseudomonas fluorescens 2-79. *Lett. Appl. Microbiol.* **2008**, *46*, 232-236.

85. Leblanc. Survival and growth of *Listeria monocytogenes* on fresh-cut apple slices and its interaction with *Glomerella cingulata* and *Penicillium expansum*. *Plant Dis*. **2000**, *84,* 177-181.
86. Liao, C. H.; Ukuku, D. K. Role of fluorescent pseudomonads and their pectolytic enzymes in spoilage of fresh and fresh-cut produce. In *Produce Degradation: Pathways and Prevention;* Lamikanra, O.; Imam, S.; Ukuku, D. K., Eds.; CRC Press: Boca Raton, FL, 2005; pp. 483-504.

Chapter 13

CARVER+Shock: Food Defense Software Tool

Phillip Pohl[1], Eric Lindgren[1], Cecelia Williams[1], Malynda Aragon[1], Jeffrey Danneels[1], Robert Browitt[1], Madison Link[1], Regina Hunter[1], Don Kautter[2], Jon Woody[2], Amy Barringer[2], Dave Acheson[2], Cory Bryant[3], Fred Shank[3], Sarah Davis[3]

[1]Sandia National Laboratories, Albuquerque, NM, USA 87185,
[2]Food and Drug Administration, College Park, MD
[3]Institute of Food Technologists, Washington, DC

The CARVER acronym represents the steps in a threat analysis exercise: Criticality, Accessibility, Recognizability, Vulnerability, Effect, and Recuperability. The CARVER+Shock software was developed as an easy-to-use tool for defending food production against malevolent attacks. Shock is added to incorporate the intangible focus of a terrorist in frightening a targeted group. The software uses the CARVER targeting methodology to identify components in a production process that are best suited for security improvements and recommends mitigative steps to improve defense. The food defense software is described and demonstrated on hypothetical yogurt and apple product food-production processes. Two of the algorithms used to generate scores in the analysis are presented and the assumptions listed. The calculated scores are compared and discussed. Version 1.0 of the software is available free from the U.S. Food and Drug Administration website. Software upgrades for pre-harvest agriculture and for retail/restaurants analyses are being developed for release in early 2009.

CARVER Background

The U.S. Food and Drug Administration (FDA) is tasked with protecting the nation's food supply (*1*). The CARVER acronym represents the steps in a threat analysis exercise: Criticality, Accessibility, Recognizability, Vulnerability, Effect, and Recuperability. The CARVER + Shock methodology was employed by the FDA and the U.S. Department of Agriculture (USDA) to assist in defending food-production systems from malevolent acts. Shock is added to incorporate the intangible focus of a terrorist in frightening a targeted group. The method recently was incorporated into stand-alone software, by Sandia National Laboratories, that is user-friendly and is designed to remove the biases that can occur by group execution of the CARVER+Shock methodology. The algorithms that give rise to scores for two of these properties are described below, followed by a case study. The group execution of the CARVER+Shock methodology, a part of the Strategic Partnership Program Agroterrorism (SPPA) typically takes 2 to 3 days' work by 15 to 30 experts (*2*).

Use of the software does not require expertise in risk assessment, chemical processing, or computer science. Rather, the goal of the software is to allow food-production personnel to execute the CARVER+Shock method in a few hours. With the software, a user can modify production design specifications as needed and can evaluate various options as a function of security.

Algorithms

A CARVER+Shock user session starts by gathering information about the process, facility security, and safety of the product being dealt with. The three steps in the session are building a process flow diagram, answering questions regarding the process nodes, and evaluating the results. The major challenge in designing the software is to ensure that the questions, subsequent answers, and reported scores adequately reflect the results of the SPPAs, of which over twenty have been done to date. To do this, algorithms were developed to depict the information flow from the user answers to preliminary calculated variables and finally to a score for each node in a process-flow diagram. Figures 1 and 2 show the algorithms for Criticality and Accessibility. The complexity of each gives the reader an idea of how many questions may be required to determine the score for each property. Note that scores from one property may be used in other scoring algorithms.

Test Processes

These case studies consider two idealized processes: apple packing and yogurt production. Apple packing was chosen because of its simplicity and lack of food processing steps or ingredient additions. Yogurt production was chosen because it includes steps for simple food processing and ingredient addition. Each process is examined on three levels, a small scale representing a local provider, a medium scale representing a regional provider, and a large scale representing a national provider. The batch size for each increment of scale increases by a factor of ten. Each process was also examined under the assumptions of best-case and worst-case security practices. The CARVER+Shock score was calculated using CARVER + Shock Version 1.0.

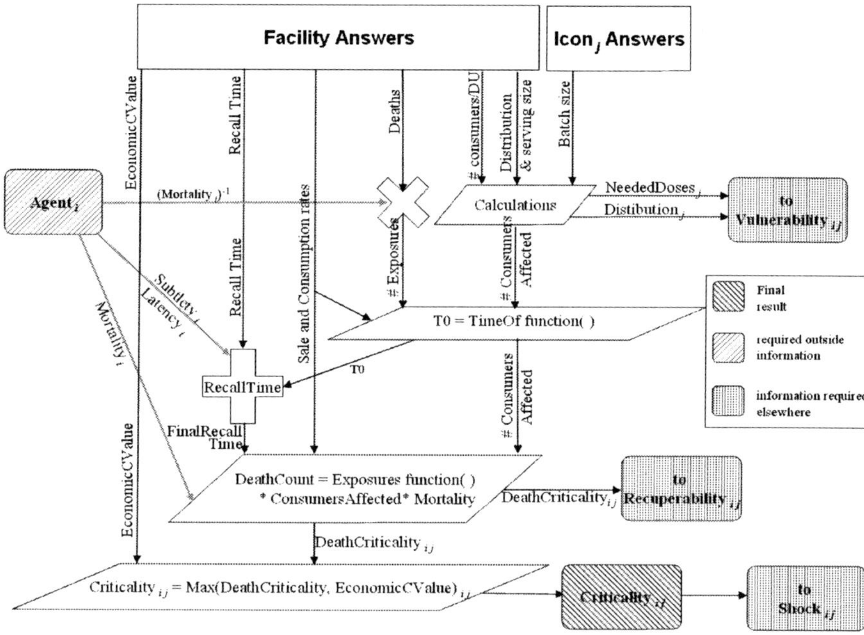

Figure 1 Algorithm for Criticality
(See page 1 of color insert.)

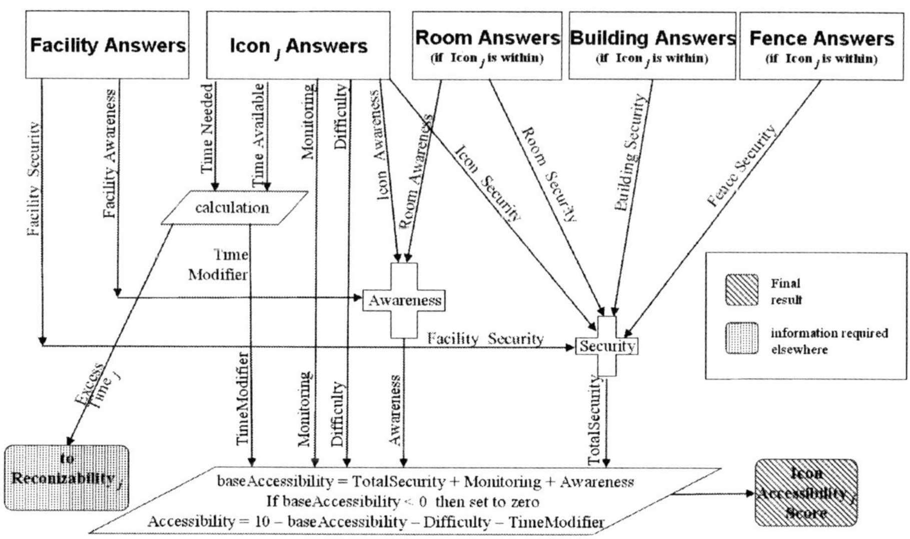

Figure 2 Algorithm for Accessibility
(See page 1 of color insert.)

The process flow diagram for apple packing is shown in Figure 3. The process is very simple, with no added ingredients. The fruit is sorted for size and quality, packed, and placed in storage until delivery to the retailer. The local supplier is assumed to have no refrigerated storage; therefore packed apples are moved directly to the truck for delivery. Contamination would not easily taint an entire batch of apples so it is assumed that a contamination attempt would affect only 1% to 10% of the batch.

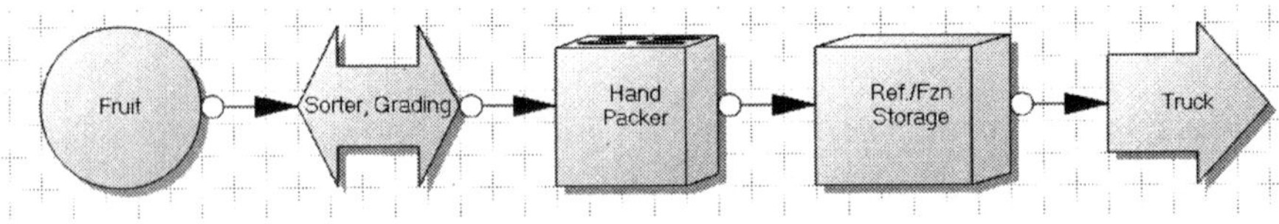

Figure 3. Apple packing process.

The process flow diagram for simplified yogurt production is shown in Figure 4. Raw milk is filtered, chilled, and then pasteurized, which raises the temperature to 85°C for 30 minutes. This heat treatment is much more severe than regular milk pasteurization. The pasteurized milk is cooled and transferred to the culturing tank, where the yogurt culture is added, and the mixture is held at 43°C for 3 to 4 hours. The yogurt is packaged directly from the culturing tank with the fruit being added directly to the containers at the time of packaging. The packaged yogurt is placed in refrigerated storage before distribution. Since yogurt is a fluid product, it is assumed that a contamination event prior to packaging would be uniformly distributed in the entire batch. A contamination event after packaging is assumed to affect only 1% to 10% of the batch based on the educated guess of the CARVER+Shock user.

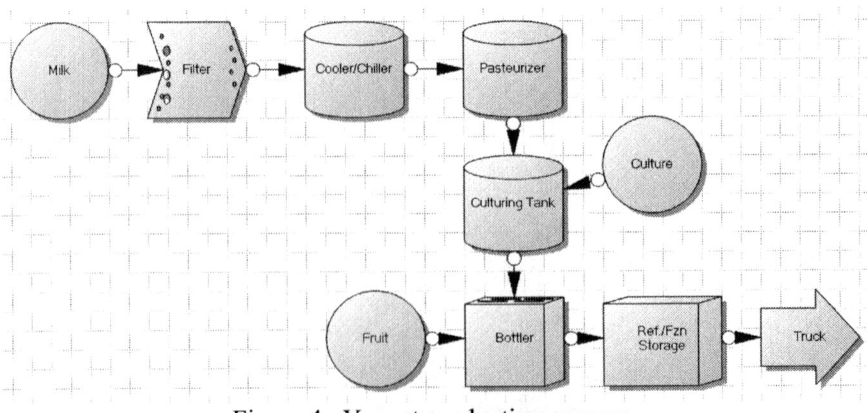

Figure 4. Yogurt production process.

Production Scale Attributes

Both apple packing and yogurt production were analyzed at three levels of production. Table 1 summarizes the general attributes of each production level. The smallest level represents a local producer who sells to one or two stores in a single locale. This producer operates a single small production line and has negligible market share and name recognition. The local producer's batch size is assumed to be 500 pounds of apples or 1,280 fluid ounces (10 gal) of yogurt. The medium level represents a regional producer who sells to a few stores in four regionally located cities. This producer operates a single, large production line and has established a small market share, but has little name recognition. The regional producer's batch size is 5,000 pounds of apples or 12,800 fl. oz. (100 gal) of yogurt. The large-scale operation distributes to many cities nationally. This producer operates multiple, large production lines and has established name recognition and an appreciable market share. The batch size for a single production line is 50,000 pounds of apples or 128,000 fl. oz. (1,000 gal) of yogurt.

Table 1. Production attributes of the three production scales considered

Production Attribute		Production Scale		
		local	regional	national
# Cities supplied		1	4	15
# Outlets/batch		2	10	50
Market share		<1%	1 to 9%	10 to 25%
Name recognition		No	No	Yes
% Production loss		>75%	>75%	15 to 24%
Batch size:	Apples (lbs)	500	5,000	50,000
	Yogurt (fl. oz.)	1,280	12,800	128,000

Contamination Agents

The properties of the five toxic agents considered in each scenario are summarized in Table 2. All agents except Agent 1 survive the heat treatment conditions of the pasteurizer (85 C for 30 min). The relative toxicity provides a measure of the quantity of the agent required for acute poisoning. In particular, the relatively low toxicity of Agent 4 limits the batch size that reasonably can be contaminated.

Table 2. Summary of toxic agent properties

	Max Temp. (Degree C)	Solubility	Relative Toxicity
Agent 1	80	water	high
Agent 2	100	water & oil	medium
Agent 3	100	oil	high
Agent 4	all temps	water	low
Agent 5	all temps	water	very high

Security Practice Scenarios

All of the production scenarios were analyzed for "best" and "worst" security practices. Table 3 summarizes the essence of the two security scenarios. The best security practices scenarios included security personnel and perimeter fences, although the sophistication of the security at the perimeter increased with the size of the operation (local producers had a basic 6 foot fence; regional producers included perimeter lighting; and national producers included security patrols). In the worst security practices scenarios neither perimeter fences nor security personnel were included. The best-case security practice included operation plans—such as plans for food defense, continuity of operation, product recall, and health department coordination plans—along with employee training and practice drills. They also had tight control on shipping and receiving. The best-case practice did not publish any information about the production process or plant location on the internet and did not allow visitors on site.

Table 3. Operation summary of the best case and worst case security practice scenarios

Operation Attribute	Security Practice	
	best case	worst case
Perimeter fence	Yes	No
Security personnel	Yes	No
Plans (defense, continuity of operation, product recall, health department)	Yes	No
Training/drills (security, defense, recall)	Yes	No
Product Traceability	good	poor
Customer support line	Yes	No
Background & drug use checks	Yes	No
Uniforms required	Yes	No
Internet information published	No	Yes
Visitors allowed	No	Yes
Shipping schedule enforced	Yes	No
GPS tracking of shipments	Yes	No
Tamper resistant seals used	Yes	No
Driver ID required	Yes	No
Acceptance testing performed	Yes	No

Production Scale and Security Practice Results

The results of the CARVER+Shock analyses for apple packing and yogurt production are shown in Figures 5 and 6, respectively. These plots show the maximum CARVER+Shock score as a function of batch size for the best and worst security practices. The maximum CARVER+Shock score is taken as the greatest CARVER+Shock score of all the process icons used and all the agents considered.

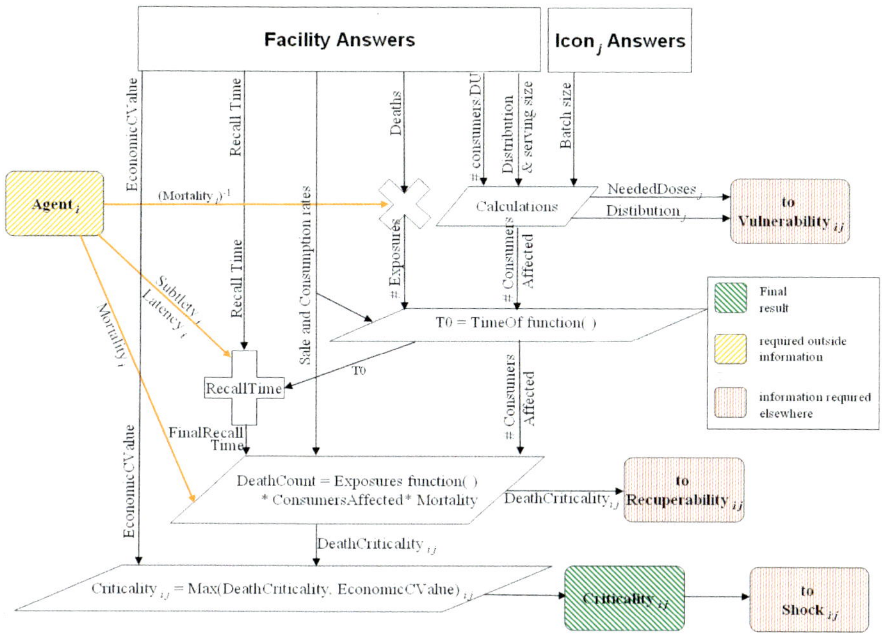

Figure 13.1 Algorithm for Criticality

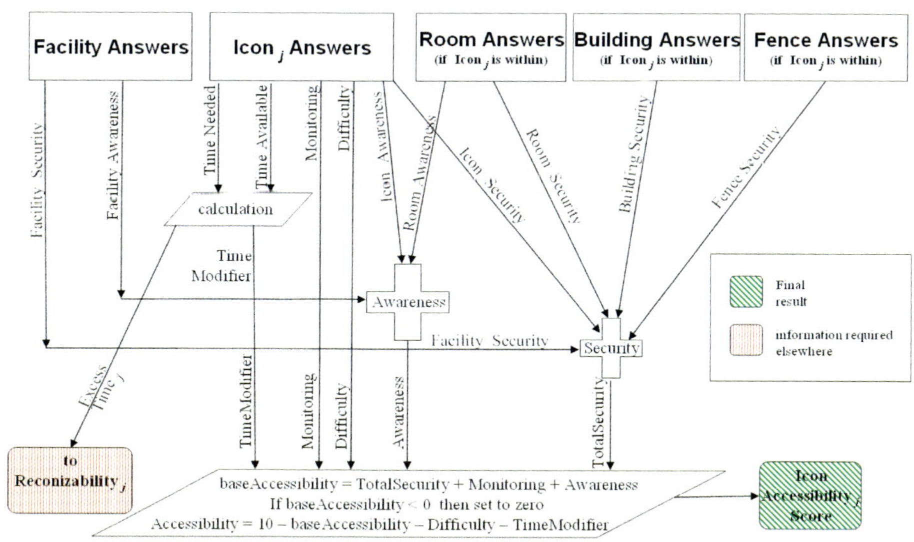

Figure 13.2 Algorithm for Accessibility

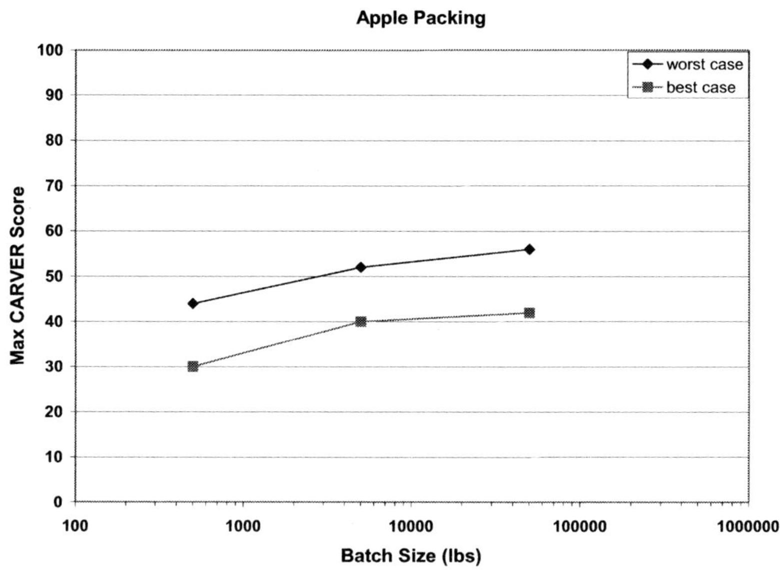

Figure 5. Batch-size dependence of total score for best/worst security practice

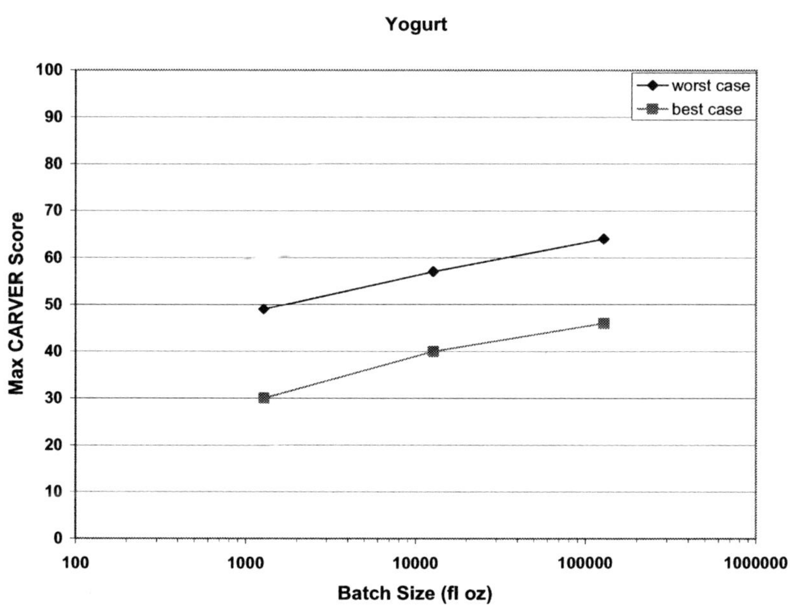

Figure 6. Batch-size dependence of total score for yogurt production security

As expected, for both apple packing and yogurt production, the CARVER scores for the worst security practices are significantly higher than for the best security practices. The best securities practices CARVER scores were all below 50, which is generally considered acceptable; however the results indicate there is possible benefit with improved security practices. The benefit gained by improved security for the yogurt operation (18 points) was greater than the benefit gained for apple packing (13 points). The difference in the improvement may be attributed to the difference in the nature of contamination spreading in apples (15 to 10%) versus yogurt (100%). For both apple packing and yogurt production, the CARVER+Shock score increased with the scale of the production. This trend is expected, because the number of affected individuals scales with the size of the contaminated volume.

Toxic Agent Effects

The effects of the different properties of the various toxic agents considered are best illustrated in the yogurt-production process. Table 4 shows scoring details for the best case scenario of the three levels of production of yogurt for toxic Agents 1, 2, and 4. As shown previously in Table 3, Agent 1 is the most temperature sensitive and would be destroyed in the pasteurizer. Agent 4 is the least toxic and is subject to dilution by large batch sizes. Agent 2 is not destroyed in the pasteurizer and is toxic enough to be effective in the batch sizes considered. The scoring trends of Agents 3 and 5 were similar to Agent 2.

The differences in the toxic agent properties are best seen in the vulnerability scores for the process icons. The effect of temperature susceptibility can be seen by comparing the vulnerability scores for Agent 2 and Agent 1. The vulnerability scores for Agent 2 are high for the culturing tank and other process steps upstream. The vulnerability scores for Agent 1 are high only for the culturing tank, which is immediately downstream from the pasteurizer. The algorithms used in CARVER+Shock look downstream for higher-temperature processing steps and adjust scores accordingly based on the properties of the toxic agent. Since the pasteurizer will destroy Agent 1, only process steps downstream from the pasteurizer have high scores.

Dilution effects are illustrated by the vulnerability scores for Agent 4. The maximum vulnerability score is 6 at the local scale, drops to a score of 4 at the regional scale, and reaches 1 at the national scale.

Table 4 Scoring details for best case yogurt production

	Agent 1								Agent 2								Agent 4							
Local Best	C	A	R	V	E	R	S	Tot	C	A	R	V	E	R	S	Tot	C	A	R	V	E	R	S	Tot
Milk	5	4	2	1	2	5	5	24	5	4	2	1	2	6	5	25	5	4	2	1	2	4	5	23
Filter	5	1	2	1	2	5	5	22	5	1	2	8	2	6	5	30	5	1	2	4	2	4	5	24
Cooler/Chiller	5	1	2	1	2	4	5	20	5	1	2	8	2	5	5	28	5	1	2	4	2	4	5	24
Pasteurizer	6	5	4	1	2	4	6	28	6	5	4	1	2	5	6	29	6	5	4	1	2	4	6	28
Culture	5	1	2	8	2	4	5	28	5	1	2	8	2	5	5	28	5	1	2	6	2	4	5	25
Culturing Tank	5	5	1	1	2	4	5	23	5	5	1	1	2	5	5	24	5	5	1	1	2	4	5	23
Fruit	5	1	2	3	2	4	5	22	5	1	2	3	2	5	5	24	5	1	2	1	2	4	5	20
Bottler	5	2	2	3	2	5	5	24	5	2	2	3	2	6	5	25	5	2	2	1	2	4	5	21
Ref./Fzn Stor.	5	1	1	1	2	5	5	20	5	1	1	1	2	6	5	21	5	1	1	1	2	4	5	19
Max Score	6	5	4	8	2	5	6	28	6	5	4	8	2	6	6	30	6	5	4	6	2	4	6	28
Regional Best	C	A	R	V	E	R	S	Tot	C	A	R	V	E	R	S	Tot	C	A	R	V	E	R	S	Tot
Milk	7	2	2	1	3	10	7	32	7	2	2	1	3	10	7	32	7	2	2	1	3	10	7	32
Filter	7	1	2	1	3	10	7	32	7	1	2	10	3	10	7	40	7	1	2	4	3	10	7	35
Cooler/Chiller	7	1	2	1	3	10	7	32	7	1	2	10	3	10	7	40	7	1	2	4	3	10	7	35
Pasteurizer	7	1	2	1	3	10	7	32	7	1	2	10	3	10	7	40	7	1	2	4	3	10	7	35
Culture	7	1	2	1	3	10	7	32	7	1	2	1	3	10	7	32	7	1	2	1	3	10	7	32
Culturing Tank	7	1	2	10	3	10	7	40	7	1	2	10	3	10	7	40	7	1	2	4	3	10	7	35
Fruit	7	1	1	1	3	10	7	30	7	1	1	1	3	10	7	30	7	1	1	1	3	10	7	30
Bottler	7	1	2	3	3	10	7	34	7	1	2	3	3	10	7	34	7	1	2	3	3	10	7	34
Ref./Fzn Stor.	7	1	2	3	3	10	7	34	7	1	2	3	3	10	7	34	7	1	2	1	3	10	7	32
Truck	7	1	2	1	3	10	7	32	7	1	2	1	3	10	7	32	7	1	2	1	3	10	7	32
Max Score	7	1	2	10	3	10	7	40	7	1	2	10	3	10	7	40	7	1	2	4	3	10	7	35
National Best	C	A	R	V	E	R	S	Tot	C	A	R	V	E	R	S	Tot	C	A	R	V	E	R	S	Tot
Milk	9	2	2	1	4	10	10	38	9	2	2	1	4	10	10	38	9	2	2	1	4	10	10	38
Filter	9	1	2	1	4	10	10	38	9	1	2	10	4	10	10	46	9	1	2	1	4	10	10	38
Cooler/Chiller	9	1	2	1	4	10	10	38	9	1	2	10	4	10	10	46	9	1	2	1	4	10	10	38
Pasteurizer	9	1	2	1	4	10	10	38	9	1	2	10	4	10	10	46	9	1	2	1	4	10	10	38
Culture	9	1	2	1	4	10	10	38	9	1	2	1	4	10	10	38	9	1	2	1	4	10	10	38
Culturing Tank	9	1	2	10	4	10	10	46	9	1	2	10	4	10	10	46	9	1	2	1	4	10	10	38
Fruit	9	1	1	1	4	10	10	36	9	1	1	1	4	10	10	36	9	1	1	1	4	10	10	36
Bottler	9	1	2	3	4	10	10	40	9	1	2	3	4	10	10	40	9	1	2	1	4	10	10	38
Ref./Fzn Stor.	9	1	2	3	4	10	10	40	9	1	2	3	4	10	10	40	9	1	2	1	4	10	10	38
Truck	9	1	2	1	4	10	10	38	9	1	2	1	4	10	10	38	9	1	2	1	4	10	10	38
Max Score	9	1	2	10	4	10	10	46	9	1	2	10	4	10	10	46	9	1	2	1	4	10	10	38

Results of CARVER + Shock Activity

In analyzing the effectiveness of the CARVER+Shock activity, we see that the software program allows users to identify the most critical, vulnerable, or accessible steps in the food-processing systems. The variation of recuperability, effect, and shock scores is negligible amongst the nodes, as is found in many of the SPPA exercises. This lack of variation suggests that modifications of the methodology or possibly the scoring mechanism should be considered.

Following identification of nodes with high scores, the program also gives mitigative information on how to reduce and even prevent potential threats (not shown).

Discussion

The CARVER+Shock scores show variability among processes and among steps within processes with regard to the seven attributes analyzed by the SPPAs. Changing the batch size can affect Criticality. Increasing security can reduce Accessibility or Recognizability. Changing the nature of process steps, if feasible, can reduce Vulnerability or Effect. Modifications of the methodology or the scoring mechanism could increase the sensitivity of Recuperability, Effect, and Shock.

In future versions of the software, the output will be connected to a database of mitigation steps that can be pursued. This database is near completion by the FDA and the scheduled release of version 2.0 of CARVER is early 2009. This version of the software will also include preharvest (horticulture and anumal husbandry) as well as retail/restaurant modules (*3*).

References

1. 1. Food defense at U.S. Food and Drug Administration http://www.cfsan.fda.gov/~dms/defterr.html
2. 2. Strategic Partnership Program Agroterrorism (SPPA) Initiative- First Year Status Report July, 2006, http://www.cfsan.fda.gov/~dms/agroter5.html; Second Year Status Report, September 2007, http://www.cfsan.fda.gov/~dms/agroter6.html.
3. Davis, SF ; McEntire, JC ; Acheson, D ; Busta, F ; Harlander, S ; Acheson, D ; Danneels, J ; Ostfield, M ; Pohl, P ; Hoffman, CJT ; Green, K ; Wankowski, D. International conference focuses on food Defense, *Food Tech*, **2007**, *61*, 125-128.

Chapter 14

Dealing with Intentional and Unintentional Contaminants in Meat and Poultry Products Regulated by the USDA/FSIS

Kerry L. Dearfield[1] and Suzanne Rigby[2]

[1]Office of Public Health Science, Food Safety and Inspection Service, U.S. Department of Agriculture, Washington, DC 20250
[2]Office of Food Defense and Emergency Response, Food Safety and Inspection Service, U.S. Department of Agriculture, Washington, DC 20250

The United States Department of Agriculture's Food Safety and Inspection Service (FSIS) monitors the intentional and unintentional additions to the meat, poultry, and egg products that it regulates. For products that have approved intentional additions (e.g., animal drug residues), FSIS ensures that consumers are not exposed to unsafe levels of these additions. Also, FSIS provides guidance and outreach materials to regulated industry to help guard against intentional additions that may be associated with threat agents. When products in commerce are found to contain unsafe contaminants (chemical and/or microbial), either through intentional or unintentional means, FSIS works with the affected establishments to recall product and informs the public of the public health risk.

Introduction to USDA's Food Safety and Inspection Service

The Food Safety and Inspection Service (FSIS) is the public health agency in the U.S. Department of Agriculture (USDA) responsible for ensuring that the nation's commercial supply of meat, poultry, and egg products is safe, wholesome, and correctly labeled and packaged. FSIS regulatory authority lies primarily with three food safety statutes that cover the products it regulates: the Federal Meat Inspection Act (FMIA), the Poultry Products Inspection Act (PPIA), and the Egg Products Inspection Act (EPIA). These statutes provide FSIS with the public health regulatory context to ensure safe products under its jurisdiction and to provide the basis for effective and practical public health decisions. FSIS is authorized to prevent products from entering commerce that are adulterated or misbranded. Key provisions for this authority can be found in FMIA Section 601(m) and (n), PPIA Section 453(g) and (h), and EPIA Section 1033(a) and (l).

Contamination of the food supply can come from either intentional and/or unintentional inputs. Food safety can be thought as the protection of food products from unintentional contamination by agents that may occur in the food supply (e.g., microbial agents as *E. coli*, *Salmonella*, *Listeria*). Food defense, on the other hand, can be thought of as the protection of food products from intentional contamination by biological, chemical, physical, or radiological agents that are not likely to occur in the food supply (i.e., threat agents such as ricin, arsenic). However, there are intentional additions to the food supply that are regulated to ensure that food doesn't become unfit to consume (e.g., tolerances for residues (pesticides, animal drugs); antimicrobial treatments (bacteriostatic or bactericidal agents like acids, sorbates, benzoates); enhancers (e.g., flavors, for "feel")). Further, other unintentional additions to food may involve levels of residues above permitted tolerances, undesirable foreign materials that adulterate food (e.g., metal, plastic), and exposure to contaminants from accidents (e.g., sewage, hydraulic fluid due to plant mishaps).

Approved Intentional Additions to the Food Supply

Certain additions to food are permitted and these are closely regulated by appropriate federal agencies. Such additions remain in food as residues after animals and plants that animals consume have been exposed. These residues can be animal drugs or pesticide residues. However, before allowing the use of such a chemical on food, a tolerance, or maximum residue limit, which is the amount of chemical residue allowed to remain in or on each treated food commodity, is established by the appropriate authority: the Food and Drug Administration's (FDA's) Center for Veterinary Medicine (CVM) for animal drugs and the Environmental Protection Agency's (EPA's) Office of Pesticide Programs (OPP) for pesticides.

In establishing a tolerance, a "safety" finding is made by considering several factors, including the toxicity of the chemical and its break-down products, how much of the chemical is used and how often, and how much of the chemical (i.e., the residue) remains in or on food by the time it is marketed and prepared. The tolerance, therefore, is the residue level that triggers enforcement actions when necessary.

The FSIS role is to monitor the meat, poultry, and egg products for the amount of residue in these foods via its National Residue Program (NRP). The foundation of the NRP was a government action to control the occurrence of toxic chemicals in the food supply that resulted from the agricultural and industrial use of new chemicals. It was established in 1967 and in 1986, a Memorandum of Understanding (MOU) set the working relationships among FSIS, USDA's Agricultural Marketing Service (AMS), FDA, and EPA. The MOU promotes coordinated federal regulatory activities concerning chemical residues that may adulterate food (i.e., setting of tolerances and monitoring the food supply).

The NRP enforces residue control with two approaches: random, scheduled sampling of healthy appearing animals and egg products; and, targeted, inspector-generated sampling of suspect animals (animals an inspector has determined may not be healthy or has reason to believe residues may be an issue). When monitored residue levels are at or below tolerance levels, these intentional additions do not warrant regulatory action. However, when a residue exceeds its tolerance value, it becomes an unintentional contaminant (i.e., adulterant) and subject to regulatory action (e.g., see recall section later). The NRP publishes its annual findings in a publication, the National Residue Program Data – Red Book (*1*), and its plans for sampling in the coming year in the National Residue Program Sampling Plan - Blue Book" (*2*).

Food Defense of Intentional Contaminants

Relatively few published reports of intentional food contamination are available (*3*, *4*, *5*). However, there have been numerous incidents globally where food has been contaminated with the intention of producing deaths, making people sick, for economic gain, or to create economic loss. FSIS believes that the best defense against intentional contamination is for food industries to build protection around the most vulnerable points in the food farm-to-table continuum as well as monitoring for suspicious activity. To do so, FSIS collaborates with other federal agencies and with industry to identify vulnerable points and possible mitigation(s). The FSIS food defense activities focus on four categories – preparedness, surveillance, response, and recovery.

Preparedness

The Homeland Security Presidential Directive-9 (HSPD-9) requires the food and agriculture sector specific agencies to conduct vulnerability assessments on the farm-to-table continuum (*6*). FSIS relies on the vulnerability assessments results to inform where to concentrate efforts to provide the most resource-effective protection against intentional contamination of food. Results guide the FSIS in establishing policy on food defense activities for personnel, in developing surveillance systems and data gathering, and in developing industry guidance and outreach (*7*).

A vulnerability assessment involves developing a description of the system under study by breaking it down into subsystems, components and finally into the smallest points. Each point is examined for risk by applying a pre-established threat scenario and scores are assigned using pre-established criteria. For the food and agriculture sector, vulnerability assessments are conducted using an offensive tool used by the military – CARVER + Shock (*8*). The tool considers seven factors that affect the desirability of a target where an intentional contaminant can be introduced into food. The scoring uses metrics expressed in mortality and economic loss. Criticality measures the public health and economic impacts to achieve the attacker's intent. Accessibility considers the physical access to the target. Recuperability measures the ability of the system to recover from the attack. Vulnerability is the ease of accomplishing the attack. Effect measures the direct loss from an attack. Recognizability scores the ease of identifying a target. And Shock takes into account the psychological effect of an attack. Some vulnerability assessments FSIS has conducted involve legal and illegal imports, the ground beef production system, hot dogs/deli meat, liquid eggs, breaded chicken nuggets, meals ready-to-eat, and the water supply as an ingredient in food. Per HSPD-9, all vulnerability assessments are updated every two years to consider any new system procedures or new threat agents.

In its work with the intelligence community and research entities, as well as results from vulnerability assessments, FSIS identifies research on specific agents and their properties in FSIS regulated food matrices at different stages of production. The results of the research provide updates to the FSIS maintained list of agents of concern. Agents from this list are used to set the parameters in the vulnerability assessments.

Surveillance

FSIS conducts random laboratory testing of product samples for agents of interest. Procedures are in place to increase the number of laboratory samples tested should the Department of Homeland Security raise the threat level with a specific threat to food and agriculture. FSIS also monitors existing data bases (such as the FSIS Performance Based Inspection System; the FSIS In-Commerce System; and the National Targeting Center's Automated Targeting System for Imports) that collect information on irregularities and anomalies that may identify a potential contaminated food.

Response

FSIS worked with other Federal agencies to develop the National Response Framework outlining what agencies are responsible for ensuring the safety and security of the food supply in the event of a national incident. The response is coordinated with States and local authorities. To assist FSIS in their role, an

Emergency Management Committee (EMC) was established to manage non-routine incidents. The committee, comprised of senior agency officials who are on-call 24/7, is activated when surveillance or a report indicates that a non-routine incident might have occurred that affects meat, poultry, or egg products. The committee, in conjunction with other federal, state, and local agencies, manages FSIS' response to the incident, including recovery through a web-based system, the Non-Routine Incident Management System (NRIMS).

The Food Emergency Response Network (FERN) is another important aspect of the FSIS' response capabilities (*9*). Through that network, FSIS has access to other federal and state food-testing laboratories to handle any surge capacity that might be needed in responding to an intentional contamination event.

Recovery

FSIS is committed to assisting industry to recover and return to operation in an expedited manner. The document Food Disposal and Facility Decontamination Guidelines for Industry, jointly developed with the Food and Drug Administration and the Environmental Protection Agency, was created to provide guidance in recovery operations (*10*). The guidance informs industry on proper disposal for food contaminated with threat agents as well as decontamination guidelines for food processing facilities contaminated with threat agents. FERN also assists by enhancing the ability to restore confidence in the food supply following a threat or an actual emergency targeting the nation's food supply.

Unintentional Contamination by Microbes in Food

The approach FSIS takes to address microbial contamination in food begins with the process of microbiological risk assessment (*11*). Microbiological risk assessment (MRA) provides a scientific process for estimating the probability of exposure to a hazard (microbe) and the resulting public health impact. Essentially, MRA is used to facilitate the application of science to policy formulation and decision making. The Codex Alimentarius risk assessment paradigm is generally followed for conducting an MRA (*12*). This risk assessment paradigm consists of four parts: *hazard identification* and *hazard characterization* (which includes a dose-response assessment) components are combined with an *exposure assessment* to provide a *risk characterization* of the consequence(s) of being exposed to a hazard.

The purpose therefore of an MRA is to inform FSIS decision makers concerning potential microbial risk as well as provide information that is needed to respond to unintentional contamination by microbes in food. MRA helps inform the establishment of industry standards for microbial contamination, allocate inspection resources, guide recall decisions, provide a basis for some trade decisions, target consumer messages, provide industry guidance, and prioritize food safety research. Importantly, MRA supports science-based policies in FSIS. Examples of FSIS risk assessments are posted on the FSIS Internet site (*13*).

Briefly, the risk assessment process in FSIS begins when a microbial food safety issue is identified. The process is planned and scoped where the purpose and scope of a risk assessment and the issues and approach(es) involved in performing the assessment are defined. Also, a clearly articulated purpose and scope provides a sound foundation for later judging the success of the risk assessment and for an effective risk characterization. As part of planning and scoping, risk managers start to formulate the issues and questions that need to be specifically addressed. A risk assessment plan is developed that summarizes the risk management questions discussed during planning and scoping, explains the public health and regulatory context for the food safety issue, and specifies the type of risk assessment to be performed based on the risk management questions and the availability of information. Specifically, it provides a brief description of the situation, product or commodity involved, information on pathways by which consumers are exposed to the hazard, possible risks associated with that exposure, consumer perceptions of the risks, and the distribution of possible risks among different segments of the population.

Once the risk assessment plan is approved, the risk assessment itself is conducted, if needed. In some circumstances, a decision can be made on the available information provided in the risk assessment plan (e.g., a recall decision that needs to be made almost immediately). For a risk assessment, basically, risk assessors analyze all relevant and available data and develop models to address the questions. If data are needed, this information is used to help point to relevant research needs and for prioritization of what research is most critical for risk assessment. The model outputs are used to inform the decision-making process, whether, for example, it is to set safe microbial levels in food commodities or provide industry guidance to keep microbial levels under standards set for microbes in food.

There are different types of risk assessments used depending upon the question(s) being addressed. A *quantitative* MRA provides an estimated risk of illness that is described numerically (e.g., a probability that a person will become sick after a certain exposure). This quantitation of risk can be used to set standards for what level of microbial contamination may be permissible in food commodities. In some instances, the presence of any detectable microbe may not be permissible based on the estimated risk and that food would be considered adulterated if microbes are detected. For example, non-intact raw beef products contaminated with *E. coli* O157:H7 are considered adulterated. This decision was supported by an MRA conducted by FSIS (*14*). Non-intact beef products include ground beef, beef that has been injected with solutions, beef that has been mechanically tenderized by needling, cubing, Frenching, or pounding devices, and beef that has been reconstructed into formed entrees. Intact raw beef products contaminated with *E. coli* O157:H7 that are intended to be processed into non-intact products are also adulterated. A quantitative MRA is considered a fairly in-depth analysis and usually requires a substantial quantity of information and data to conduct. An analysis of uncertainty in the data and models as well as of the variability of the potential responses is critical to quantitative MRA. Uncertainty and variability analyses actually are important in all risk assessments.

Other types of MRA usually don't involve an in-depth quantitation using probabilities of possible risk. Qualitative risk assessments instead use verbal descriptors of risk and severity as well as uncertainty. A *qualitative* MRA describes risk as a likelihood (e.g., high vs. low). Many assessments are actually *safety assessments*, where the amount of hazard (microbe) or level of exposure is compared to an already established safety standard. A safety assessment may be used when an unintentional exposure to a contaminant on or in food is detected above the set safety standard for that contaminant and the decision is made that the food is a high risk and therefore adulterated and subject to recall (see section below on recall activity). It should be noted that if microbial contamination is found before food enters commerce, in many instances it can be diverted to other processes, such as rendering or cooking, where the microbes of concern are killed and the risk removed.

Relative risk ranking assessments compare the relative risk among several hazards. Risk ranking assessments help establish regulatory program priorities and identify critical research needs. For example, the FDA/USDA *Listeria monocytogenes* assessment provides a relative risk ranking of *Listeria* in twenty-some food commodities (*15*). Ready-to-eat (RTE) deli meats ranked as the highest risk to consumers and thus provided FSIS with information to make reduction of *Listeria* in RTE deli meats a very high priority. *Comparative* risk assessments are similar to relative risk ranking, but here the risk of one hazard is compared against the risk of another hazard. For example, comparing the risk of exposure to microbial (e.g., *Listeria*) contamination in RTE meats that are not treated with a bacteriostatic agent to the risk from eating treated RTE product would provide information on what methods may be used to best maintain public health (*16*).

Unintentional Contamination by Chemicals in Food

Similar to dealing with microbial contamination of food, risk assessment provides the basis for dealing with unintentional contamination by chemicals in food. Chemical risk assessment provides the basis for determining tolerable levels of chemicals in the meat, poultry, and egg products regulated by FSIS (see discussion earlier on tolerances). The risk assessment paradigm for chemicals was articulated by the National Research Council (NRC) in its seminal volume *Risk Assessment in the Federal Government; Managing the Process* (*17*; aka "Red Book"). Here, risk assessment is defined as "*the qualitative or quantitative characterization of the potential health effects of particular substances on individuals or populations.*" The chemical risk assessment paradigm consists of four parts including hazard identification, dose-response assessment, exposure assessment, and risk characterization. These four parts are essentially equivalent to the four parts articulated above for MRA.

Once a chemical (e.g., pesticide or animal drug residue) is detected at a level greater than the tolerance level, the food containing that chemical is considered contaminated and therefore adulterated. Detection can occur via the monitoring conducted as part of the NRP (discussed earlier), by routine testing of product by the industry, or by routine testing of product during inspection activities by FSIS. As the food is considered adulterated, the contaminated food becomes subject to a recall decision (see section below on recall activity).

There are other instances where foreign material(s) may appear in food that do not have tolerances, but should not be in food for public health and/or safety reasons. Materials such as metal bits or plastic parts, for example, may or may not be intentionally added to food, but nonetheless make the food unfit for consumption. When these contaminants are detected in food, again, the food is considered adulterated and subject to recall. Also, food may be unintentionally contaminated by accident, for example, animal carcasses in a slaughter facility may be exposed to hydraulic fluid from leaking or ruptured lines and fluid drips or sprays onto the carcasses. While there is probably no tolerance for this type of exposure, there may be components of the fluid that may have been examined (e.g., setting of a reference dose). These components may be examined to see if the exposure was under an established "safety" value. If the food is considered adulterated, an approach to deal with it is considered (e.g., rendering, rehabilitation by removing the exposed parts of the carcass).

Another way to deal with unintentional contamination is to monitor it (e.g., FSIS Dioxin08 survey). While there is not a current "safe" level determined for dioxin in meat and poultry products, exposure to dioxins and dioxin-like compounds should be reduced as far as reasonably possible (*18*). Most human exposure to dioxins is through the diet with major inputs from animal products (meat, fish, and dairy). The Dioxin08 survey goals include obtaining statistically-valid information about current levels of dioxins in domestically-produced meat and poultry, investigating any unusual findings, and comparing these new results with those from the mid-90s and 2002-2003 surveys to determine temporal trends (*19*). When an unusual finding is uncovered, for example, a sample with a level of dioxin congener higher than the rest of the samples (statistically determined to be greater than two standard deviations from the mean of the earlier survey results), then an in-field investigation is initiated to try to determine the dioxin source so it may be eliminated. For example, in the earlier survey (2002-2003), it was discovered that a ball clay component in feed had unusually high levels of dioxin congener. Due to this discovery, the ball clay component of feed was discontinued.

Recall Activity for Unintentional Contamination

One of the major ways that FSIS deals with unintentional contamination of food is through a recall. A food recall is a voluntary action by a manufacturer or distributor to protect the public from products that may cause health problems or possible death. A recall is intended to remove food products from commerce when there is reason to believe the products may be adulterated or misbranded. Recalls are initiated by the manufacturer or distributor of the meat or poultry, sometimes at the request of FSIS. It is important to note that recalls are voluntary; however, if a company refuses to recall its products, then FSIS has the legal authority to detain and seize those products in commerce.

There are four, primary means by which contaminated or improperly labeled meat and poultry products come to the attention of FSIS:

- The company that manufactured or distributed the food informs FSIS of the potential hazard.
- Test results received by FSIS as part of its sampling program indicate that the products are adulterated, or, in some situations, misbranded.
- FSIS field inspectors and program investigators, in the course of their routine duties, discover unsafe or improperly labeled foods.
- Epidemiological data submitted by State or local public health departments, or other Federal agencies, such as the FDA or the Centers for Disease Control and Prevention (CDC) reveal unsafe, unwholesome or inaccurately labeled food.

There are three classes of recalls (Table I). FSIS notifies the public in a press release for Class I and Class II recalls, and a Recall Notification Report (RNR) for Class III recalls. The press release notifications are posted on the FSIS web site, distributed through an email listserve, and when possible, accompanied by pictures of the recalled product. FSIS recall releases include the name of the establishment recalling the meat or poultry, the reason for the recall, a description of the recalled product, any identifying product codes, the recall classification and contact information at FSIS, and the company involved. The RNR provides substantially the same information as the press release; the format is different. If the recalled product was purchased by USDA and distributed through a food distribution program, e.g., the National School Lunch Program, FSIS notifies the Federal agency responsible for the food program, and that agency will hold the product.

Table I. Recall Classes

Class I - A Class I recall involves a health hazard situation in which there is a reasonable probability that eating the food will cause health problems or death
Class II - A Class II recall involves a potential health hazard situation in which there is a remote probability of adverse health consequences from eating the food
Class III - A Class III recall involves a situation in which eating the food will not cause adverse health consequences

After a recall has been issued, FSIS field enforcement personnel conduct "effectiveness checks" to ensure that the recalling firm makes all reasonable efforts to notify the consignees of the recalled product that there is a need to remove the product from commerce. Further, FSIS personnel verify that the recalling firm has been diligent and successful in notifying its customers of the need to retrieve and control recalled products and that the customers have responded accordingly. FSIS actions may include public warnings, product detentions and seizure, or other appropriate actions.

During the recall effectiveness checks, FSIS compiles a list of subsequent recipients as the recalled products are traced through each level of distribution to the retail level. As an enhancement to the recall process, USDA in later 2008 intends to post the list of retail stores and locations receiving meat and poultry products involved in Class I recalls, the highest risk category, on the FSIS Web site within three to ten business days after the recall release is issued. Retail stores are those that sell products to the final consumer and include supermarkets or other grocery stores, convenience stores, meat markets, wholesale clubs, and supercenters. FSIS will not identify distribution centers, institutions, or restaurants, since they prepare food for immediate consumption without packaging that is identifiable or available to consumers.

Summary

FSIS plays a major role in ensuring the safety and wholesomeness of the meat, poultry, and egg products for the U.S. public. It is also responsible for guarding against possible threats to these products in the U.S. food supply. Through the statutory authority given to FSIS, there are certain additions to food that are allowable to maintain a safe and plentiful supply of food. FSIS monitors for levels of these additions to ensure they do not attain unsafe levels. Similarly, FSIS also monitors food for microbial contamination to ensure that unsafe levels are not found. Once meat, poultry, or egg product is found to be contaminated, either intentionally or unintentionally, FSIS works with other federal agencies, States, and local authorities to identify and remove the contaminated product, most usually through a recall action.

Although FSIS does not have statutory authority to mandate food defense against intentional contamination, the agency has taken an active role to protect the food supply under its purview. Guidance and outreach material that has been developed based on vulnerability assessment findings has been provided to industry to voluntary develop countermeasures for intentional contamination during food processing. Additionally, an FSIS threat agent list has been developed and periodically updated when informed by the intelligence community and by research. This list provides the bases for threat agent testing of random product samples adding to the countermeasures against intentional contamination.

References

1. USDA (U.S. Department of Agriculture). *2006 National Residue Program data - Red Book*, **2008**, URL http://www.fsis.usda.gov/Science/2006_Red_Book/index.asp.
2. USDA (U.S. Department of Agriculture). *FSIS National Residue Program Scheduled Sampling Plans - "Blue Book"*, **2008**, URL http://www.fsis.usda.gov/Science/2007_Blue_Book/index.asp.
3. CDC (Centers for Disease Control and Prevention). Nicotine poisoning after ingestion of contaminated ground beef – Michigan, 2003. MMWR Morb. Mortal. Wkly. Rep. **2003**, *52*, 413-416.
4. Kolovec, S.A.; Kimura, A.; Simons, S.L.; Slutsker, L.; Barth, S.; Haley, C.E. An intentional food contamination. *JAMA* **1997**, *278*, 396-398.
5. Torok, T.J.; Tauxe, R.V.; Wise, R.P.; Livengood, J.R.; Sokolow, R.; Mauvias, S. A large community outbreak of salmonellosis caused by intentional contamination of restaurant salad bars. *JAMA* **1997** *278*, 389-395.
6. HSPD (Homeland Security Presidential Directive) 9. *Defense of United States Agriculture and Food*, **2004**, URL http://www.whitehouse.gov/news/releases/2004/02/20040203-2.html.
7. USDA (U.S. Department of Agriculture). *Food Defense and Emergency Response Guidance Material*, **2008**, URL http://www.fsis.usda.gov/Food_Defense_&_Emergency_Response/Guidance_Materials/ index.asp.
8. USDA/FDA (U.S. Department of Agriculture/Food and Drug Administration). *An Overview of the Carver Plus Shock Method for Food Sector Vulnerability Assessments*, **2007**, URL http://www.fsis.usda.gov/PDF/CARVER.pdf.
9. FERN (Food Emergency Response Network). *The Food Emergency Response Network*, **2008**, URL http://www.fernlab.org/index.cfm.
10. USDA (U.S. Department of Agriculture). *Guidelines for the disposal of intentionally adulterated food products and the decontamination of food processing facilities*, **2006**, URL http://www.fsis.usda.gov/PDF/Disposal_Decontamination_Guidelines.pdf.
11. USDA (U.S. Department of Agriculture). *Risk analysis at FSIS: standard operating procedures, USDA, Food Safety and Inspection Service (FSIS)*, **2003**, URL http://www.fsis.usda.gov/OPPDE/rdad/FRPubs/RASOPs.pdf.
12. Codex (Codex Alimentarius Commission). *Principles and guidelines for the conduct of microbial risk assessment, CAC/GL-30 (1999)*, **1999**, URL http://www.who.int/foodsafety/publications/micro/ cac1999/en/.
13. USDA (U.S. Department of Agriculture). *Science: Risk Assessments*, **2008**, URL http://www.fsis.usda.gov/Science/Risk_Assessments/index.asp.
14. USDA (U.S. Department of Agriculture). *Risk assessment of the public health impact of* Escherichia coli O157:H7 *in ground beef*, **2001**, URL http://www.fsis.usda.gov/OPPDE/rdad/FRPubs/00-023N/00-023NReport.pdf.
15. FDA/FSIS (Food and Drug Administration, Food Safety and Inspection Service). *Draft assessment of the relative risk to public health from foodborne* Listeria monocytogenes *among selected categories of ready-to-*

eat foods, FDA Docket No. 99N-1168 and FSIS Docket No. 00-048N, **2001**, URL http://www.foodsafety.gov/~dms/lmrisk.html.

16. USDA (U.S. Department of Agriculture). *FSIS Risk Assessment for* Listeria monocytogenes *in Deli Meats*, **2003**, URL http://www.fsis.usda.gov/PDF/Lm_Deli_Risk_Assess_Final_2003.pdf.
17. *Risk Assessment in the Federal Government: Managing the Process*; NRC (National Research Council); National Academy Press: Washington, DC, 1983.
18. Codex (Codex Alimentarius Commission). *Code of practice for the prevention and reduction of dioxin and dioxin-like PCB contamination in foods and feeds*, CAC/RCP 62-2006, **2006**, URL http://www.codexalimentarius.net/download/standards/10693/CXP_062e.pdf.
19. Hoffman, M.; Huwe, J.; Deyrup, C.; Lorentzsen, M.; Zaylskie, R.; Clinch, N.; Saunders, P.; Sutton, W. Statistically designed survey of polychlorinated dibenzo-p-dioxins, polychlorinated dibenzofurans, and co-planar polychlorinated biphenyls in U. S. meat and poultry, 2002-2003: results, trends, and implications. *Environ. Sci. Technol.* **2006**, *40*, 5340-5346.

Chapter 15

Lead in Food

The Neo-Classical Contaminant

Michael E. Kashtock

Center for Food Safety and Applied Nutrition, U.S. Food and Drug Administration, College Park, MD 20740

This chapter traces the history of U.S. efforts to control lead contamination in food from the early twentieth century to the present. It draws liberally from the earliest records of the Food and Drug Administration and its predecessor agencies. It reviews the multitude of sources of lead's introduction into food that have resulted from the extensive use of lead in the country's industry and agriculture. It also notes key advances in analytical science and the understanding of lead's harmful effects and details how these have impacted efforts to limit lead exposure from food. Written from a regulator's viewpoint, it ascribes due weight to the impact of the law and changes in the law over time. The chapter documents how the challenges of the twentieth century were successfully met and sounds the call of new challenges in the new century as the global food trade brings to the U.S. an increasing flow of products from countries that have yet to effectively control the hazards of lead in their food production systems.

The history of lead and food is a long one. Historians have debated the extent to which plumbism, i.e., lead poisoning, existed due to the use of lead in food and food vessels in the Roman Empire. The subject of this chapter is a comparatively brief portion of that history; it will focus on experiences with lead in food in the United States. In the U.S., efforts addressing lead in food span over 100 years. The ways in which lead has been introduced into foods historically in the U.S. are almost too numerous to list ranging from the wide scale spraying of apples and pears until after World War II with lead arsenate, which was necessary to control economic devastation wrought by the codling moth, to blunders such as the accidental use of commercial lead acetate for making pie meringue, that caused an outbreak of food poisoning in Texas in 1931. Of the latter example, an FDA report stated, "The fact that this poison is commonly known as sugar of lead may furnish a clue to how such a blunder was made *(1)*."

Lead contamination in food was one of the first priorities of the United States Department of Agriculture's Bureau of Chemistry, a predecessor agency to the Food and Drug Administration (FDA) in the aftermath of the passage of the first national food safety law in 1906. As the twentieth century progressed FDA and its predecessors addressed lead contamination in foods from a wide range of sources reflecting the broad use of lead in industry and agriculture. Well before the twentieth century drew to a close it was apparent that the major sources of lead in domestically produced foods had either been eliminated or were strictly controlled so as to pose no significant risk.

However, in short order, the globalization of world food trade brought the challenge of lead in foods to the forefront anew. As food imports into the U.S. have increased, we are discovering that other parts of the world have not necessarily learned what we did decades ago about the need to eliminate or strictly control industrial, agricultural or food processing practices that may contaminate food with lead. While it has been known that there is no established "safe" level of lead intake, the most recent research is showing that subtle adverse developmental effects from this toxin can occur at blood lead levels near those of the population norm. These issues now frame the challenges of addressing lead in food in the current century.

The Pure Food and Drugs Act of 1906: Where it Began

The Pure Food and Drugs Act of 1906 was the first law that forbade the introduction of adulterated food into U.S. commerce. Armed with the authority of this new law, FDA's (this term as used in this Chapter may refer to the FDA proper or to any of its predecessor agencies) three major areas of pursuit in its early years were the patent medicine industry, traffic in filthy and decomposed foods and the contamination of foods with heavy metals due to the use of impure ingredients or utensils in their manufacture.

Early archives of the Bureau of Chemistry document in 1908 an "investigation of the polishing and coating of green coffees with certain colors, which led, in one instance, to the confiscation and destruction of 84 bags of such product coated with lead chromate *(2)*."

Investigations also documented the contamination of some common food ingredients during their manufacture. A 1912 Bureau report stated "During the last year attention has been given to the addition of arsenic and lead to food incidentally in the method of manufacture and without intention or knowledge of the makers. ... Lead is almost universally contained in the tartaric acid on the market, and a study was made of the manufacture of tartaric acid and cream of tartar. The presence of lead was found to be due to the use of lead receptacles and pipes and lead-lined vacuum pans, and to the fact that in the attempt to remove it in the manufacturing establishments the solution to which hydrogen sulphide is added is too concentrated and at too high a temperature *(3)*."

Similarly, a 1913 report stated "The investigation of gelatin has been continued. Results obtained show the presence in gelatin of such metallic impurities as zinc, copper, arsenic, and lead. The cause of these impurities has been shown to be the action of sulphurous acids on the zinc and copper containers during manufacture. The work will enable the manufacturers to avoid introducing these impurities into gelatin *(4)*."

The Bureau worked with the U.S. Army during World War I in studying enamelware and glazed earthenware for their potential to contaminate foods with lead and other metals. Some concern was expressed about the use of lead in some foils used for wrapping food. One Bureau report in 1919 noted that "there are no Federal laws to protect the public against the presence of poisonous substances in articles of common use in the household," however "many foreign countries have long had such legislation on their statute books *(5)*." As noted below, this would not change until 1958.

Lead Arsenate Spray Residue on Fruit: Herald of the Modern Era

Though it seems unfathomable today, for much of the first half of the twentieth century nearly all apples and many other fruits grown in the U.S. were sprayed with lead arsenate to control insects. Lead arsenate was the insecticide of choice for codling moth control in apple orchards. A 1915 FDA report cites annual usage of di-lead arsenate as "several thousand tons *(6)*." After harvest, apples and other fruits were treated with a hydrochloric acid wash process that effectively reduced residues of both lead and arsenic. The consumption of treated but unwashed apples could have fatal consequences. In 1934, it was reported that the death of a boy in West Virginia who consumed sprayed apples off the ground while playing in an orchard was deemed to be due to lead and arsenic poisoning *(7)*.

FDA enforced an administrative tolerance (an informal regulatory level not established via the rulemaking process) for arsenic, but not lead prior to 1933 because rapid and accurate methods of analysis for low levels of lead on fruit did not exist at the time. In addition, the acid wash procedure proportionately removed residues of both lead and arsenic so washed apples that contained acceptable levels of arsenic were not considered to pose a risk for excessive lead exposure. By enforcing the tolerance for arsenic on fruit, FDA could also protect the public against residues of lead.

In 1933, FDA reported that some apple packers had, in the previous year, used other processes for removing arsenic residues that did not concurrently remove residues of lead *(8)*. The changes in fruit washing processes compelled FDA to develop rapid and accurate methods for low levels of lead in fruits. FDA initially published an electrolytic method that could accurately detect as little as 0.05 milligrams of lead *(9)* and soon thereafter published a dithizone colorimetric method capable of detecting a few thousandths of a milligram *(10)*. These developments replaced inaccurate and cumbersome procedures and reduced the time required for analysis to 30 minutes. The availability of suitable methodology also enabled FDA to establish an administrative tolerance for lead in fruit on June 22, 1933 of 0.02 grain of lead per pound of fruit (about 2.85 parts per million) *(11)*. That tolerance would subsequently be reduced to as low as 0.018 in 1935 and then raised to 0.025 in 1938 and finally to 0.05 in 1940. FDA carried out extensive monitoring of apples and apple products such as jams, jellies and apple butter for lead until newer insecticides such as DDT largely replaced the use of lead bearing sprays on apple foliage after World War II. FDA also enforced the tolerance for lead in fruit as a general administrative tolerance for other foods during this period. The era of trace analysis of lead in foods had begun.

The Era of Trace Analysis

With the newly developed analytical methods in hand, FDA began developing data on lead levels in a broad variety of foods for the first time in the 1930s. By 1935, FDA reported that it had tested over 2000 foods from all categories of the diet for lead, with emphasis placed on foods such as milk, candy, chocolate products and jam, which are consumed extensively by children *(12)*. This enabled FDA to develop an understanding of what foods were subject to lead contamination and subsequent investigations could then establish the cause of the lead addition.

Cacao Products

Early studies on cacao products showed that some products, e.g., cocoa beans, cocoa, and chocolate candy, could contain extremely high levels of lead, even in excess of the tolerance for lead in fruit, while many products had insignificant lead levels *(13)*. Investigations of the sources of lead in cacao products determined that in one case soldered joints in manufacturing equipment were likely responsible for lead contamination of chocolate liquor *(14)*. It was also discovered that lead seals attached to bags of cocoa beans before shipment from Africa could become commingled in the process stream and be ground with product during milling *(15)*. Only one 40 grain lead seal (about 2.6 g) could contaminate a one ton lot of cocoa beans in excess of the tolerance for lead in fruit. Shippers quickly replaced the lead seals with ones made from tin or iron.

Sardines

Early studies also showed that while canned sardines in general, did not contain excessive levels of lead, some samples of imported sardines from Portugal, Spain and France had high levels of lead contamination. FDA detained numerous shipments of imported sardines in the 1930s. By 1935, FDA reported that investigations revealed two sources of lead. One was a metallic grid used in processing that contained a lead-tin coating. Speculation had also centered on solder used in certain types of cans used to pack sardines domestically and abroad. Although it was thought that lead levels in the food might be proportional to the amount of exposed solder in the can seams, irregardless of other factors, subsequent investigation revealed that an oleic acid based flux used in soldering the can seams extracted lead from the solder and that residues of the flux contaminated the sardines with excessive amounts of lead *(16)*. The industry corrected the manufacturing practices that gave rise to the contamination and by 1940 FDA found no tested entries of imported sardines to contain excessive lead, although the problem resurfaced to some degree later in the 1940s.

Tea

FDA also reported in 1935 that imported tea wrapped in lead foil or packaged in lead lined containers was subject to lead contamination if the tea was not protected by a barrier of paper or another material *(16)*. One shipment had a lead level above 7 parts per million. Many tea shipments were detained by FDA during this period *(17)*.

Maple Products

In 1937 FDA reported on investigations to establish the causes of lead contamination of maple syrup and maple products, a problem that had been known for some time. The studies determined that the use of teme plate buckets (made with a tin-lead coating) for sap collection and the use of lead paint and lead containing solder on equipment and utensils such as evaporating pans were responsible for the contamination. In particular, if lead containing sap buckets were not frequently emptied, the product would ferment and increase in acidity, which would increase the leaching of lead into the product *(18)*. This problem was not easily addressed by small farmers who collected the sap because most of them were not financially equipped to replace their equipment and utensils.

Because maple products were subject to the general tolerance for lead (e.g., in 1938, FDA reported 22 domestic seizures and 41 import refusals of maple products) much of the syrup purchased from cooperatives by major firms was treated before or after purchase with a process that removed the lead by adsorption onto fine aluminum particles. FDA reported in 1938 that some central collectors of maple products were inspecting their suppliers' operations to provide assurance that their products would not require the costly lead removal treatment *(19)*. It was even reported that some tobacco companies that used maple sugar to treat cigarette and chewing tobacco products required that the sugar comply with FDA requirements *(20)*.

State authorities in maple product producing states carried out educational efforts during the 1930s for farmers on how to replace suspect equipment and utensils. States also offered assistance to farmers and cooperatives in the testing of their products and treatment to reduce lead levels. However, FDA continued to report seizures of domestic maple products into the 1940s as well as detentions of imported products from Canada.

The Foundation of a Policy for Lead in Foods

Some significant things were learned during FDA's first time surveying of foods for lead levels in the 1930s. First it was found that most foods actually showed no indication of problematic amounts of lead contamination. However, the finding of small amounts of lead in many foods led FDA to conclude that "Absolute freedom from lead is impossible of attainment in civilized and perhaps even primitive society because of the widespread occurrence in natural products of minute though appreciable amounts of this metal in the order of a few thousandths of a grain per pound *(21)*."

Thus, FDA recognized from its earliest experiences measuring lead levels in food that food could not be reasonably expected to be free of lead. FDA's policies consequently focused on the following two objectives which continue to be the basis of FDA policy to this day:

1. Preventing the avoidable introduction of lead into food.
2. Controlling the introduction of lead into food where its presence can not be avoided.

For example, because there was no acceptable alternative to the use of arsenicals, e.g., lead arsenate, for insect control on apple foliage, such usage allowed, but the public's exposure to lead and arsenic was limited by washing the fruit. Because alternatives to lead bearing insecticides were available for vegetables, FDA viewed the use of lead bearing insecticides on vegetables as unacceptable.

A key factor to bear in mind is that the multitude of historical sources of lead in food, e.g., insecticides, solder, seals for cocoa bean bags, metal foils, etc. reflects the broad utility of lead in agricultural, food processing and food packaging applications. Although the U.S. would in later years ban or phase out the uses of lead in such applications as new technologies became available to industry, a comparable degree of technological evolution has not necessarily occurred in some parts of the world that are now participating in global food trade. To the extent that lead may still be used in some countries' agriculture and food processing and packaging industries, the food they export may be subject to lead contamination from a variety of sources.

The Boundaries of the Law

While FDA had learned that food contact and packaging materials such as ceramic ware, solder used in processing equipment and metal foil could be sources of significant lead contamination of food, FDA's did not have authority to regulate food contact and packaging materials under the 1906 law and this would remain so until 1958. While FDA was authorized to seize or detain shipments of contaminated food per se, it could not establish specifications for the materials used in food contact applications. What FDA could and did do, was to issue notices to food contact product manufacturers and request that they manufacture products from materials of such quality that they would not contaminate food. This approach was fully consistent with FDA's goals and policies for protecting the public from the hazards of lead in food.

However, one legal change brought about by the 1938 amendments to the 1906 law that established the Federal Food, Drug and Cosmetic Act, was the addition of legal authority permitting FDA to prohibit the sale of adulterated medical devices. Using this new authority, FDA carried out many seizures of lead containing nipple shields for nursing infants beginning in 1939 based upon evidence that they posed a danger of lead poisoning to infants.

Metal Foils Used With Food

Illustrative of FDA's willingness to reach outside of its direct legal authority in the interest of protecting the public, here are portions of the text of an FDA notice to metal foil manufacturers issued in 1936 *(22)*:

- "We have recently collected all of the information which we have secured on the use of metal foil wrappings on foods. While the Food and Drugs Act does not cover containers per se, it does hold to be adulterated any food contaminated with lead from lead-bearing foil as well as from any other sources."
- "Of some 150 samples of metal foil examined, there were not more than 20 per cent showing lead in greater amounts than are found in tin of the highest quality such as is used for tin-plate containers. The lead content of this 20 per cent ranged all the way from a few tenths of 1 percent up to practically pure lead. In a few instances the pure lead foil carried a thin coating of tin. Nearly always the food was protected from the foil by a layer of paper. In the case of dry foods, such as tea and coffee, such protection would probably be fairly effective, unless the protecting paper were inadvertently omitted. In the latter case we know already that tea can take up serious amounts of contaminating lead by attrition, and doubtless coffee would act in the same manner."
- "By far the greater portion of the lead-bearing foils were employed, however, on moist or salty products, principally cheese and cheese mixtures. In a few instances they were employed on confectionary and butter. Careful analyses of the outer and inner portions of the cheese gave occasional evidence of contamination from the foil in spite of the protective paper wrapping. While no contamination of a really serious character has been noted, we believe that you will agree with us in the conclusion that these preliminary findings indicate the wisdom of reducing the lead content of foil food wrappings to the lowest point consistent with the utmost care in the selection of metals for this purpose."
- "We appreciate that your influence can be exerted only to the extent of endeavoring to persuade your food customers against the purchase of impure foils, but this alone will exert a great corrective influence."

Ceramic Teacups

FDA similarly addressed the ceramicware industry by letter in 1938 after it learned of individuals who became ill after consuming tea from poorly manufactured teacups. The glaze on the teacups was shown by testing to leach large amounts of lead and copper into tea (an emetic dose of copper was the likely cause of the illnesses). A 1938 FDA report states "A complaint of illness from the consumption of tea made in cups with a colored glaze led to an investigation of the pottery in question. It was found that tea with added lemon juice dissolved from 15 to 25 parts per million of lead from these cups, besides an amount of copper sufficient to affect the color and flavor of the beverage. While the Food and Drug Act provides no direct jurisdiction over this condition, the matter was taken up with the organized ceramic industry with the result that immediate steps were taken to adopt a form of glaze that would be free from objection on this ground *(23)*."

In the letter itself, FDA stated: "The calling of this matter to your attention is in line with our well-established policy. Similar letters were, in the past, addressed to the manufacturers of metal food containers, and to dealers in second-hand wooden or other food containers, as soon as the danger of contamination of foods from such sources was evident *(24)*."

In summarizing FDA's efforts to regulate lead in food during the first half of the 20th century, it can be said that FDA's actions stemmed from the legal authority vested by the 1906 Pure Food and Drugs Act and were greatly spurred by advances in analytical science during the early 1930s which enabled FDA to effectively monitor the broad food supply for lead levels for the purpose of taking enforcement action against contaminated foods that posed a threat to human health. Figure 1 presents the significant events marking the regulation of lead in food during the first half of the twentieth century in the U.S.

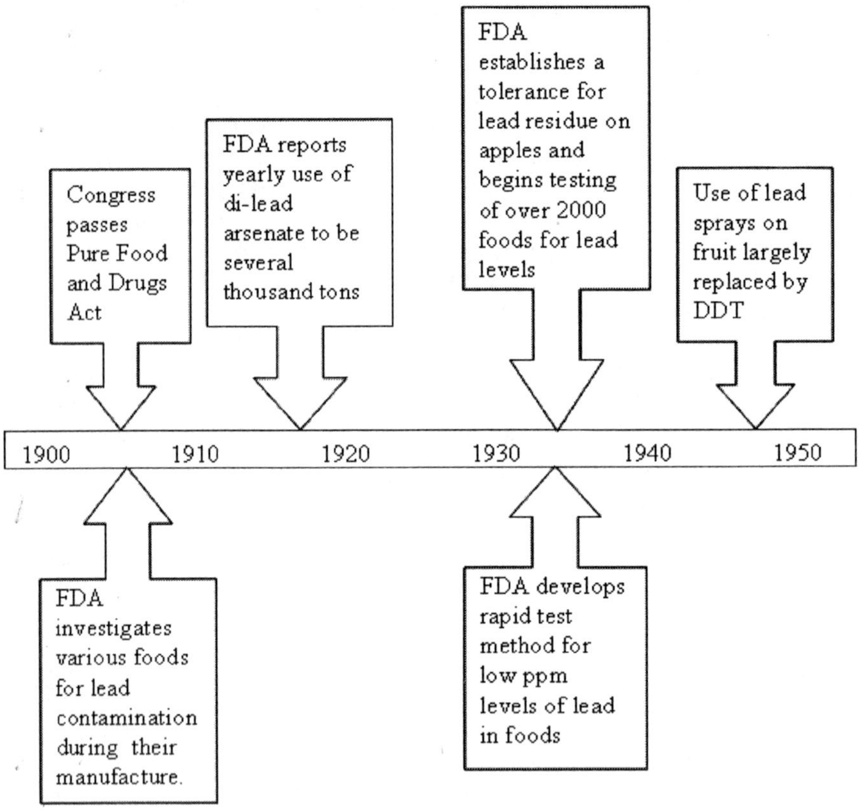

Figure 1. Significant events in the regulation of lead in food in the U.S. from 1900 to 1950.

Latter Half of the 20th Century: The Law and Toxicological Science Become the Key Drivers

In the latter half of the 20th century, significant changes in the U.S. laws and advances in the understanding of lead's adverse effects, particularly in children would become the key drivers of broad efforts to limit the population's exposure to lead, including lead in food.

As noted above, over the years, FDA had addressed numerous matters involving lead contamination of food originating from various types of food contact surfaces, e.g., soldered food cans, soldered seams in processing equipment, metal foil. Because it lacked authority to directly regulate food contact materials, FDA relied on notices to the manufacturers requesting voluntary actions on their part. Amendments to the Federal Food, Drug and Cosmetic Act in 1958 (the 1958 Amendments) for the first time gave FDA the authority to directly regulate the composition of food contact materials. FDA now had the authority to ban the use of food contact materials that were "unsafe" within the meaning of the law. It could also establish specifications for materials to govern their purity and to limit the migration, or leaching, of substances into food.

The latter part of the twentieth century also brought important new findings about the adverse effects of lead. Acute lead poisoning due to exposure from food was not a significant public health problem in the U.S in the twentieth century. Until the 1970s, FDA's policies and programs addressing lead in food were predicated upon protecting the public, and particularly children from the hazards of symptomatic chronic lead poisoning. For example, in reviewing published data on chronic lead poisoning in 1933 *(25)*, FDA noted that persons ingesting as little as 0.1 milligram daily of lead over an average period of 8-1/4 years, were found to have exhibited gross symptoms of lead poisoning, namely, a blue line about the gums, stippled blood cells, and in many instances, pallor, weakness, and digestive disturbances. FDA noted that one moderately sprayed apple might readily contain more than 0.1 milligram of lead unless subjected to appropriate cleansing treatment *(26)*.

Throughout much of the twentieth century, when FDA carried out seizures of shipments of food because of lead contamination, it brought forth expert testimony as necessary to show the court that such actions were justified under the law to protect the public from the danger of symptomatic chronic lead poisoning, i.e., the accumulation of lead in the body over a period of time ultimately resulting in manifestations observable in the clinical setting. In 1933 the head of the FDA stated "We are advised that, in enforcing the Food and Drug Act, the Food and Drug Administration has continuously secured the advice of qualified specialists on matters pertaining to the poisonous effects of insecticidal residues on food products. Through the information and testimony supplied by these experts – including such men as Doctor Carlson of the University of Chicago, Doctor Voegtlin of the U. S. Public Health Service, and the late Doctor Loevenhart of the University of Wisconsin – they have fully substantiated the necessity of the regulations promulgated under the Act and have successfully defended court cases contesting their legality *(26)*."

While it was historically understood that children were particularly susceptible to lead poisoning, by the 1970s, concerns about children at risk increased as the peculiar vulnerabilities of children to lead poisoning became better understood, and more was learned about the effects of lead on behavior and learning in children. In addition, concerns emerged over the potential problems associated with children living in deteriorating dwellings with lead paint. A government estimate in the early 1970s, at which time a blood lead level of 40 μg/dL was considered to be the concern level, indicated that hundreds of thousands of children had elevated lead body burdens or unduly high blood lead levels *(27)*. During the early 1970s, the government initiated several actions aimed at reducing the population's lead exposure. In 1971, the Surgeon General expressed the need to shift the focus of intervention from identifying poisoned children to primary prevention *(28)*. That same year, Congress passed the Lead-Based Paint Poisoning Prevention Act emphasizing prevention of exposure to lead-based paint in housing. Most importantly with respect to lead exposure from food, EPA initiated a phase-out of leaded gasoline and FDA initiated actions to reduce the addition of lead to food from solder used in food cans.

By the 1990s, studies had demonstrated that blood lead levels well below previous concern levels could cause subclinical, i.e., non-symptomatic behavioral and performance deficits, e.g., lower IQ scores, in studied populations of children *(29)*. Other research demonstrated that exposure to low levels of lead in utero could adversely affect the neurobehavioral development of the fetus *(29)*. These findings of subclinical adverse effects in populations have been shown to persist into adulthood *(29)*. Although the Centers for Disease Control in 1991 established a blood lead level of 10 μg/dL as the level of concern, no studies contain evidence of a threshold and recent studies have strengthened the evidence of effects at even lower blood lead levels *(29, 30)*.

The findings on the effects of lead that emerged beginning in the 1970s re-defined the public health objective of regulatory efforts in the U.S. to protect the public from lead exposure from food and other sources. The objective is no longer only to protect the public from chronic lead exposure that might eventually manifest symptomatic lead poisoning e.g., gastrointestinal, and kidney problems, anemia, insomnia and lethargy. The objective now includes protecting the public from lead exposure that could adversely affect a young individual's behavioral and cognitive development, e.g., IQ and learning ability, through the effect of lead upon the developing nervous systems of the fetus and small children. Figure 2 presents the significant events marking the regulation of lead in food during the second half of the twentieth century in the U.S.

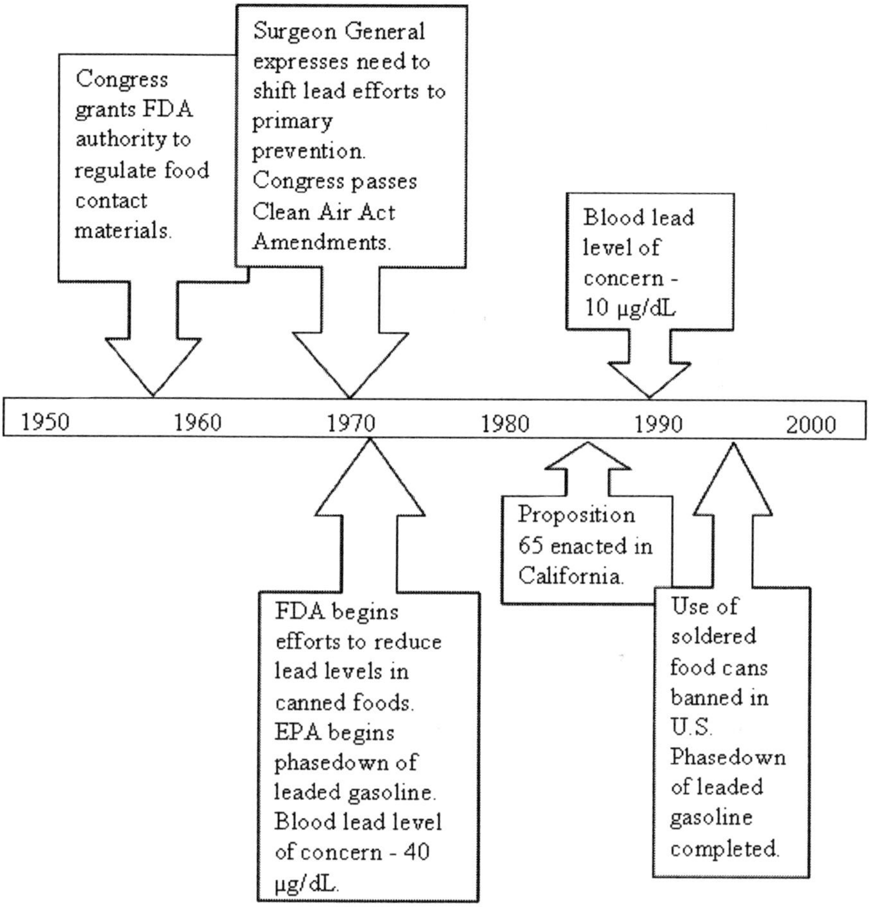

Figure 2. Significant events in the regulation of lead in food in the United States from 1950 to 2000.

Impact of 1958 Amendments: Food Contact Substances Become a Regulatory Focus

Ceramic Dinnerware

The 1970 FDA annual report noted that one of the oldest problems of food safety—lead poisoning from cooking and serving vessels, had recurred with reports in Good Housekeeping magazine of illness traced to lead glazes used on some types of Mexican pottery. Furthermore, subsequent investigations had revealed similar glazes on pottery from other countries *(31)*:

These events led to the establishment of programs in the early 1970s to monitor imported and domestic pottery (ceramic dinnerware) for excessive levels of leachable lead and to remove ceramicware from commerce or refuse its entry into the U.S. if it leached excessive levels of lead that were indicative of inadequate firing of the glaze or the use of improper glaze materials. FDA worked with ceramic dinnerware manufacturers to eliminate manufacturing practices that could lead to excessive leaching of lead. FDA also identified lead containing decals used in ceramicware manufacture as a source of excessive lead leaching and worked with the industry to eliminate this problem. Under FDA's monitoring programs for leachable lead, testing is done by contacting the ware with an acidic leaching solution for 24 hours and then measuring the amount of lead that leaches into the solution expressed as micrograms of lead per milliliter of leach solution (μg/mL). The original limit established in 1970 was 7 μg/mL. That limit has been subsequently lowered on multiple occasions and now ranges from 0.5 to 3.0 μg/mL depending upon the type of ceramic piece being tested *(32)*.

The agency's ability to take enforcement action against non-compliant food contact articles reflects a significant advance in its ability to protect the public afforded by the 1958 Amendments. An important case decision by a Federal District court upheld FDA's authority to regulate ceramicware, and by implication of other food contact articles. In discussing the case FDA's 1974 Annual Report stated "A District Court decided that pottery dinnerware (consisting of plates, bowls, cups, and saucers (intended to be used in the service of foods for human consumption that contains a lead substance which may migrate from the dinnerware to the food being served, and thus cause harm, is subject to the "food" provisions of the Act. The Court ruled that the legislative history of the Food Additives Amendment of 1958 showed a clear Congressional intent that substances which are subject to being ingested by human beings because of migration are "food additives" and thus "foods" within the meaning of the Act *(33)*."

FDA during this time period also established regulatory programs to limit leachable lead from other types of tableware items such as enamelware, pewter, silver plated hollowware. These programs also rely on the authority granted to the agency under the 1958 Amendments.

Food Cans

Throughout much of the twentieth century, two types of food cans were commonly used in the U.S., the vent-hole can (used for evaporated milk) and the sanitary can (the common "tin can"). Both utilized lead solder to either fabricate the can body, apply the can ends or seal the can's opening. The composition of solder used to manufacture food cans was even modified during World War II to contain more lead and less tin, which was needed for the war effort. FDA did not turn its interest toward reducing the migration of lead from soldered cans into food until the 1970s because prior to then it was believed that levels of lead in the food supply were of no safety consequence.

In the 1970s, when it began efforts to reduce the lead content of canned foods, FDA believed that there might be no alternatives to the use of soldered cans in the packing of food. Thus, FDA worked with can manufacturers and food packers to identify improvements in soldered can manufacturing technology to reduce lead levels in canned foods, with foods consumed by infants and children being those primarily targeted.

A quality assurance program initiated with the evaporated milk industry helped to reduce lead levels from 0.52 ppm in 1972 to 0.10 ppm in 1979. The average lead content of canned infant formula concentrate was reduced from 0.1 ppm to about 0.06 ppm during this period. The average lead level in canned juice was reduced from 0.30 ppm in 1972 to 0.05 ppm in 1979 *(34)*. By 1979, the entire infant juice industry had converted its packaging from cans to glass containers. The industry also continued to seek lead reduction through improvements in raw material selection.

During the 1980s, the U.S. can manufacturing industry implemented major changes in can manufacturing technology that brought about the elimination of the use of lead solder in the manufacture of food cans. Foods formerly packed in lead soldered cans were now packed either in cans with welded seams, in two-piece cans without side seams, or in other packaging media such as glass jars or flexible pouches. In 1991, U.S. can manufacturers and processors of canned foods announced that lead soldered food cans were no longer produced in the United States. These remarkable developments occurred less than 20 years after FDA had proposed in 1974 to set a tolerance for evaporated milk and evaporated skim milk in lead soldered cans and had stated in 1979 that it was considering establishing lead limits for other canned foods consumed by infants *(35, 36)*.

By 1992, FDA decided that the use of lead solder should be eliminated from all food cans because it is not required and can be avoided. FDA determined that if all food cans were soldered, the dietary lead intakes for infants, children and women of childbearing age would place infants and children at risk due to the contribution of lead from the solder. FDA initiated a sequence of somewhat complicated proceedings necessitated by the "prior sanctioned," i.e., legally grandfathered, status of soldered food cans, that ultimately resulted in the prohibition of the use of lead solder in food cans in the U.S. in 1995 *(37)*. At the time of its proposal to ban lead soldered food cans, FDA sent letters to over 65 countries alerting them of its intention in this regard. FDA received responses from several countries, e.g., Brazil, Guatemala, Poland and Hungary that indicated that the use of lead soldered cans in those countries would also be discontinued during the 1991-1993 time period.

Other Packaging Materials

As part of a broad agency effort to reduce lead levels in food during the 1990s, FDA also banned the use of tin coated lead foil capsules used for covering the cork and neck area on wine bottles *(38)*. Studies by the United Kingdom's Ministry of Agriculture, Fisheries and Food (MAFF) had demonstrated that significant lead contamination of wine by corrosion products of the capsules that were present on the bottle mouth could occur during pouring of the wine. The MAFF's findings were confirmed in U.S. studies by the Bureau of Alcohol Tobacco and Firearms (BATF). FDA's ban was only applied to wine bottles to which capsules were applied after its February 1996 effective date.

In 1995, FDA learned of a child who was found to have an elevated blood lead level that was attributed to candy he frequently consumed that had become contaminated with lead from a poorly designed candy wrapper that contained a lead based pigment in its outer paper portion. FDA issued a letter to manufacturers, importers and distributors of imported candy and candy wrappers strongly discouraging this industry from using lead in its packaging materials *(39)*. Under the agency's policy as outlined in the letter, if the design or construction of the package permits the migration of intentionally added lead components into the food product, the package is deemed to be "unsafe" and in violation of the law and the food is subject to FDA regulatory action.

Impact of the Phase-out of Leaded Gasoline

In 1971, the Administrator of the newly established Environmental Protection Agency (EPA) recognizing the emerging consensus on the effects of lead stated that "Only very recently have scientists been able to prove that low-level lead exposure resulting from automobile emissions is harmful to human health in general, but especially to the health of children and pregnant women *(40)*."

EPA published a position paper on this issue in 1973 *(41)*. In December of 1973, under the Clean Air Amendments of 1970, EPA issued initial regulations calling for a gradual phased own in the lead content of the total gasoline pool. Though the phase down was not completed until 1996, by 1986, the EPA requirements had eliminated 98 percent of the lead from gasoline compared to 1970 usage *(42)*. The completion of the phase-out was announced in an EPA statement in 1996. In recounting the history of the effort the agency stated that the average lead content in gasoline in 1973 was 2-3 grams per gallon or about 200,000 tons of lead a year. Beginning in 1975, passenger cars and light trucks were manufactured with a more elaborate emission control system which included a catalytic converter that required lead-free fuel. By 1995, leaded fuel accounted for only 0.6 percent of total gasoline sales and less than 2,000 tons of lead per year *(43)*."

The phase-out of leaded gasoline significantly impacted one of the major sources of lead in food. In the mid-1980s, it was estimated that anthropogenic lead aerosols accounted for 40% of the lead in food. Lead contamination from this route affected numerous food commodities. Studies had shown that the lead content of crops correlated with their proximity to anthropogenic sources of lead and that lead levels on the surfaces of vegetation were proportional to air lead concentrations. Other studies had shown that lead concentrations were the highest in leafy crops (that have high surface to volume ratios), i.e., that are more subject to atmospheric deposition. Conversely, the lowest concentrations were found in root crops which grow in the soil and are not subject to surface deposition. Grazing animals were also contaminated with lead from the consumption of forage and feed contaminated by atmospheric lead deposits *(44)*.

Broader Lead Reduction Efforts in the 1990s and Beyond

As studies continued to demonstrate adverse effects of lead on the developing nervous system at lower lead body burdens, FDA in the 1990s undertook additional efforts to reduce lead levels in food. FDA tightened its lead leaching limits for ceramic dinnerware in 1992. It lowered the allowable level for lead in bottled water in 1995 *(45)*. FDA also imposed limits on lead in wine in 1991 and issued guidance to the states on the public health significance of lead levels in shellfish in 1993 *(46, 47)*. After initially addressing concerns about lead in imported candy and candy wrappers in a 1995 letter to the industry, FDA subsequently tightened its limit for lead in candy in a guidance document issued in 2006 *(48)*. While not directed at sources that posed a potential for significant lead exposure to the broad population, these additional actions did prohibit as a matter of law, certain uses of lead that were no longer necessary, e.g., tin coated lead foil seals for wine bottles, and they also established feasibility based limits that ensured that other occurrences of lead in food would be governed by strict limitations in keeping with the best available technology, e.g., ceramic dinnerware. Actions addressing imported candy may have had a significant impact on lowering the dietary lead intake of population subgroups such as Hispanic children who frequently consumed imported candy from Mexico which was prone to lead contamination problems.

Tracking the Progress of Dietary Lead Reduction Efforts

FDA conducts a Total Diet Study (TDS), sometimes called the market basket study. The TDS is an ongoing FDA program that determines levels of various analytes in foods. From this information, dietary intakes of those analytes by the U.S. population are estimated for various age/gender groups. Since its inception in 1961 as a program to monitor radioactive contamination of foods, the TDS has grown to encompass pesticide residues, industrial chemicals, and toxic and nutrient elements. A unique aspect of the TDS is that foods are prepared as they would be consumed (table-ready) prior to analysis, so the analytical results provide the basis for realistic estimates of the dietary intake of these analytes.

The TDS design allows for the identification of trends in dietary lead intake reduction corresponding with the dietary lead reduction efforts that began in the 1970s. In 1996, FDA published TDS data illustrating a marked reduction in dietary lead intakes for 14-16 year old males from 38 µg/day in 1982-1984 to 3.2 µg/day in 1990-1991 *(49)*. FDA stated that in the decade prior to 1982, the TDS dietary intake of lead by teenage males, measured under a somewhat different scheme, was 60 to 90 µg/day. FDA highlighted the data for teenage males because this group had the longest continuous reporting in the TDS and the data illustrated a remarkable reduction in dietary lead intake during the time period corresponding to the phase down of leaded gasoline and the elimination of lead soldered food cans in the U.S.

FDA stated that similar reductions in dietary lead have been seen in infants, young children or toddlers and women of different age groups and that in absolute terms, the dietary lead intake of these groups had been reduced from 34 to 44 μg/day in 1980 to 2 μg/day by 1990-1991. These reductions were stated to have leveled off, or continued at a reduced rate through the mid-1990s. Indeed, FDA TDS data published in 2002 indicated that during the 1991-1996 period 14-16 year old males had a dietary lead intake of 4.0 μg/day, virtually unchanged from the 1990-1991 value *(50)*. The dietary lead intakes for other population groups likewise by 1991-1996 indicated that changes were minimal after 1990-1991.

These data indicate that the efforts to reduce general population exposures to lead from gasoline and lead soldered cans undertaken in the 1970s and 1980s effectively targeted the significant addressable sources of dietary lead exposure for the broad population. Additional initiatives since then have brought about incremental reductions in dietary sources of lead that were not necessarily significant sources for the general population, e.g., ceramicware, bottled water, wine. Current dietary lead exposure in the general population likely reflects lead present in food at background levels due to its ubiquitous presence in the environment. It is a major accomplishment of the U.S. federal agencies and the regulated industries that by the mid-1990s, the dietary lead intake of all population groups in the TDS was only 1-5% of the provisional tolerable daily intake (PTDI) for lead. The PTDI is a level of dietary lead intake that would account for a 1 μg/dL rise in an individual's blood lead level *(51)*.

However, as will be discussed below, certain groups, e.g., users of poorly made Mexican pottery, can still be subject to significant dietary lead intakes and lead poisoning, if because of cultural or other preferences they largely consume certain types of foods or use food contact articles that carry a higher risk of lead contamination. This risk is not borne by the general population.

The 21st Century and the Challenge of Global Food Trade:

The U.S. successfully confronted many challenges in protecting the public from lead in food during the 20th century. The three major avenues of lead exposure from food were ultimately brought to an end in large part by technological advances that were instituted in their respective industries. The use of arsenicals for insect control purposes on fruit trees was replaced after World War II by newer insecticides. The use of lead soldered cans was supplanted by new types of food cans and other forms of packaging in the 1980s and the use of leaded gasoline was brought to an end as the automotive industry developed emission control technologies that necessitated the elimination of lead from gasoline. While government efforts limited exposures to lead in food from uses of lead that were necessary or unavoidable at times in history, technological progress ultimately provided alternatives that eliminated these uses of lead altogether in the U.S.

Likewise, other applications of lead in U.S food related industries, e.g., in printing inks, in glazed ceramicware, and in construction of food processing equipment, have either been replaced or subjected to strict controls during the latter part of the 20th century.

Nonetheless we are now in an era of global food trade. The U.S. Food Protection Plan states that between 1997 and 2007, the number of food entry lines, i.e., food shipments listed on import documentation, has tripled *(52)*. Furthermore, increasing amounts of the food we consume can be expected to come from nations who's governmental, agricultural and food production systems may not have incorporated the advances in lead controls experienced in the U.S. Consider that since the mid-1990s:

- A study published in 2000 reported finding high quantities of lead in printing inks used on numerous cellophane type candy wrappers for candies imported into the U.S. from Mexico *(53)*.
- Recent studies have shown that Mexican style candy products containing chili powder contained elevated levels of lead due to the common practice of not washing the chili peppers used in the production of chili powder in Mexico, allowing for soil-borne lead on the peppers to contaminate the powder and the finished candy product *(54)*.
- Articles of poorly made pottery with lead glazes continue to enter the U.S. from Mexico, many through informal means, such as personal carriage. Such pottery is capable of contaminating food with lead and inducing substantial increases in blood lead levels after only brief usage *(55)*. Cases on record indicate that long term exposure to high lead leaching tableware can cause symptomatic lead poisoning *(56)*.
- In 1998 nine members of an extended family including 2 small children in Michigan were found to have elevated blood lead levels ranging from 25 to 84 μg/dL. An investigation established that lozeena, a spice used in Iraq to color food, had been brought into the U.S. by a relative and used to prepare food consumed by the group. The spice was found to contain 7.8-8.9% lead *(57)*.
- From 1993-1997, nine children and one adult member of an extended family were found to have elevated blood levels ranging from 26 to 59 μg/dL. Review of the serial blood lead levels established that the increases coincided with the return of a maternal aunt from visits to Mexico. Further questioning of the family revealed that the aunt personally carried Mexican tamarind candy jam products packed in ceramic jars into the U.S and gave the candy to the children *(57)*. FDA subsequently learned that poorly manufactured glazed ceramic jars could contaminate candy with high amounts of lead *(58)*.
- In 1995-96, FDA found 15 of 22 samples of raisins from Turkey to contain excessive lead levels, i.e., greater than 0.25 ppm, and ultimately instituted import restrictions on all shipments of raisins from Turkey *(59)*. The lead contamination problem was also reported to affect raisins from Turkey shipped to Canada and the United Kingdom. The problem was ultimately traced to the widespread use of a copper fungicide on grapes, which contained high concentrations of lead *(60)*.
- In 1998, it was reported that a 4 month old child was found to have a blood lead level of 46 μg/dL when tested to establish a baseline blood lead level prior to the family's reoccupancy of an apartment after its lead abatement. The child's mother had a blood lead level of 29 μg/dL. An investigation

ultimately established the source of lead to be a samovar, i.e., urn recently purchased by the family in Iran and used for boiling water to prepare infant formula for the child and tea for the mother. The urn contained lead solder at locations including the base and handles and was found by testing to leach extremely high levels of lead into water placed therein *(61)*.

- Findings of elevated lead levels in maple syrup have resulted in renewed efforts by U.S. and Canadian institutions and authorities to encourage syrup producers to eliminate potential sources of lead contamination from their operations *(62, 63)*. These include soldered equipment, galvanized equipment and bronze and brass fittings manufactured as recently as the 1990s. Producers have been encouraged to replace the problematic equipment e.g., evaporators with items made from stainless steel.

Confronting a Different Type of Challenge

The challenge of addressing lead in food is a different type of challenge today than it was as recently as the 1970s. Therefore the means of addressing this challenge will need to be different. The at risk population is no longer the general population of the U.S., but rather certain sub-populations, e.g., ethnic groups who preferentially consume products subject to a greater risk of contamination. The industries that manufacture the products of concern are no longer mainstream U.S. industries such as can manufacturers, who possess technological resources to pursue solutions and who can partner with government in this pursuit, but may be technologically unsophisticated small producers in foreign countries who practice trades such as traditional pottery manufacture, using crude or outdated manufacturing practices that may result in products that pose risk.

Another point of concern is that some countries have not imposed technologically feasible control measures on uses of lead within their borders. A recent study noted the absence of controls on the lead content of paint available in Nigeria, China, India and Malaysia *(64),* stating "When it comes to public awareness of lead and its detrimental health effects, Nigeria and many other large, developing countries are 25 years behind." Given the potential for lead to be used in a broad variety of agricultural and industrial applications, the prospect of increasing food imports from countries that have yet to control lead exposures for their own populations suggests that U.S. consumers may be at risk for some time to come.

Unlike the circumstances in 1970, when two well understood sources of lead in food were of most concern, we do not know precisely where we will need to turn our efforts to confront the new challenges we may face. But as has been shown time and again, when lead is broadly used in a country's industry and agriculture, we can expect incidents of lead contaminated food to occur. When problems do occur, we will have to resort to different means of addressing them than we have used in the past.

Education and Outreach

The issues associated with substandard Mexican pottery are useful to consider. Poorly made low fired lead glazed pottery, sometimes made in wood burning kilns is produced by indigenous potters throughout Mexico. This type of pottery, typically fails FDA testing at import if it is formally offered for entry into the U.S. However, it can be brought into the U.S. in small quantities that are not declared as commercial entries or carried into the country by individuals. The items may become available for sale or may be used by the families of the travelers. Such substandard pottery, particularly items used for cooking food and holding beverages, can substantially contaminate food and give rise to increases in blood lead levels well beyond thresholds established for intervention by public health authorities.

Because it is not always possible to interdict these items at the time of importation, agencies have undertaken educational and outreach efforts directed at population groups likely to use this type of pottery. In 2007, a coalition of federal and state agencies led by The Office of Binational Border Health in the California Department of Public Health and the U.S.-Mexico Border Health Commission, California Outreach Office developed and implemented a health risk communication project primary intended to reach Latino populations in California, Arizona, New Mexico, and Texas, and secondarily, U.S. tourists in Mexico. The project's focus was the potential hazard posed by lead in traditional pottery, with the objectives being: (1) to raise awareness in at-risk populations about lead in traditional pottery; and (2) to provide information on how to prevent lead exposure *(65)*.

Subsequently, the coalition, with the assistance of a communications firm, graphic artist, and media production service produced five bilingual (English-Spanish) products: a poster, a flyer, a brochure and two public service announcements (15-second and 30-second). The following messages are incorporated into the products, which will be posted on the sponsoring agencies' websites pending the availability of funding which has been delayed at the time of publication of this chapter *(66)*:

- Lead is dangerous for everyone, especially for children under 6 years of age.
- Lead is in some traditional pottery
- Lead can cause serious health problems
- Lead can get into food and drinks that are made, served or stored in pottery.
- If you don't know if your pottery has lead, do not use it for cooking, serving or storing food or drinks. You can still use it for decoration.
- A child with lead poisoning may not look or act sick. Ask your doctor to test your child for lead.
- To learn more about how to keep your family safe from lead, contact: ....

Once publically available, these products will be adaptable for use by government and community organizations situated to reach populations who may be at risk.

This type of effort serves as an example of how authorities can address a source of lead exposure at its point of use because it is not readily addressable at its origin. Low fired pottery manufacture using lead glazes continues to be widespread in Mexico even though efforts to encourage potters to use non-lead glazes developed for low-fire usage are being undertaken by the Mexican government. The proximity of the U.S. and Mexico, and the large subpopulation of residents of Mexican origin in the U.S. virtually ensure that some quantities of substandard pottery from Mexico will find their way into the U.S. in spite of government efforts to the contrary. In the face of this reality, carefully conceived risk communication outreach to the user community can be an important tool for reaching a subpopulation at risk.

Understanding Foreign Production Practices

When the U.S. began finding elevated lead levels in Mexican candy, it was initially thought that the use of lead containing printing inks on the candy wrappers might be the source of the contamination. Although a few instances of candy contamination were ultimately found to be due to the use of such inks, as noted above, it was ultimately shown that chili powder, a common ingredient in Mexican-style candies was the source of the most pervasive contamination and that the powder contained high levels of lead because the common method of powder manufacture in Mexico did not include washing the peppers to remove soil particles (and the lead bound to them). This source of lead was not known or suspected initially in part because of the limited understanding in the U.S. of foreign production practices for minimally processed ingredients such as chili powder.

As global food trade increases, we can expect to see increasing volumes of foreign produced foods enter the U.S for which U.S. authorities likewise have a limited understanding of their ingredients and processing. It should be noted in this regard that the FDA does not routinely inspect large numbers of food processing facilities abroad. Thus, some foreign produced food products subject to low level lead contamination due to substandard agricultural or processing practices not well known or understood by U.S. authorities may reach the U.S. The ability of the FDA to anticipate these types of problems will increasingly hinge on its ability to develop an understanding of growing and production practices in parts of the world in which it has no working history. It is interesting in this regard to note that FDA is currently moving forward with plans to establish a permanent office in China *(67)*.

The Need to Involve Third Parties

In 2006, the FDA and the Certification and Accreditation Administration of the People's Republic of China renewed a Memorandum of Understanding

(MOU) that established a certification system for factories manufacturing daily use ceramicware in China to be offered for import into the U.S *(68)*. The certification is based upon Chinese government inspection of the ceramicware production facilities and periodic analysis of samples of finished product for compliance with FDA's action levels for leachable lead from ceramicware. The agreement is intended to increase the likelihood that daily use ceramicware manufactured in China and offered for import into the U.S. will comply with FDA's requirements. The agreement is also intended to provide a basis whereby FDA can reduce its frequency of sampling and testing of the subject ceramicware at the time of import in accordance with the effectiveness of the factory certification system.

Because of FDA's limited resources for monitoring food imports and food production operations in foreign countries, the function of a competent third party in certifying the safety attributes of a foreign manufacturer's product can provide FDA with a measure of confidence that a foreign made product was manufactured to comply with the applicable standards of FDA regarding lead safety. Such an arrangement can free up more FDA resources for monitoring products that are more likely to pose risk. The value of a pre-certification arrangement to foreign manufacturers is that it provides a means by which they can establish that their products are not likely to pose a significant risk of contamination and thus do not warrant more restrictive FDA procedures upon import.

The Negative Public Perception Factor for Lead

Lead has also been a focus of efforts to require labeling information, i.e., warning statements in labels or labeling, about its presence in products, even though the products meet all applicable safety standards. The best known of these efforts is California's "Safe Drinking Water and Toxic Enforcement Act of 1986" known as Proposition 65, a law established by public referendum. These types of laws are distinctly different from the historical laws that prohibit the sale of food that does not meet an applicable standard of safety. These types of laws are based upon the premise that a consumer has a right to know that a product, i.e., ceramic dinnerware, contains a toxic substance, e.g., lead, even if the product meets federal (and state) safety standards and is acceptable for sale to the public *(69)*.

Proposition 65 has proved to be a powerful incentive for industry to reformulate products to avoid the use of lead. Non-lead glazed ceramicware, labeled as such, is commonly sold today. The incentive is the prospect of consumer rejection of the product if it bears a warning label. Because California comprises such a large share of the U.S. marketplace, companies, e.g., ceramicware retailers, typically source California compliant non-lead containing products for their entire U.S. inventory.

The California law also includes provisions for large monetary penalties for violations. A lawsuit can be brought against a violator by a private individual of by the state of jointly. Settlement monies can be awarded to the State if the state participates in a legal action and can be used to fund ongoing efforts to enforce the requirements of the law.

This law has been the basis of legal actions in California against ceramic dinnerware and lead crystal. The compositions of many products sold in the U.S. today have been impacted because of Proposition 65. Irregardless of whether this law has brought about significant risk reduction, it has provided a means by which a powerful negative public perception about lead in consumer products has been expressed. That perception would likely give rise to a sharp expression of public dissatisfaction if a significant episode of lead contamination of an imported food product was to occur.

Conclusion

The challenge of protecting the public from lead in food was successfully met in the U.S. in the 20th century. That challenge was once posed by a food supply that was largely domestic in origin and subject to lead contamination predominantly traceable to only a few industrial uses of lead. The challenge has now reappeared in a different form as a globally sourced food supply of infinite variety and origins poses the potential for lead contaminated food to reach the U.S. While the risk may be minimal to the general population, it may be considerable to certain sub-populations, e.g., ethnic groups who use products at greater risk of contamination. New means of government action will be required to meet this challenge as the source industries are not readily identifiable, much less prepared to step forward and partner with the government in the lead reduction effort. The challenge posed by lead in food, once thought to be vanquished in the twentieth century, may in fact be with us well into the twenty first.

The author acknowledges Paul Peterson's (FDA retired) most valuable work in compiling information on lead related actions from historical documents of the FDA.

References

1. *Report of the Food and Drug Administration,* August 29, 1931, pp 11-12.
2. *Report of the Bureau of Chemistry*, September 14, 1908, pp 15-16.
3. *Report of the Bureau of Chemistry,* September 2, 1912, p 27.
4. *Report of the Bureau of Chemistry,* November 24, 1913, p 8.
5. *Report of the Bureau of Chemistry,* October 11, 1919, p 20.
6. *Report of the Bureau of Chemistry* October 1, 1915, pp 193-195
7. *Food and Drug Review*, **1934,** *18,* p 254.
8. *Report of the Food and Drug Administration*, August 31, 1933, p 3.
9. *Report of the Food and Drug Administration*, August 29, 1934, p 13.
10. *Report of the Food and Drug Administration*, August 29, 1935, p 20.
11. *Report of the Food and Drug Administration*, August 31, 1933, p 3.
12. FDA notice, December 14, 1935.

13. Letter from W. B. White, Chief, Food Division, to Chiefs of Districts, October 19, 1934
14. *Food and Drug Review*, **1935,** *19,* p 87.
15. *Food and Drug Review*, **1935,** *19*, p 91.
16. *Report of the Food and Drug Administration*, August 29, 1935, pp 3-6.
17. *Food and Drug Review*, **1935,** *19*, p 90.
18. *Food and Drug Review*, **1937,** *21*, p 65.
19. *Report of the Food and Drug Administration,* August 30, 1938, pp 5, 8, 9, 17.
20. *Food and Drug Review*, **1937,** *21,* p 191.
21. *Report of the Food and Drug Administration*, August 29, 1935, pp 3-6.
22. FDA notice, May 8, 1936.
23. *Report of the Food and Drug Administration*, August 30, 1938, pp 5, 8, 9, 17.
24. *Food and Drug Review*, **1938**, *22,* pp 101-104.
25. Wright, W.; Sappington, C.O.; Rantoul, E. *J Indust. Hyg.* **1928,** *10*, 234.
26. Letter from S. A. Rohwer, Acting Chief of Bureau of Entomology, to Entomologists of the Bureau and Cooperating States, February 13, 1933.
27. U.S. Code of Federal Regulations, Title 16, Sec. 1303.5.
28. United States Congress Public Law 91-695. 91st Congress. H.R. 19172, January 13, 1971, Lead-Based Paint Poisoning Prevention Act, Washington, U.S. Government Printing Office, 1971, p 3.
29. Carrington, C.D.; Bolger, P.M. *Regul. Toxicol. Pharm.* **1992,** *16,* 266.
30. *MMWR. Morb. Mortal. Wkly. Rep.* November 2, 2007, pp 1-14.
31. *Food and Drug Administration 1970 Annual Report*, p 266.
32. FDA Compliance Policy Guide, Sec. 545.450. Available: http://www.fda.gov/ora/compliance_ref/cpg/cpgfod/cpg545-450.html. [Accessed August 22, 2008].
33. *Food and Drug Administration 1974 Annual Report*, p 132.
34. *Federal Register,* **1979,** *44,* p 51236.
35. *Federal Register,* **1974,** *39,* p 42740.
36. *Federal Register,* **1979,** *44,* p 51240.
37. *Federal Register,* **1995,** *60,* pp 33106-09.
38. *Federal Register,* **1996,** *61,* pp 4816-20.
39. Letter from Fred R. Shank, Director, Center for Food Safety and Applied Nutrition to Manufacturers, Importers, and Distributors of Imported Candy and Candy Wrappers, June 13, 1995. Available: http://www.cfsan.fda.gov/~dms/pbguid.html. [Accessed August 21, 2008].
40. Lewis, J.: *EPA Journal,* **1985,** May. Available: http://www.epa.gov/history/topics/perspect/lead.htm. [Accessed August 22, 2008].
41. EPA Health Document: EPA's Position on the Health Implications of Airborne Lead. November 28, 1973.

42. Lewis, J.: *EPA Journal,* **1985,** May. Available: http://www.epa.gov/history/topics/perspect/lead.htm. [Accessed August 22, 2008].
43. EPA press release, Jan 29, 1996. Available: http://www.epa.gov/history/topics/lead/02.htm. [Accessed August 22, 2008].
44. Flegal, A.R.; Smith, D.R.; Elias R. *Environ. Sci. Technol.* **1990,** *23,* 96.
45. *Federal Register,* **1994,** *59,* p 26933.
46. Kaufrmann, A. *Food Addit. Contam.* **1998,** *15,* 437-445.
47. Food and Drug Administration. Guidance Document for Lead in Shellfish. 1993. Available: http://vm.cfsan.fda.gov/~frf/guid-pb.html. [Accessed August 22, 2008].
48. FDA Guidance Document: Lead in Candy Likely to be Consumed Frequently by Small Children: Recommended Maximum Level and Enforcement Policy. November 2006. Available: http://www.cfsan.fda.gov/~dms/pbguid3.html [Accessed August 22, 2008].
49. Bolger, P.M.; Yess, N.J.; Gunderson, E.L.; Troxell, T.C.; Carrington, C.D. *Food Addit. Contam.* **1996,** *13,* 53-60.
50. Egan, S.K.; Tao, S.S.-H.; Pennington, J.A.T.; Bolger, P.M. *Food Addict. Contac.* **2002,** *19,* 103-125.
51. Carrington, C.D.; Bolger, M.P. *Regul. Toxicol. Pharm.* ***1992,*** *16,* 265-272.
52. Department of Health and Human Services and Food and Drug Administration: Food Protection Plan: An Integrated Strategy for Protecting the Nation's Food Supply. November 2007. Available http://www.fda.gov/oc/initiatives/advance/food/plan.html [Accessed August 22, 2008].
53. Fortes, L.; Bauer, E. *Vêt. Hum. Toxicol.* **2000,** *42,* 41-42.
54. FDA Supporting Document: Recommended Maximum Level of Lead in Candy Likely to be Consumed Frequently by Small Children. November 2006. Available: http://www.cfsan.fda.gov/~dms/pbcandy2.html [Accessed August 21, 2008].
55. Matte, T.D.; Props, D.; Palazuelos, E.; Graef, J.; Mauricio Hernandez, A. *Lancet (London),* **1994,** *344,* 1064-1065.
56. Scarlett, J.D.; Hodges, J.; Romain, P.R.; Bowlen, C. *Med. J. Aust.* **1995,** *163,* 589-590.
57. *MMWR. Morb. Mortal. Wkly. Rep.* December 11, 1998, pp 1041-1043.
58. Food and Drug Administration Import Alert 33-10. Available: http://www.fda.gov/ora/fiars/ora_import_ia3310.html. [Accessed August 22, 2008].
59. Food and Drug Administration 20-03. Available http://www.fda.gov/ora/fiars/ora_import_ia2003.html. [Accessed August 22, 2008].
60. Dabeka, R.W.; McKenzie, A.D.; Pepper, K. *Food Addit. Contam.* **2002,** *19,* 47-54.
61. Shannon, M. *Environ. Health Perspect.* **1998,** *106,* 313-316.

62. Ontario Ministry of Agriculture, Food and Rural Affairs. Available: http://www.omafra.gov.on.ca/english/food/inspection/maple/pdf/maple_lead_fact.htm. [Accessed August 22, 2008].
63. University of Maine Cooperative Extension. Available: http://www.umaine.edu/umext/maplesyrupproduction/reducinglead.htm. [Accessed August 22, 2008].
64. *Science Daily*. October 2, 2007. Available: http://www.sciencedaily.com/releases/2007/09/070927113155.htm. [Accessed August 21, 2008].
65. United States – Mexico Border Health Commission. Available: http://www.borderhealth.org/lead.php?curr=bhc_initiatives, [Accessed September 26, 2008].
66. Personal Communication, Dr. Maura Mack, Office of Binational Border Health, California Department of Public Health
67. FDA News, March 14, 2008. Available: http://www.fda.gov/bbs/topics/NEWS/2008/NEW01806.html. [Accessed August 22, 2008].
68. *Federal Register,* **2006,** *71,* pp 15188-15210.
69. State of California, Office of Environmental Health Hazard Assessment. Available: http://www.oehha.ca.gov/prop65/law/P65law72003.html. [Accessed October 1, 2008].

Chapter 16

Achieving Total Food Protection: Benefits From Integrating Food Safety and Food Defense Programs

David K. Park, President, Food-Defense, LLC

Opportunities exist to introduce biological agents, industrial chemicals, toxins, or radiological agents into foods and packaging which could go undetected in traditionally-structured HACCP-based food safety systems. Select pathogens, heat-resistant microbial toxins, poisonous, odorless and tasteless chemicals and radiological contaminants are examples of agents that can often avoid normally-employed HACCP detection methods. It is clear that to proactively engage the evolving threats to our food supply, a fully-integrated food safety *and* food defense strategic platform helps rationalize significant U.S. public and private investments in achieving total food protection goals for the industry and consumer. In helping to protect agriculture and food as a U.S. critical infrastructure, George W. Bush signed Homeland Security Presidential Directives HSPD-7 and HSPD-9. These mandates are anchored in the U.S. Food and Drug Administration (FDA) adoption of an "all-hazards" risk/threat management approach designed to enhance protection of our nations' food supply, sourced domestically and increasingly from overseas, as presented in its 2007 *Food Protection Plan.* This chapter suggests the use of a Hazard Analysis and Critical Control Points (HACCP)-like core risk assessment and management methodology to the identification and mitigation of food defense vulnerabilities. Hazard/threat identification methodology involving food safety and food defense are fundamentally the same; only the perspectives in identifying and managing these risks are different. It is also imperative that the food industry shifts to the use of more comprehensive total food protection facility assessments and not rely solely on current third-party auditing structures to identify vulnerabilities. Food defense technologies must correctly detect (ideally identify) and mitigate potential hazards that, while reasonably unlikely to occur, could result in a catastrophic business and/or industry outcome (low probability/high consequence). These must run parallel to detection and mitigation of traditional food safety hazards which are "reasonably likely to occur". Adopting a common technology platform using developments such as rapid chem/bio sensors, trace-back and product identity preservation technologies, tamper-proof and tamper-evident packaging, surveillance technologies and risk decision support and human factor assessment methods are examples of food safety and food defense integration opportunities which complement total *food protection* objectives to best and most efficiently protect our critical infrastructure, industry and the consumer. This chapter makes practical recommendations on how the U.S. food industry can best equip itself to efficiently address the challenges and opportunities in identifying and mitigating product integrity vulnerabilities before catastrophic consequences are allowed to occur.

Introduction

Years have now passed since fuel-laden commercial airliners were used in strategic coordinated terrorist attacks against multiple targets as weapons of mass destruction. No major terrorist attack has occurred on United States (U.S.) soil since September 11, 2001. Unfortunately, Americans' collective memory of September 11th has dimmed over time, as there have been no subsequent terrorist attacks on our shores. While some have argued that governmental responses to the attacks have made us less vulnerable than we were before September 11, the absence of a major attack may be merely attributable to luck. Many experts concur with this hypothesis. A second national catastrophe, Hurricane Katrina, ravaged the City of New Orleans in 2005 with floodwaters from dramatic levee breaches that inundated the sub-sea-level city. The chaos that ensued resulted in devastating loss of life and property. While one of these events was intentional and the other a natural disaster, both catastrophic outcomes were aided by system failures, principally caused by humans, when threats were not taken seriously and the aftermath's responses being awkwardly unprepared and poorly executed.

In the intervening years, Americans have become more acutely knowledgeable about their vulnerabilities, as well as the complexity of the threats they must attempt to mitigate. Interestingly, both September 11th and the Katrina scenarios were contemplated by experts in advance of the event. Each disaster provided compelling new "lessons learned," accompanied by trite vows to "make sure that this never happens again." One of those lessons is that we are likely to fail again in preventing future disasters despite our best intentions. The real problem is that most of our efforts are spent fixing problems that have already happened, and we fail to select decision support tools that would best help us solve problems pre-emptively. The best approach to risk management is to use lessons learned from the past to predict future events.

The U.S. food supply is known to be on terrorist target lists, and a future attack by an adaptive adversary is probable. Imagine the outcome of a carefully directed multi-city lethal attack using a contaminated branded product with a label recognized around the world. At the very least, the economic consequences could be catastrophic for whichever global business is targeted. Food is central to human social behavior. Such an attack might not only cause tragic loss of life, but would reduce consumer confidence in the food supply and alter market demand. American lives and their shopping habits would be significantly affected. Residents of the metropolitan Washington, DC. area greatly altered their daily habits in the face of snipers who attacked random innocence in 2002. People feared for friends, family and total strangers as they filled up gas tanks, parked in shopping malls, escorted their children to school or adjusted to the disappointment of cancelled sporting events. Might the reaction be the same from a coordinated food attack?

As U.S. Health Secretary Tommy Thompson made his resignation speech on Friday, December 3, 2004, he invoked the unsettling fears of many within the food industry with his now infamous words: "For the life of me, I cannot understand why the terrorists have not attacked our food supply because it is so easy to do" *(1). The* public, the media, and government officials voiced strong criticism of his comment, chastising Thompson for publicly revealing this previously unapparent national security vulnerability. This vulnerability remains under-valued by consumers, the private sector and public officials, and requires new and creative solutions to protect the food supply.

Protecting the American food supply poses one of the most monumental national security challenges, yet this threat was initially overlooked in simulated national terrorism exercises post-September 11. Our U.S. food supply is one of the most unique and complex systems of interdependencies operating within highly differentiated, efficient, and dynamic international markets. From farming, harvesting, processing, transportation, wholesaling, and retailing, there are myriad global supply chain challenges in protecting industry assets associated with these system interdependencies. Food safety systems alone are insufficient; Americans also need a parallel system of food defense.

Does Our Food Industry Stand Ready?

Years after September 11 and awakening to a shocking new threat environment, the food industry acknowledged the threat to the US food supply, claiming to have assessed its vulnerabilities, identified risks, developed a food defense plan, tested scenarios, worked with the government, cooperated with audits, and responded to customer demands. The industry response was largely defensive, citing that there had been ample opportunities for our food supply to be attacked, and that there wasn't much more that the food industry could do.

But there is much that the food industry can do and needs to do to be prepared for a potential catastrophe. Reaching this new and needed preparedness level will require reassessing the situation, learning from past mistakes and successes, and improving upon the already-traveled course. Answers for our government and industry are not easy, but complacency in the face of this threat will only increase Americans' vulnerability. Regulatory officials claim that American consumers have one of the safest food supplies in the world. These words resonate today in the food sector, but the truth is that we are still not properly positioned to defend our food supply, and the American consumer is growing less and less confident.

Food industry professionals have inadvertently complicated the development of solutions to these issues by failing to connect basic food safety and food defense elements into a common risk-based "total food protection" platform, by listening to the opinions of third-party auditors who often lack the combined food safety and food defense expertise required to assess vulnerabilities, and by relying upon commercial software to mass-produce food defense plans for businesses based upon pre-scripted, largely subjective inputs. The food industry must make up for these missteps by strategically managing the body of combined food safety and food defense knowledge and converting that knowledge into cost- and resource-effective, implementable industry actions.

Misguided By The Past

Immediately after September 11, the food industry was misled by the loudest voices in the food and agricultural communities. The predominant message conveyed to the food sector immediately post-September 11 by industry spokespersons was that food security (a.k.a. food defense) is not food safety. Industry was carefully instructed to purposely segregate and compartmentalize food safety and food defense activities into two separate risk management components. This "divide and conquer approach" that persists today is fundamentally flawed. At the time, the justification for this dichotomy was that *food safety* requires us to look at conventional, symmetrically occurring threats to food. In other words, food scientists should "think like scientists with the science they know" in managing food safety risks. Admittedly, food safety scientists are not typically trained to anticipate malevolent interference with the food supply. Yet these scientists are increasingly tasked with considering these non-conventional, asymmetrically occurring threats to foods, the essence of *food defense.* Food defense hazards are more complex and adaptive than those encountered in the context of a food safety environment. However, the principal premise presented in this chapter is that food safety and food defense should never be addressed with independent risk management approaches, as this strategy heightens the probability of "system failure" in protecting our food supply.

Food safety and food defense are approached and managed in a traditionally different context, Small- to medium-size companies are well-versed in food safety, but because food defense is often approached and managed at a different level, companies often find it difficult to develop and implement a comprehensive food defense plan, given perceived resource constraints (finances, personnel) and confusion about multiple strategic options. Large companies, too, may balk at dedicating full-time staff, extra resources and attention to manage food defense and vulnerability issues, particularly if they perceive the terrorist threat to be a distant, external potentiality over which they have little control, compared to internal product safety standards which they can easily control using proven process management protocols. Simply put, the objectives, justifications and returns on food defense investment are easily blurred against "speculative" threats. In addition, asset management, crisis management and business continuity plans are often embraced as stand-alone un-integrated components dangling precipitously outside of the context of a *total food protection* plan.

Widely used food defense audits employ bolt-on standard checklists to prerequisite food safety audits that likely overestimate the effectiveness of mitigations in place. Unfortunately, this approach generates dangerous overconfidence in actual preparedness. Today's vendor verification and audit systems are non-standardized across the industry. Food defense audits are often performed by experienced food safety auditors, but with inexperienced food defense perspectives. These auditors rarely understand the complex, adaptive and opportunistic strategies of the criminal or terrorist mind to breach and penetrate protective barriers surrounding food production and processing establishments. Some of this confusion stems from misconceptions about how food defense audits and food defense vulnerability assessments are conducted and applied.

A food defense vulnerability assessment is a risk-based evaluation of a site's or system's hazard control strengths that could cause failure in achieving the set standard or documented process within a food defense environment. The process involves the identification and classification of the primary vulnerabilities that may impact site or system function. A food defense audit consists of an evaluation of the business (or businesses) specific systems, processes and controls and is performed against an already established set of standard or documented processes previously developed using the results of vulnerability assessments. Audits are designed to provide an independent evaluation of system processes and controls using personnel with expert knowledge about such systems and processes. An audit also provides a gap analysis of the operating effectiveness of the internal controls in meeting a system or control requirement. Unlike a vulnerability assessment, the auditor provides limited feedback about how to mitigate the system gap. Because the purposes of audits and vulnerability assessments are not the same, very different outcomes can result from that are used to develop or verify the placement of critical elements in a food defense plan. A thorough food defense vulnerability assessment must *precede* an audit. Unfortunately, many businesses only conduct an audit. If they do conduct both, the audit often improperly precedes a vulnerability assessment.

There are other problems in understanding and effectively addressing today's food defense issues. Post-September 11th, the reality is that food businesses have other operational priorities and cannot effectively focus on the high capital and human resource demands needed to continuously assess vulnerabilities and identify and mitigate business threats. Also, when food safety systems are dysfunctional, it is also likely that any food defense system, if it exists at all, is also dysfunctional. This is particularly true of many foreign countries that export their products to the U.S.

We know that the Food and Drug Administration (FDA) foreign firm inspection data show that a staggeringly high percentage of firms fail to comply with mandatory provisions of U.S. food safety regulations. Presumably, most FDA-uninspected foreign firms would also fail to comply. Ironically, as long as these uninspected firms were registered under the provisions of the Bioterrorism Act of 2002, and comply with Prior Notice requirements of the Act registration (in some cases, manufacturing process registration), their products will likely be accepted into the U.S., even if produced and/or manufactured under conditions violating the Good Manufacturing Practices (GMPs) and effective Hazard Analysis Critical Control Points (HACCP) programs. Due to the complexity of proper oversight, FDA Registration and Prior Notice compliance with the Bioterrorism Act often provide little assurance that uninspected foreign firms have actual food safety and/or food defense safeguards in place. Inadequacies in product surveillance at U.S. Ports of Entry and insufficient international regulatory inspections compound the problem. Could these vulnerabilities be the Achilles heel in U.S. food supply protections?

The U.S. has been witnessing annual import growth rates averaging 15%, which the FDA must adapt to manage. There have been dramatic changes in the volume, variety, and complexity of FDA-regulated products arriving at U.S. ports. The United States trades with over 150 countries and territories, with products entering more than 300 U.S. Ports of Entry. More than 65% of FDA goods received at U.S. Ports of Entry are food goods. In addition to an increased volume of imports, the nature of these imports has changed. Traditionally, the bulk of FDA products consisted of unprocessed food ingredients. Today, prepared, ready-to-eat food products, and fresh produce account for an increasing proportion of all FDA regulated imported food products. The $65 billion in food goods imported each year make up a total of 15-20% of the U.S. food supply, and as much as 50-60% of some products such as fresh fruits. According to the U.S. FDA, almost 10 million lines of food entries constituted at least five times as many as they did in 1994 in 2004; this figure has tripled in the last decade alone *(2)*. The vast majority of imported foods are unsampled, unexamined and untested. Today, and until further requested funding is in place, FDA currently estimates that it can only conduct border inspections on approximately 1% of the food that it regulates (i.e. vegetables, fruit, seafood, grains, dairy and animal feed) at the border since 2007, a decrease from prior years, due to the surge in entries.

From 1995-2005, imports of seafood have increased by 33%; produce by 50%; confections by 61%; and pet foods by 45%. Total food imports have increased nearly 40% since the North American Free Trade Agreement and World Trade Organization went into effect in the mid-1990s *(2)*.

A system-wide, global impact case in point is the crisis in 2007 involving animal deaths from wheat gluten, a protein enhancer used in pet food formulations. The FDA received thousands of reports of pet illness that owners suspected were connected with the consumption of contaminated pet food. After an investigation begun in March 2007, the FDA's Dr. David Acheson, Associate Commissioner of Foods, announced that FDA investigations had concluded that the animal deaths were traceable to imported pet food ingredients contaminated with the industrial chemical melamine and melamine analogs. Melamine-contaminated production waste from the pet food manufacturing process had been used as an ingredient in animal feed for hogs and chickens. FDA identified the supplier of the contaminated wheat gluten as a Chinese firm, Xuzhou Anying Biologic Technology Development Company, and issued an import alert providing for detention of all wheat gluten imported from that firm to assure that contaminated product did not enter U.S. commerce. The import alert was broad, covering all vegetable protein products entering the U.S. from China. All entries from China were detained by FDA upon arrival into the U.S. and were not released into domestic commerce unless third-party analysis demonstrated the entry was not contaminated with melamine or melamine analogs. According to the FDA, the response required was intense including mobilization of employees, increased inspections, increased lab analyses, intense consumer communications, dispatch of a foreign investigational team and activation of the Center for Food Safety and Applied Nutrition Emergency Operations Center (CFSAN).

What if the contaminated pet food ingredients had been part of a complex scheme to launch an intentional terrorist attack on the U.S.? Could the situation have been different or worse? Imagine if the pathogenic *E. coli* and *Salmonella Saintpaul* fresh produce outbreak events in recent months were associated with an intentional attack on our food supply. What if there had been more than one food attack including several products, several brands, several cities, and several agents? What would be different about the outcomes, the way in which the event would be handled, the impact upon the food industry and the ability of the state, federal government and public health institutions to mange laboratory, law enforcement and public health surge requirements in such an event? What would be required to maintain consumer confidence in our nation's food supply? The solutions needed to protect our food supply depend on collective and decisive action.

The U.S. government is concerned primarily with preparing itself to respond to terrorist or criminal attacks when such attacks occur. Yet if there were a successful terrorist attack on our food supply, Congress would likely promulgate tightened food defense regulations. Industry should engage in more proactive food defense efforts than these to avert economic and life losses. But despite real and known threats, local, state, and federal governments have expressed no near-term intention of providing any direct government subsidies to the food industry sector for food defense preparedness. Recent heightened bi-partisan political interests in food safety are promising indications that greater food defense industry oversight by the government may emerge under a new administration.

A New "Total Food Protection" Model

There is little argument that more specialization is needed in the food defense discipline, including the assessment skills of physical security specialists, counterterrorism experts and criminal investigators. However, despite these specialized security-oriented tasks, the basic risk assessment methodology for both food safety and food defense is the same. This methodology includes identifying hazards, assigning risks, analyzing risk controls, making risk control decisions, implementing controls against the risks and system vulnerabilities, and supervising and reviewing the process (e.g. Operational Risk Management, or ORM). Other assessment methods, such as the "CARVER + Shock" method, use specific metrics (in this example, Criticality, Accessibility, Recuperability, Vulnerability, Effect, Recognizability, and Shock) to help assessors refine the definition of risk into economic and psychological terms. Other more sophisticated assessment tools that have been developed are most often used in the context of national security. The food industry has historically focused on safety and the quality of the products that America grows, processes and sells. From a food safety perspective, the emphasis since the 1970s has been a preventive philosophy, anticipating bad things before they happen. This is presented in the guiding philosophy of the HACCP approach to risk management *(3)*. This new and constantly changing risk/threat environment requires leveraging existing lessons learned from managing the highly effective food safety processes and procedures along with traditional physical security and criminal threat and risk approaches. Due to its lack of expertise with the latter, as well as only recent awareness of the need for these security and criminal risk approaches, the food industry has had only limited success in the sphere of food defense. Rather than the traditional food defense "react-and-respond" culture, the American food industry would be well served to adopt an "anticipate-and-prevent" approach as part of a strategy of *total food protection*.

Simple and effective solutions are possible, and need not pit food safety against food defense. Food defense plans should be developed and taught using the lessons learned from the prerequisite framework of HACCP, the globally accepted and practiced "core" food safety platform. This process, which includes a risk management methodology, is widely used globally to manage and respond to biological, chemical and physical hazards in foods.

Understanding the concept of "risk" and appropriately evaluating it is central to the establishment of a total food protection platform and process management evaluation. The predecessor food safety-directed HACCP risk management approach can be modified to accommodate the food defense process, offering a logical, practical and workable solution. The food defense requirements should not be forced directly into existing HACCP plans to make them universal. However, the HACCP process can be emulated within a developmental systems approach for food defense, and can be evaluated alongside evolving HACCP plan requirements. Combined with food safety, managing food defense risks on a total food protection platform offers an efficient, practical industry application. Food defense risk assessment objectives and methods used today, including ORM and CARVER + Shock, and traditional food safety risk assessments using HACCP hazard identification and risk control measures are fundamentally alike. The risk process is the same and the application is the same, but the specific risk assessment perspectives are different.

To justify combining different perspectives of both food safety and food defense, whether assessing a microbiological, chemical, physical, radiological, or explosive hazard, the following formula defines risk:

Risk = Vulnerabilities + Threat + Consequence

Therefore, we can conclude that this risk determination equation can be simply and elegantly applied to both food safety and food defense.

The real risk determination difference between food safety and food defense is a mere modification of this formula rather than a new formula. To define food safety and food defense, the basic risk equation can be refined as:

Food Safety*...*Risk = Vulnerabilities + Unintentional (Hazards) + Consequence

Food Defense...Risk = Vulnerabilities + Intentional (Threats)* + Consequence

Threat* = Capability + Intent

The concept of *intent* applies only to food defense, demarcating the terrorist or criminal forces at play. Food protection threats can be distinguished from hazards based solely "intent". While intent requires adoption of a different and broader perspective, assessing intent does not require a unique risk process. Combining the two risk equations depicts the components of *total food protection*:

***Total Food Protection*... Risk = *All* Vulnerabilities + *All* Threats & Hazards + Consequence**

To simplify an otherwise complex risk management approach to simultaneously address both food safety and food defense requirements requires expansion upon and application of existing expertise in safeguarding the American food supply. Food defense is a newer concept compared to the accomplishing food safety objectives. Relating the food defense process development and implementation requirement to HACCP demystifies the novelty of food defense and places it into a more comprehensible industry context. Adapting methodologies and terms such as HACCP, Critical Control Points (CCP) and Control Points (CP), the food industry can develop parallel concepts and terms to include food defense concepts of risk, such as "Threat Analysis Critical Defense Control Points" (TACDCP), "Critical Defense Control Points" (CDCPs) and "Defense Control Points" (DCPs). Rather than introducing entirely new strategies and terms, CDCPs and DCPs would reflect alterations of current concepts and terms to reflect the integrated approach needed to identify and manage new and different hazards around all critical nodes of food plant operation. Additionally, HACCP could be redefined using Threat Analysis Critical Safety Control Point (TACSCP) with Critical Safety Control Points (CSCP) and Safety Control Points (SCP), if needed. TACDCP could be overlaid directly onto current HACCP systems, emulating the HACCP process steps and performing the risk assessment for all hazards and all threats. Mitigating these threats would take the form of a parallel, *a priori* decision-making process for all identified vulnerabilities at once. This integrated process approach provides for an "all hazards, all threats," *total food protection* approach.

For example, a change in the food safety HACCP plan would signal a requirement to reevaluate its impact upon a corresponding change to the food defense TACDCP plan and vice-versa. With the fundamental process objectives of risk/threat assessment unchanged, the change in perspective can be accomplished with a diverse assessment team and critical decision-path facilitation. When food safety and food defense risk processes are purposefully and carefully integrated, the results are both synergistic and distinct from one another. Food defense elements can remain securely "partitioned" from general view with related documents protected from employee view, thus restricting sensitive information access on a "need-to-know" basis. A select few trusted, knowledgeable and qualified employees can gain access to both HACCP and TACDCP plans to perform total food protection plan required maintenance.

The underlying risk process should be understood by all individuals involved, regardless of the full range of perspectives and subject matter input embedded in the plan. When changes in a HACCP plan might occur, the TACDCP plan may also be directly affected, and vice versa. This would indicate a required review on whether or not the HACCP change would mitigate an existing vulnerability, create another, or have no effect at all on the TACDCP critical defense points. HACCP and TACDCP programs must, then, evolve together for food protection to be assured. Changes in HACCP may impact TACDCP because critical process nodes may be shared in a total food protection program. Developing and implementing a new integrated total food protection system must have different perspectives contributed by different subject matter experts (SMEs), which is key to positive process and risk/threat management outcomes.

Traditional HACCP encourages hazard identification and hazard control as far upstream in the food and packaging supply chain as is practical. In an intentional threat environment, processing and packaging access points are open and vulnerable to an intentional, malevolent attack, either upstream or downstream in the process. This requires the TACDCP approach, which must consider opportunities and probabilities for the introduction of biological, chemical, or radiological agents into foods or packaging at any point in the process stream. This is especially important for breaches that may go undetected in the context of presently developed and industry-practiced HACCP plans.

For example, select pathogens, heat-resistant microbial toxins, poisonous, odorless and tasteless toxic chemicals, and radiological contaminants are examples of hazards that can go undetected under currently configured HACCP plans. For this reason, food defense requirements must be evaluated on a customized basis for the purpose of best selecting, positioning, and deploying "dual-utilization" food safety and food defense detection devices or procedures to be used in mitigation of all hazards in the food production or processing application. Dual-utilization technologies must now also correctly detect (ideally, identify) and mitigate all potential hazards, even those that are reasonably unlikely to occur, which could contribute to a high-impact, or catastrophic, business failure or national security incident.

Managing the risk of product diversion, theft and counterfeiting is of principal concern. These incidents often involve higher market-value products marketed to high-risk populations. These products have been traditionally involved in economic adulteration and often find their way back into the food supply. In most cases, criminal, not terrorist, activity is implicated in diversions of these products. However, these incidents do indicate a food defense concern that can be mitigated by available technology countermeasures. Sensor technology development; rapid microbial, chemical and radiological detection; supply-chain trace-back and product identity preservation technologies; tamper-proof and tamper-evident packaging; and surveillance technologies are examples of what works in food safety risk management. These technologies are equally applicable to food defense threat management, as they offer enhanced food protection and dual-mitigation utilization. Industry is now beginning to learn that adopting investment strategies that meet both food safety and food defense objectives rationalizes the investment required to meet growing needs in managing all food threats. Radio-frequency identification (RFID) and enhanced bar code technology investments, for example, bring supply chain efficiency gain as well as provide an added layer of food protection. Enhanced passive and real-time sensor capability, with global positioning systems (GPS) and geographic information systems (GIS) technology support, provide industry with the ability to consistently scan the environment for food safety and food defense "indicators and warnings" which are built around "critical food protection nodes" within the entire supply chain.

New Simulation Tools Are Available

Perhaps the most exciting advanced technology developments for food defense are in the area of computer simulation, visualization tools and complex decision-making tools. These new products offer far greater insight to better define sector risk, visualize the consequences of an intentional attack on the U.S. food supply, and to stimulate ways of thinking differently about threats to the U.S. food supply in complex and adaptive environments.

Purdue University's Center for Computational Homeland Security (CCHS) and the Purdue Food Science Food Bio-security Simulation have teamed up to conduct agroterrorism simulations in synthetic environments. The alliance has produced an agent-based simulation used by Purdue's partners in government and business to help them prepare for and train to prevent and respond to terrorist events in the food supply chain. The simulation develops a virtual society that mirrors essential socio-demographic and epidemiologic characteristics of the U.S., mapped onto a simulated geography and infrastructure of the U.S. The simulator introduces as many as one million artificial agents with variable attributes and behaviors to indicate position, mobility, susceptibility to infection, and well-being of the citizenry based on real data. The artificial agent population is tied to different locations, and attributes of individual well-being draw upon paradigms from economics and psychology *(4)*.

Another example of new risk evaluation tools for the food industry is the Consequence Management System and Crisis Management and Response System developed by BT Safety, LLC, with the support of the U.S. FDA and Homeland Security's National Center for Food Protection and Defense (NCFPD). The Consequence Management System simulates and estimates the consequences of a food contamination event, guiding the user through scenarios that depict the human and economic impact of the contamination event in light of different intervention strategies *(5)*.

The FDA Food Protection Plan (FPP)

Since September 11th, Food Defense, LLC has been publicly supporting the integration of food safety and food defense into a common "all hazards/threats" platform, when FDA announced that it too would look at managing food safety and food defense hazards as part of an integrated food protection plan.

In November 2007, the FDA released the *Food Protection Plan,* which provides a framework to identify and counter potential hazards in both domestic and imported food. Achieving the food safety enhancements identified by these plans will require the involvement of all food safety partners, including federal, state, local, tribal, and foreign governments; industry; academia; consumers; and Congress. The approach was to build in safety measures across a product's life cycle, from the time a food is produced to the time it is distributed and consumed. They encompass three core elements: prevention, intervention, and response *(2)*.

The FDA *Food Protection Plan* identified ten legislative authorities necessary for achieving full implementation. Hopefully, Congress will fully support FDA with these these authorities, which would:

- Allow FDA to require preventive controls against intentional adulteration at points of high vulnerability in the food chain;
- Authorize FDA to issue additional preventive controls for certain high-risk foods;
- Require food facilities to renew their FDA registrations at least every two years and allow FDA to modify the current food product categories for purposes of registration;
- Authorize FDA to accredit highly-qualified third parties for voluntary food inspections;
- Require a new re-inspection fee from facilities that fail to meet current Good Manufacturing Practice (cGMPs) requirements;
- Empower FDA to require electronic import certificates for shipments of designated high-risk products from countries with which FDA has concluded an agreement on a certification program that provides a level of safety sufficient to meet FDA standards;
- Allow FDA to charge export certification fees for food and animal feed to improve the ability of U.S. firms to export their products;
- Authorize FDA to refuse admission of imported food if FDA inspection access is delayed, limited or denied;
- Empower FDA to issue a mandatory recall of food products if voluntary recalls are not effective; and
- Give FDA enhanced access to food records during emergencies *(2)*.

In June 2008, HHS Secretary Michael Leavitt announced that the Bush administration was increasing its Fiscal Year (FY) 2009 budget request for the FDA by $275 million. This increase brings the Administration's total proposed increase in the FDA's budget, including user fees, for FY 2009 to $406.3 million, a 17.9% increase over FY 2008. A large portion of this increase ($125 million) was earmarked for food safety and will allow the FDA to intensify actions to implement the Food Protection Plan. This increased funding is in addition to the $42.2 million increase proposed for food protection in the budget announced in February 2008 *(6)*.

On June 30 2008, the President signed the FY 2008 Supplemental Appropriation into law. This appropriation act provided $150 million for the FDA, and these resources will allow the FDA to accelerate its transformation of its regulatory strategies to meet the challenges of the evolving global marketplace for food and medical products. The funds in the supplemental appropriations act will allow the FDA to further implement the Food Protection Plan, the Action Plan for Import Safety, and important medical product priorities. It will specifically allow the FDA to expand its food safety activities, such as increasing inspections, performing research on mechanisms of food contamination, establishing offices overseas to build capacity with our foreign partners, developing and validating more rapid detection tools, enhancing its information technology systems to support interoperable databases, and enhancing the FDA's ability to identify and target the greatest threats from intentional and unintentional contamination *(7)*.

As part of its Food Protection Plan, the FDA has now adopted a policy of more proactive engagement in food-related human illness outbreaks, including prevention, intervention and response upon implementation. This strategy intends to link food safety with food defense. To date, this approach has fallen short. However, additional integration of risk management efforts based on the HACCP approach when formulating and industry-implementing food defense plans.

What the Food Industry Needs to Do

Unparalleled economic damage and preventable losses of human life were caused by the recent *E-coli* O157:H7 contamination of fresh-bagged spinach and nationwide recall. Americans have also been warned in recent months to stop consuming fresh red tomatoes, jalapeno peppers, and peanut butter products for fear of salmonella contamination. How the food industry alongside the U.S. government respond to these food safety lessons lends insight into the national response—or failure to respond—should the U.S. suffer an intentional attack contaminating its food supply. Why might this sequence of events be any different from an intentional attack on our food supply, and how might we expect to respond to these probable challenges in the future?

If the food industry and regulators can change their focus from "reaction and response" to "anticipation and prevention," while recognizing that criminal and terrorist tactics are adaptive, the landscape of food defense becomes clearer. The American food industry is poised to lead the way globally, by using new guidance with established protocols and tools to radically enhance the protection of America's critical infrastructure system.

Rather than rationally and comprehensively defining the range of potential threats to the U.S. food supply and by extension, the American people, the food industry and the U.S. government together tend to focus on the most recent crisis. The law of large numbers rules the day. For example, at nuclear weapons facilities across the U.S., this type of thinking has led to a phenomenon called the "design basis threat." As an illustration of this weakness, the U.S. government spends millions of dollars to design, engineer and build the most sophisticated security systems in the world; however, if the design basis threat does not provide for the potentiality that terrorists could use toxic gas, then security guards won't be issued gas masks. Security systems based on designs of the past can easily fall victim to a new generation of technologically adaptive, inventive, and evasive terrorists. Professionals and regulators in the food industry face the same threats; they too may be operating based on a myopic design basis threat.

The changing nature of today's threat and risk management picture call for greater investment in early detection and preventing intentional contamination, theft or destruction of the U.S. food supply. As both September 11th and the U.S. government response to Hurricane Katrina illustrated, by the time tragedy strikes, much damage has already been done. If a carefully executed food supply attack happens, it may already be too late to save a business, a brand, or consumer confidence. An ounce of prevention is worth a pound of cure: in the absence of effective prevention and response measures, disaster quickly turns into catastrophe.

It is important that every American business in the food industry and those in the agriculture sector review the Homeland Security Presidential Directives (HSPD)-7 "Critical Infrastructure Identification, Prioritization, and Protection." This document establishes a national policy for federal departments and agencies to identify and prioritize United States critical infrastructure and key resources and to protect them from terrorist attacks *(8)*. An additional directive, "Defense of United States Agriculture and Food" (HSPD-9), defines the national policy to defend the agriculture and food system against terrorist attacks, major disasters, and other emergencies *(9)*.

The National Response Framework (NRF) replaced the National Response Plan on January 22, 2008. This document presents the guiding principles that enable all response partners to prepare for and provide a unified national response to disasters and emergencies, from the smallest incident to the largest catastrophe. The NRF establishes a comprehensive, national, all-hazards approach to domestic incident response *(10)*.

The Department of Homeland Security (DHS) National Infrastructure Protection Plan (NIPP) and supporting Sector-Specific Plans (SSPs) provide a coordinated approach to Critical Infrastructure and Key Resources (CIKR) protection roles and responsibilities for federal, state, local, tribal, and private sector security partners *(11)*. The NIPP sets national priorities, goals, and requirements for effective distribution of funding and resources to help ensure that our government, economy, and public services continue in the event of a terrorist attack or other disaster.

The plan is based on the following:

- Strong public-private partnerships to foster relationships and facilitate coordination within and across CIKR sectors.
- Robust multi-directional information sharing which will enhance the ability to assess risks, makes prudent security investments, and takes protective action.

- Risk management framework establishing processes for combining consequence, vulnerability, and threat information to produce a comprehensive, systematic, and rational assessment of national or sector risk *(11)*.

The DHS Federal Emergency Management Agency (FEMA) conducted a final review of the National Incident Management System (NIMS) in conjunction with the recent release of the National Response Framework (NRF) from 2006 until 2008 *(12)*.

To connect the complex requirements that the food and agricultural sector must manage from both an individual business and a national security sector perspective, one entrepreneurial company in Frederick, MD, ThoughtQuest, LLC, is reassessing critical infrastructures and the relationship between security and safety, including food defense responsibilities *(13)*. One of the top priorities of ThoughtQuest is to help protect the nation's food supply; to this end this company has developed a new generation of "smart software learning knowledge bases" and "decision support products." These products identify and continuously monitor the early warning signs of natural and man-made events, as well as evolution and change in terrorist tactics. ThoughtQuest is engaged in developing proactively pre-conceived event sequences and better decision support tools around a system of organizing and managing complex inputs and adaptive changes. This approach is far-reaching, and can be applied by any U.S. industry, including the food industry, as well as government decision-makers to better understand and manage the threat environment. One module designed specifically for the agriculture/food infrastructure, FoodDefenseTQ™, establishes both food safety and food defense prevention and response metrics using an integrated systems approach. Using the collaborative private sector partnering with federal and state governments and academic partners (e.g. Argonne National Laboratories, the Indiana National Guard, and the University of Maryland), core knowledge bases are introduced from other non-food critical infrastructures. The Complexity Systems Management (CSM)™ software:

1. projects what events might happen;
2. sequences the specific pathway in how the event would actually occur;
3. works backwards from hypothetical disaster scenarios to identify potential strategies for prevention of given events;
4. establishes threat quotients (TQ) to deter, detect and prevent a projected event by computing a measured response against compliance standards and best practices; and
5. identifies mitigation options to gain necessary compliance and better conform to best practices *(13)*.

America has made significant progress in food protection. To continue the progress made, the food industry, government, and other stakeholders should systematically examine the potential threats to our food supply—and likely consequences if these threats are not dealt with effectively—before a catastrophic event takes place. Studying hypothetical and real catastrophes, there are lessons to be learned about the information needed to prevented disaster or mitigate its consequences. A working food safety and food defense framework bound together into a total food protection solution will help all professionals, regulators, and consumer advocates in the food supply chain to understand and accomplish their compatible, shared goals. We have to think about prevention and response in a different way. The most effective responses to complex events are not those that identify a perpetrator (intentional or not) to bring about public punishment. Rather, the most effective process is to deter, detect, and prevent threats. Total food protection requires a combination of proven best food safety and defense practices, procedures and compliance standards that incorporate the best science that experts can offer. Success is measured in effective prevention and mitigation, in reduced outbreaks, in documenting threats successfully detected and dealt with. Total food protection requires sustained commitment to outpacing the adaptation of criminal and terrorist strategies, using novel technologies, new ways of thinking, and effective communication between all stakeholders.

References

1. Branigin, W., Allen, M., & Mintz, J. "Tommy Thompson Resigns From HHS." *The Washington Post,* December 3, 2004.
2. Department of Health and Human Services & the U.S. Food and Drug Administration. *Food Protection Plan: An integrated strategy for protecting the nation's food supply.* Rockville, MD: DHHS & FDA, November, 2007. http://www.fda.gov/oc/initiatives/advance/food/plan.pdf.
3. National Advisory Committee on Microbiological Criteria for Foods. *Hazard Analysis and Critical Control Point Principles and Application Guidelines. Adopted 1997 by the U.S. Food and Drug Administration and United States Department of Agriculture.* Rockville, MD: U.S. Food and Drug Administration, August 14, 1997. http://www.cfsan.fda.gov/~comm/nacmcfp.html
4. Chaturvedi, A. "Live and computational experimentation in bio-terror response." Presentation at the Carlson School of Management, University of Minnesota, January 27, 2006.
5. BTSafety, LLC. http://www.btsafety.com.
6. U.S. Food and Drug Administration. "Statement of David Acheson, M.D., F.R.C.P, Associate Commissioner for Foods, Food and Drug Administration before the Committee on Agriculture Subcommittee on Horticulture and Organic Agriculture, United States House of Representatives, July 30, 2008." (*Salmonella Saintpaul* outbreak investigation) http://www.fda.gov/ola/2008/traceability073008
7. U.S. Food and Drug Administration. *Food Protection Plan Funding to Protect American Food Supply.* Rockville, MD: Food and Drug Administration. http://www.fda.gov/ola/2008/fpp091708.html
8. Department of Homeland Security. *Homeland Security Presidential Directive 7.* Washington, DC: Department of Homeland Security, December 17, 2003.
9. Department of Homeland Security. *Homeland Security Presidential Directive 9.* Washington, DC: Department of Homeland Security, January 30, 2004.
10. Department of Homeland Security. National Infrastructure Protection Plan (NIPP) 2006. http://www.fema.gov/pdf/emergency/nims/NIMS_core.pdf.
11. Federal Emergency Management Administration (FEMA). *National Response Framework.* Washington, DC: Federal Emergency Management Administration (FEMA), January, 2008. http://www.fema.gov/pdf/emergency/nrf/nrf-core.pdf.
12. Federal Emergency Management Administration. *National Incident Management System (NIMS),* Federal Emergency Management Administration (FEMA), December 2008.
13. ThoughtQuest, LLC. http://www.thoughtquest.com.

Indexes

Author Index

Subject Index

Q

R

S

T

U

V

W

Y